APPLIED STRENGTH
OF MATERIALS

APPLIED STRENGTH
OF MATERIALS

ROBERT L. MOTT, P.E.
Associate Professor and Chairman
Mechanical Engineering Technology
University of Dayton
Dayton, Ohio

PRENTICE-HALL, INC., *Englewood Cliffs, New Jersey* *07632*

Library of Congress Cataloging in Publication Data

Mott, Robert L
   Applied strength of materials.

   Includes index.
   1. Strength of materials. I. Title.
TA405.M883   620.1'12   77-26914
ISBN   0-13-043299-7

Printed in the United States of America

10 9 8 7 6

PRENTICE-HALL INTERNATIONAL, INC., *London*
PRENTICE-HALL OF AUSTRALIA PTY. LIMITED, *Sydney*
PRENTICE-HALL OF CANADA, LTD., *Toronto*
PRENTICE-HALL OF INDIA PRIVATE LIMITED, *New Delhi*
PRENTICE-HALL OF JAPAN, INC., *Tokyo*
PRENTICE-HALL OF SOUTHEAST ASIA PTE. LTD., *Singapore*
WHITEHALL BOOKS LIMITED, *Wellington, New Zealand*

To my wife MARGE
To our children LYNNÉ, ROBERT JR., and STEPHEN
And to my MOTHER and FATHER.

CONTENTS

PREFACE

OBJECTIVE

The objective of this book is to present the principles of strength of materials in a way which will facilitate their application to practical problems in several fields of engineering technology. Applications are shown particularly in the areas of mechanical design, construction technology, and civil engineering technology. However, it is recognized that many other disciplines require an understanding of strength of materials.

PREREQUISITES

The book is directed primarily to students in college level engineering technology programs. It is expected that the student will have competence in analytic geometry, algebra, and trigonometry. Calculus is used in selected topics where its use is considered important to the understanding of a concept or where problem solution procedures are facilitated. The arrangement is such that those sections in which calculus is employed may be left out without impeding the logical coverage of material.

Those using the book are also expected to have an understanding of physics mechanics and statics. The concepts of forces, moments, resultants, static equilibrium, free-body diagrams, and support reactions should be familiar to the student to the extent that problems in statics can be solved. It would also be helpful to have a previous knowledge of centroids and moments of inertia of areas. However, these topics *are* presented in the book.

APPROACH

A balance is attempted between consideration of problems in *analysis* of members and problems requiring *design*. After completing this book, a student should have confidence in his or her ability to do original design of load-carrying machine parts and structural members. Extensive use is made of tables of material properties and properties of structural shapes. The reasons for choosing a particular shape for a part or for specifying a certain material are often discussed.

UNITS

The International System of Units, generally known as SI, is used as the primary unit system of the book. However, its use is combined with that of the conventional English Gravitational Unit System. This dual approach is taken because of the applied nature of the book. Persons entering the field of engineering technology in the next several years will encounter with increasing frequency the use of SI in their work. New products in many fields are being made to metric dimensions. However, these same people must work with other equipment designed in the English system. They must also deal with people who may retain the use of the English system. Thus, it seems mandatory to use a dual approach in teaching applied strength of materials for the next several years.

The basic reference for implementing the use of SI in this book is Standard E380, *Metric Practice Guide*, issued by the American Society for Testing and Materials. Other references which have been helpful are:

International Standard ISO 1000, SI Units and Recommendations for the Use of Their Multiples and of Certain Other Units, International Organization for Standardization.

ASME Orientation and Guide for Use of SI (Metric) Units, 5th ed., The American Society of Mechanical Engineers.

SI Units in Strength of Materials, The American Society of Mechanical Engineers.

AISI Metric Practice Guide, The American Iron and Steel Institute.

Reference Manual for SI (Metric), Inland Steel Company, Steel Division.

Many example problems are worked in complete detail, including the manipulation of units in equations. In the more complex examples, a programmed instruction format is used in which the student is asked to perform a small segment of the solution before being shown the correct result. The programs are of the linear type in which one panel presents a concept and then either poses a question or asks that a certain operation be performed. The following panel gives the correct result and the details of how it was obtained. The program then continues.

DESIGN EXAMPLES

Throughout the book, a number of design examples are worked which summarize the concepts presented in a particular chapter or a number of chapters. The examples are written in the form of design projects for which there are a variety of possible solutions. One solution is worked out in detail, illustrating the kind of thought process which might be used to solve the problem. The given solution can be treated as a case study by the instructor with the possibility of asking the students to generate their own unique solutions. Other projects are included in the problem section of some chapters for the students to work completely independently.

EXPECTED RESULTS

The entire book has the objective of helping the student become confident in his or her competence to use the principles of strength of materials in real problems. For the topics included in the book, a person who successfully completes the study should have an understanding of the fundamental concepts and of the applications of the concepts to analysis and design problems.

ROBERT L. MOTT

1

BASIC CONCEPTS
IN STRENGTH
OF MATERIALS

1-1 NEED FOR STRENGTH ANALYSIS

Insuring that a product, a machine, or a structure is safe for the expected conditions under which it will be used is the most important requirement in design. An analysis of the design according to the principles of strength of materials is the way safety can be assured.

Consider the automobile jack stand shown in Figure 1-1. This is an obvious example of a mechanical device which must support a load safely. A person servicing the underside of a car depends on the jack stand to carry the force exerted by the car. Of course, the stand must also be designed for rigidity and stability. The designer of the jack stand must answer the questions:

What size and shape should the main support shaft be?

What size and shape should the feet of the

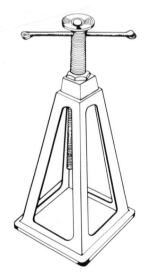

Figure 1-1

1

stand be in order to rest properly on concrete, asphalt, wood, or gravel surfaces?

What material should all the parts be made of?

Should bolts be used to fasten the frame members?

Should welding be used for joining parts? If so, how should the welded joint be made?

What are the expected loads on the jack stand?

What are the maximum conceivable loads that could be placed on the jack stand, even if it is being used improperly?

The designer must understand the principles of strength of materials in order to answer these questions satisfactorily.

The crane shown in Figure 1-2 is rated by the manufacturer to provide a lifting force of 4000 pounds (lb). In order to insure safety, the designer had to analyze all parts of the crane to determine what load each one carries when the crane itself holds 4000 lb. Each part was then designed to withstand its unique loading. This illustrates the strong interdependence between the study of strength of materials and the use of the principles of physics mechanics and statics. A load analysis must precede the strength analysis. The concepts

Figure 1-2   Two Ton Shop Crane

of static equilibrium, determination of reactions due to applied loads, free-body diagram analysis, and resultants of forces will be used continually throughout the book.

The building structure shown in Figure 1-3 contains numerous standard

Figure 1-3   (*Courtesy of* B. G. Danis Co., Dayton, Ohio)

structural members for its columns, beams, and roof supports. Each was specified to carry a design load determined by the use to which the building will be put and the environmental conditions it must withstand. Design data included such things as:

> Amount of load on each floor and how it is distributed.
> How much the floors can deflect under load.
> Expected winds on the building.
> Amount of snow expected.
> Likelihood of earthquakes occurring at the building site.

These data are required *before* the strength analysis is performed.

These three examples point out some of the applications for the principles of strength of materials. Certainly there are many others. Each design problem presents a different requirement for analysis. The designer must be capable of recognizing what approach should be used for each problem. A firm foundation based on the concepts presented in this book will enable one to perform the proper analysis and design safe products, machines, and structures.

1-2 THE CONCEPT OF STRESS

The objective of any strength analysis is to insure safety. Accomplishing this requires that the *stress* produced in the member being analyzed is below a certain *allowable stress* for the material from which the member is made. Understanding the meaning of stress is of utmost importance in studying strength of materials.

Stress is the internal resistance offered by a unit area of the material from which a member is made to an externally applied load.

We are concerned about what happens *inside* a load-carrying member. The effect of the internal resistance on the material determines whether the member is safe or not. Different materials react differently to loads and thus have different strengths. Therefore, the selection of a material is an important part of the design process. However, it will be shown in later chapters that the *shape* as well as the *size* of a load-carrying member affects its strength. Actually, then, it might be more nearly correct to refer to the study of the "strength of load-carrying members" rather than the "strength of materials."

Stress is a measure of the amount of force exerted on a specific area of the material in a part. In some cases the magnitude of the stress is uniform over a large section of the part. In such cases stress can be computed by simply dividing the total force by the area of the part which resists the force. This concept can be expressed mathematically as

$$\text{Stress} = \frac{\text{force}}{\text{area}}$$

For example, one of the loaded components of the jack stand shown in Figure 1-1 is the main support shaft. The force exerted by the car pushes down on the shaft. That means that all parts of the shaft must resist the load of the car. This is shown in Figure 1-4. On the left is shown the complete shaft, with the load of the car labeled "Applied force." In order to hold the shaft in position, a reaction force is provided by other parts of the jack stand. But what about a section inside the shaft? It too must resist the applied force. Figure 1-4(b) shows only a portion of the shaft as if it was cut at section

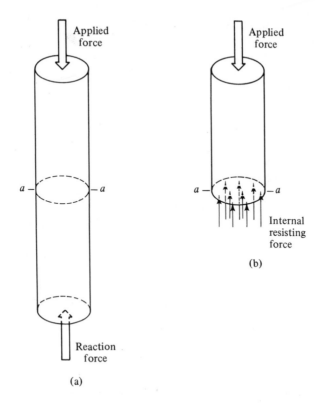

Figure 1-4

a-a. The entire area across the shaft is available to resist the applied force. The arrows at section *a-a* indicate that the material inside the shaft is exerting a force equal in magnitude and opposite in direction to the applied force. From this observation it can then be said:

$$\text{Stress on section } a\text{-}a = \frac{\text{applied force}}{\text{area of section } a\text{-}a}$$

In use, a reasonable load on the jack stand would be 2000 lb. If the shaft is circular with a diameter of 1.0 inch (in.), the area of section *a-a* is

$$\text{Area} = \frac{\pi}{4}(\text{diameter})^2 = \frac{\pi}{4}(1.0)^2 \text{ in.}^2 = 0.785 \text{ in.}^2$$

Then

$$\text{Stress on section } a\text{-}a = \frac{2000 \text{ lb}}{0.785 \text{ in.}^2} = 2548 \text{ lb/in.}^2$$

Stress will always be expressed in the units of force per unit area.

The example of the jack-stand shaft was used to illustrate the concept of stress. It must be understood that this is a simple example of stress due to a direct compressive load. There are other kinds of loads that produce stress, which will be explained in later chapters.

1-3   THE CONCEPT OF STRAIN

The previous section deals with stress, the calculation of which is necessary to evaluate the strength of a part to determine if it is safe. In some cases, the total deformation of a load-carrying member is also important.

In building construction, long beams tend to sag or deflect downward under their own weight and because of applied forces. In order to insure a satisfactory appearance and proper fit of mating members of the structure, the designer must be able to calculate the magnitude of the expected deflection. Under certain conditions, the choice of the size and shape of the beam may depend more on limiting deflection than on keeping stresses to a safe level.

An important application of deflection analysis occurs in machine design. Figure 1-5 shows two parallel rotating shafts in a machine carrying gears.

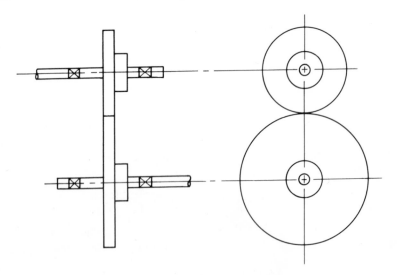

Figure 1-5

As the gears transmit power, forces are developed which act in a direction which tends to separate the two shafts. Since the proper operation of precision gears requires that the mating gear teeth be carefully positioned relative to each other, deflection of the shafts must be limited to only about 0.005 in. Deflection of beams, including rotating shafts, is covered in Chapter 12.

Whenever anything is subjected to a load, it deforms. A simple example is shown in Figure 1-6. An aluminum bar having a diameter (dia.) of 0.750 in. is subjected to an axial tensile force of 10 000 lb. While carrying the load, the length of the bar, which was originally 10.000 in., elongates by 0.023 in. The elongation is referred to as the *total deformation*.

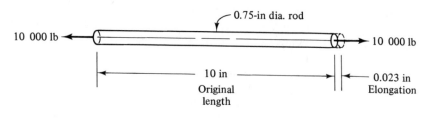

Figure 1-6

Dividing the total deformation by the original length of the bar produces a *unit deformation* called *strain*. Expressing this concept mathematically gives

$$\text{Strain} = \frac{\text{total deformation}}{\text{original length}}$$

For the case of the aluminum bar,

$$\text{Strain} = \frac{0.023 \text{ in.}}{10.000 \text{ in.}} = 0.0023 \text{ in./in.}$$

More will be said about strain in later chapters. However, you should now have an understanding of the physical concepts of stress and strain, which is what strength of materials is all about.

1-4   UNITS IN STRENGTH OF MATERIALS

Computations required in the application of strength of materials involve the manipulation of several sets of units in equations. For numerical accuracy it is of great importance to insure that consistent units are used in the equations. Throughout this book units will be carried with the applicable numbers.

Because of the present transition in the United States from the traditional English units to metric units, both are used in this book. It is expected that persons entering or continuing an industrial career within the next several years will be faced with having to be familiar with both systems. On the one hand, many new products such as automobiles and business machines are being built using metric dimensions. Thus components and manufacturing equipment will be specified in those units. However, the transition is not being made uniformly in all areas. Designers will continue to deal with such items as structural steel, aluminum, and wood whose properties and dimensions are given in English units in standard references. Also, designers, sales and service people, and those in manufacturing must work with equipment which has already been installed and which was made to English system dimensions. Therefore, it seems logical that persons now working in industry should now be capable of working *and thinking* in both unit systems.

The complete name for the English system is the *English Gravitational Unit System* (EGU). The metric system which has been adopted internationally is the *International System of Units*. The common abbreviation for this system is SI. The following sections will describe the main features of each system.

It should be noted that problems will be worked in *either* the EGU system or the SI system of units. It is most undesirable to mix unit systems in any analysis. Conversion of all data to one system should be done prior to performing the computations. Results, then, can be converted from one system to the other. Appendix tables are included as an aid in performing conversions.

The quantities usually encountered in strength of materials are listed below and described in terms of their relationship to the basic quantities of *length, time, force,* and *mass.*

Distance	length
Area	$(\text{length})^2$
Energy or work	length-force
Stress or pressure	force/area
Power	length-force/time or work/time
Torque or moment	force-length
Section modulus	$(\text{length})^3$
Moment of inertia	$(\text{length})^4$

In addition, use will be made of angular measurement, rotational speed, and temperature measurement.

1-5 THE ENGLISH GRAVITATIONAL UNIT SYSTEM

The units for the basic quantities in the English Gravitational Unit System (EGU) are:

$$\text{Length} = \text{foot (ft)}$$
$$\text{Time} = \text{second (sec)}$$
$$\text{Force} = \text{pound (lb)}$$
$$\text{Mass} = \text{slug (lb-sec}^2/\text{ft)}$$

In this system, the unit for mass is derived from the other three basic units from Newton's law,

$$\text{Force} = \text{mass} \times \text{acceleration}$$

$$F = ma$$

where a is acceleration, having the units of ft/sec². Therefore, the derived unit for mass is:

$$m = \frac{F}{a} = \frac{\text{lb}}{\text{ft/sec}^2} = \frac{\text{lb-sec}^2}{\text{ft}} = \text{slug}$$

The slug is used very infrequently in strength of materials.

The list below shows the principle quantities encountered in strength of materials with their EGU system units and other accepted units.

Distance	ft or in.
Area	ft² or in.²
Energy or work	ft-lb
Stress or pressure	lb/ft² or lb/in.² (psi)
Power	ft-lb/sec or horsepower (hp)
Torque or moment	ft-lb or in.-lb
Section modulus	ft³ or in.³ (usual unit)
Moment of inertia	ft⁴ or in.⁴ (usual unit)
Angle	degree (deg)
Rotational speed	rev/min or rpm
Temperature	°F

Many times, large magnitudes of forces are found to act on machine parts and structures. For this reason, the unit *kip* is defined as equal to 1000 lb. For example: 12 000 lb = 12 kips.

1-6   THE INTERNATIONAL SYSTEM OF UNITS

The base units in the International System of Units (SI) of interest in strength of materials are listed in the following table.

Quantity	Unit Name	SI Symbol	Other Permissible Units
Length	metre	m	—
Mass	kilogram	kg	—
Time	second	s	—
Temperature	kelvin	K	°C (Celsius)
Angle	radian	rad	degree (deg)

Units for other quantities are derived from these base units. In some cases special names and SI symbols have been developed for derived quantities. The following table lists the quantities of importance in strength of materials.

Quantity	Unit	SI Symbol	Formula
Distance	metre	m	—
Area	square metre	—	m^2
Force	newton	N	$kg \cdot m/s^2$
Energy or work	joule	J	$N \cdot m$
Stress or pressure	pascal	Pa	N/m^2
Power	watt	W	J/s or $N \cdot m/s$
Torque or moment	newton-metre	—	$N \cdot m$
Section modulus	metre cubed	—	m^3
Moment of inertia	metre to the fourth power	—	m^4
Rotational speed	radian per second	—	rad/s

In the SI system, prefixes should be used to indicate orders of magnitude, thus eliminating insignificant digits and providing a convenient substitute for writing powers of 10, as generally preferred for computation. Prefixes representing steps of 1000 are recommended. Those usually encountered in strength of materials are listed in the following table.

Prefix	SI Symbol	Factor	
giga	G	10^9	$= 1\ 000\ 000\ 000$
mega	M	10^6	$= 1\ 000\ 000$
kilo	k	10^3	$= 1\ 000$
milli	m	10^{-3}	$= 0.001$
micro	μ	10^{-6}	$= 0.000\ 001$

It should be noted that in the SI system of units there are separate and distinct units for force and mass. Force is in newtons and mass is in kilograms. The term *weight* will be used in this book to denote the attractive force of the earth on a body due to the acceleration of gravity. Thus it may be said that a beam weighs 1760 N. However, if the physical quantity of the substance of a body is referred to, its *mass* will be given. For example, it may be said that a hoist can lift 425 kg of concrete. Then in order to determine what force is exerted on the hoist by the concrete, Newton's law of gravitation ($W = mg$) should be used.

$$W = mg = \text{mass times acceleration of gravity}$$

But

$$g = \text{acceleration of gravity} = 9.81 \text{ m/s}^2$$

Then

$$W = 425 \text{ kg} \cdot 9.81 \text{ m/s}^2 = 4170 \text{ kg} \cdot \text{m/s}^2 = 4170 \text{ N}$$

Thus, 425 kg of concrete weighs 4170 N.

PROBLEMS

1-1   Define stress.

1-2   Define strain.

1-3   Define total deformation.

1-4   State the units in the English Gravitational Unit System for the following quantities:

    (a) Length.   (f) Energy.

    (b) Time.   (g) Stress.

    (c) Force.   (h) Power.

    (d) Mass.   (i) Temperature.

    (e) Area.   (j) Moment.

1-5   State the units in the SI system of units for the quantities listed in Problem 1-4.

1-6   A truck carries 1800 kg of gravel. What is the weight of the gravel in newtons?

1-7   A four-wheeled truck having a total mass of 4000 kg is sitting on a bridge. If 60 percent of the weight is on the rear wheels and 40 percent is on the front wheels, compute the force exerted on the bridge at each wheel.

1-8   A total of 6800 kg of a bulk fertilizer is stored in a flat-bottom bin having side dimensions 5.0 m × 3.5 m. Compute the loading on the floor in newtons per square metre, or pascals.

1-9   A mass of 25 kg is suspended by a spring which has a spring scale of 4500 N/m. How much will the spring be stretched?

1-10   Measure the length, width, and thickness of this book in millimetres.

1-11   Determine your own weight in newtons and your mass in kilograms.

1-12   Express the weight found in Problem 1-6 in pounds.

1-13   Express the forces found in Problem 1-7 in pounds.

1-14   Express the loading in Problem 1-8 in pounds per square foot.

1-15   For the data in Problem 1-9, compute the weight of the mass in pounds, the spring scale in pounds per inch, and the stretch of the spring in inches.

1-16   Express the power rating of a 7.5-hp motor in watts.

1-17   A pressure vessel contains a gas at 1200 psi. Express the pressure in pascals.

1-18   A structural steel has an allowable stress of 21 600 psi. Express this in pascals.

1-19   The stress at which a material will break under a direct tensile load is called the *ultimate strength*. The range of ultimate strengths for aluminum alloys ranges from about 14 000 psi to 76 000 psi. Express this range in pascals.

1-20   An electric motor shaft rotates at 1750 rpm. Express the rotational speed in radians per second.

1-21   Express an area of 14.1 in.² in the units of square millimetres.

1-22 An allowable deformation of a certain beam is 0.080 in. Express the deformation in millimetres.

1-23 A base for a building column measures 18.0 in. by 18.0 in. on a side and 12.0 in. high. Compute the cross-sectional area in both square inches and square millimetres. Compute the volume in cubic inches, cubic feet, cubic millimetres, and cubic metres.

1-24 Compute the area of a rod having a diameter of 0.505 in. in square inches. Then convert the result to square millimetres.

2

DESIGN PROPERTIES
OF MATERIALS

2-1 METALS IN ENGINEERING DESIGN

The study of strength of materials requires a knowledge of how external forces and moments affect the stresses and deformations developed in the material of a load-carrying member. In order to put this knowledge to practical use, however, a designer needs to know how such stresses and deformations can be withstood safely by the material. Thus, material properties as they relate to design must be understood along with the analysis required to determine the magnitude of stresses and deformations.

Metals are most widely used for load-carrying members in buildings, bridges, machines, and a wide variety of consumer products. Beams and columns in commercial buildings are made of structural steel or aluminum. In automobiles, a large number of steels are used, including plain carbon sheet for body panels, free-cutting alloys for machined parts, and high strength alloys for gears and heavily loaded parts. Cast iron is used in engine blocks, brake drums, and cylinder heads. Tools, springs, and other parts requiring high hardness and wear resistance are made from steel alloys containing a large amount of carbon. Stainless steels are used in transportation equipment, chemical plant products, and kitchen equipment where resistance to corrosion is required.

Aluminum sees many of the same applications as steel. Aluminum is used in many architectural products and frames for mobile equipment. Its corrosion resistance allows its use in chemical storage tanks, cooking utensils, marine equipment, and products such as highway signposts. Automotive pistons, trim, and die-cast housings for pumps and alternators are made of aluminum. Aircraft structures, engine parts, and sheet-metal skins use aluminum because of its high strength-to-weight ratio.

Copper and its alloys such as brass and bronze are used in electric conductors, heat exchangers, springs, bushings, marine hardware, and switch parts. Magnesium is often cast into truck parts, wheels, and appliance parts. Zinc sees similar service and may also be forged into machinery components and industrial hardware. Titanium has a high strength-to-weight ratio and good corrosion resistance, and thus is used in aircraft parts, pressure vessels, and chemical equipment.

Material selection requires consideration of many factors. Generally, strength, stiffness, ductility, corrosion resistance, machinability, workability, weldability, weight, appearance, cost, and availability must all be evaluated. Relative to the study of strength of materials, the first three of these factors are most important: strength, stiffness, and ductility.

Figure 2-1 Universal Testing Machine for Obtaining Stress-Strain Data for Materials. (*Source:* Tinius-Olsen Testing Machine Co., Inc., Willow Grove, Pa.)

Strength. Reference data listing the mechanical properties of metals will almost always include the *ultimate tensile strength* and *yield strength* of the metal. Comparison of the actual stresses in a part with the ultimate or yield strength of the material from which the part is made is the usual method of evaluating the suitability of the material to carry the applied loads safely. More is said about the details of stress analysis in Chapter 3 and subsequent chapters.

The ultimate tensile strength and yield strength are determined by testing a sample of the material in a tensile-testing machine such as that shown in Figure 2-1. A round bar or flat strip is placed in the upper and lower jaws. A pulling force is applied slowly and steadily to the sample, stretching it until it breaks. During the test, a graph is made which shows the relationship between the stress in the sample and the strain or unit deformation. A typical stress-strain diagram for a low carbon steel is shown in Figure 2-2. It can be seen that, during the first phase of loading, the plot of stress versus strain is a straight line, indicating that stress is directly proportional to strain. After point *A* on the diagram, the curve is no longer a straight line. This point is called the *proportional limit*. As the load on the sample is continually increased, a point called the *elastic limit* is reached, marked *B* in Figure 2-2.

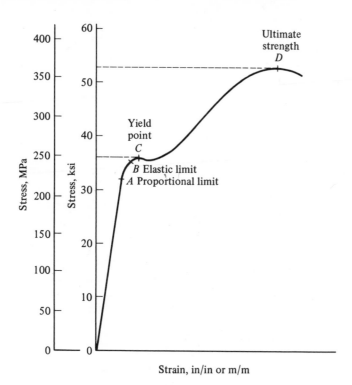

Figure 2-2 Typical Stress-Strain Curve for Steel

At stresses below this point, the material will return to its original size and shape if the load is removed. At higher stresses, the material is permanently deformed. The *yield point* is the stress at which a noticeable elongation of the sample occurs with no apparent increase in load. The yield point is at *C* in Figure 2-2, about 36 000 psi (248 MPa). Applying still higher loads after the yield point has been reached causes the curve to rise again. After reaching a peak, the curve drops somewhat until finally the sample breaks, terminating the plot. The highest apparent stress taken from the stress-strain diagram is called the *ultimate strength*. In Figure 2-2, the ultimate strength would be about 53 000 psi (365 MPa).

In most designs, a part is considered to have failed if it breaks or if it deforms enough so that it cannot perform its intended function. Therefore, the yield strength and ultimate strength are the two most important strength properties of materials.

Many metals do not exhibit a well-defined yield point like that in Figure 2-2. Some examples are high strength alloy steels, aluminum, and titanium.

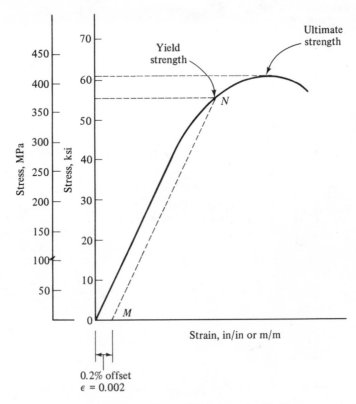

Figure 2-3 Typical Stress-Strain Curve for Aluminum

However, these materials do in fact yield in the sense of deforming a sizeable amount before fracture actually occurs. For these materials, a typical stress-strain diagram would look as shown in Figure 2-3. The curve is smooth with no pronounced yield point. For such materials, the yield strength is defined by a line like *M-N* drawn parallel to the straight-line portion of the test curve. Point *M* is usually determined by finding that point on the strain axis representing a strain of 0.002 in./in. This point is also called the point of 0.2 percent offset. The point *N*, where the offset line intersects the curve, defines the yield strength of the material, about 55 000 psi in Figure 2-3. The ultimate strength is at the peak of the curve, as was described before. *Yield strength* is used in place of yield point for these materials.

Stiffness. It is frequently necessary to determine how much a part will deform under load in order to insure that excessive deformation does not destroy the usefulness of the part. This can occur at stresses well below the yield strength of the material, especially in very long members or in high precision devices. Stiffness of a material is a function of its *modulus of elasticity*, sometimes called Young's modulus. The symbol *E* is used for modulus of elasticity, which is measured by determining the slope of the stress-strain curve during the first part, where it is a straight line. This can be stated mathematically as

$$E = \frac{\text{stress}}{\text{strain}} = \frac{s}{\epsilon} \tag{2-1}$$

Therefore, a material having a steeper slope on its stress-strain curve will be stiffer and will deform less under load than a material having a less steep slope. Figure 2-4 illustrates this concept by showing the straight-line portions of the stress-strain curves for steel, titanium, aluminum, and magnesium. It can be seen that if two otherwise identical parts were made of steel and aluminum, respectively, the aluminum part would deform about three times as much when subjected to the same load.

Expressing Equation (2-1) as $s = E\epsilon$, we can say that stress is proportional to strain as long as the stress is below the proportional limit of the material. This is called *Hooke's law*. The assumption that stress is proportional to strain is used in developing many of the formulas in strength of materials.

Ductility. When metals break, their fracture can be classified as either ductile or brittle. A ductile material will stretch and yield prior to actual fracture, causing a noticeable decrease in the cross-sectional area at the fractured section. Conversely, a brittle material will fracture suddenly with little or no change in the area at the fractured section. Ductile materials are preferred for parts which carry repeated loads or are subjected to impact loading because they are usually more resistant to fatigue failure and because they are better at absorbing impact energy.

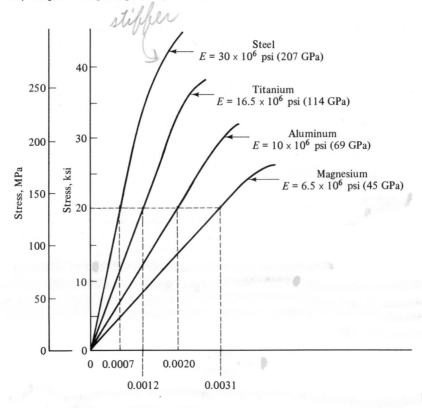

stiffer

Figure 2-4   Modulus of Elasticity for Different Metals

Ductility in metals is usually measured during the tensile test by noting how much the material has elongated before fracturing. At the start of the test, a set of gauge marks is scribed on the test sample. Most tests use 2.0 in. between the gauge marks. After the sample has been pulled to failure, the broken parts are fitted back together and the distance between the marks is again measured. From these data a *percent elongation* is computed. A high value of percent elongation indicates a highly ductile material. A metal is considered to be ductile if its percent elongation is greater than about 5 percent. Virtually all steels and aluminum alloys are considered to be ductile. Gray cast iron and some highly hardened steel alloys are examples of brittle materials.

The following sections will discuss the typical properties of metals which affect their load-carrying performances. Then nonmetals such as wood, plastics, and concrete are described.

2-2 STEEL

The term *steel* refers to alloys of iron and carbon and, in many cases, other elements. Because of the large number of steels available, they will be classified in this section as *plain carbon steels, alloy steels, stainless steels, high strength steels*, and *structural steels*.

For *plain carbon steels* and *alloy steels*, a four-digit designation code is used to define each alloy. Figure 2-5 shows the significance of each digit.

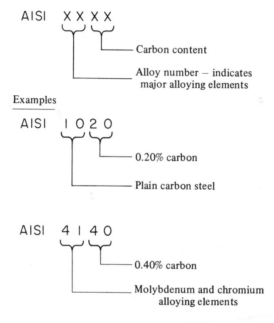

Figure 2-5 Steel Designation

The four digits would be the same for steels classified by the American Iron and Steel Institute (AISI) and the Society of Automotive Engineers (SAE). Classification by the American Society for Testing and Materials (ASTM) will be discussed later.

Usually the first two digits in a four-digit designation for steel will denote the major alloying elements, other than carbon, in the steel. The last two digits denote the average percent (or points) of carbon in the steel. For example, if the last two digits are 40, the steel would have about 0.4 percent carbon content. Carbon is given such a prominent place in the alloy designation because, in general, as carbon content increases, the strength and hardness of the steel also increases. Carbon content usually ranges from a low of

19

0.1 percent to about 1.0 percent. It should be noted that while strength increases with increasing carbon content, the steel also becomes more brittle.

Table 2-1 shows the major alloying elements which correspond to the first two digits of the steel designation. Table 2-2 lists some common alloys along with the principal uses for each.

TABLE 2-1   Major Alloying Elements in Steel Alloys

Steel AISI No.	Alloying Elements	Steel AISI No.	Alloying Elements
10xx	Plain carbon	46xx	Molybdenum-nickel
11xx	Sulphur (free-cutting)	47xx	Molybdenum-nickel-chromium
13xx	Manganese	48xx	Molybdenum-nickel
14xx	Boron	5xxx	Chromium
2xxx	Nickel	6xxx	Chromium-vanadium
3xxx	Nickel-chromium	8xxx	Nickel-chromium-molybdenum
4xxx	Molybdenum	9xxx	Nickel-chromium-molybdenum (except 92xx)
41xx	Molybdenum-chromium	92xx	Silicon-manganese
43xx	Molybdenum-chromium-nickel		

TABLE 2-2   Common Steel Alloys and Typical Uses

Steel AISI No.	Typical Uses
1020	Structural steel, bars, plate
1040	Machinery parts, shafts
1050	Machinery parts
1095	Tools, springs
1137	Shafts, screw machine parts (free-cutting alloy)
1141	Shafts, machined parts
4130	General purpose, high strength steel; shafts, gears, pins
4140	Same as 4130
4150	Same as 4130
5160	High strength gears, bolts
8760	Tools, springs, chisels

The mechanical properties of plain carbon steel are also sensitive to the manner in which they are formed and to heat-treating processes. Table A-1 lists the ultimate strength, yield strength, and percent elongation for a variety of steels. Plain carbon steels having a low carbon content are not usually heat-treated. Their properties vary, however, depending on whether they have been rolled while hot or drawn through a die while cold. Generally, the more severely a steel is worked, the higher its strength will be. This can be seen in Table A-1 by comparing the strength for the same alloy, such as AISI 1040, in the hot-rolled and the cold-drawn condition.

A graphical picture of the effect of cold working on the properties of a steel can be seen in Figure 2-6. Music wire (ASTM A228) and oil-tempered ASTM A229 wire are both high strength alloys used for springs. The desired diameter is achieved by drawing the wire through progressively smaller dies.

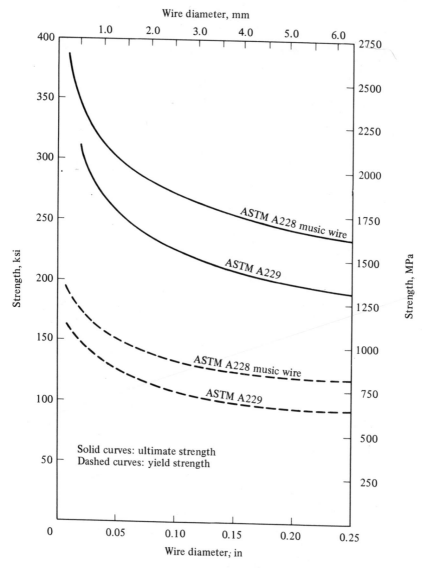

Figure 2-6 Strength of Drawn Wire

For music wire the ultimate strength increases from about 235 000 psi (1620 MPa) for a 0.25-in. diameter to 375 000 psi (2585 MPa) for a 0.010-in. diameter, due to the greater working of the material.

Alloy steels are usually heat-treated to develop specified properties. Heat treatment involves raising the temperature of the steel to above about 1450°F to 1650°F (depending on the alloy) and then cooling it rapidly by quenching in either water or oil. After quenching, the steel has a high strength and hardness, but it may also be brittle. For this reason a subsequent treatment called *tempering* (or *drawing*) is usually performed. The steel is reheated to a temperature in the range of 400°F to 1300°F and then cooled. The effect

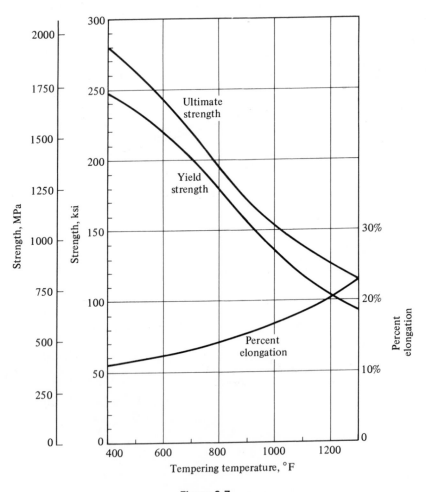

Figure 2-7

of tempering an alloy steel can be seen by referring to Figure 2-7. Thus, the properties of a heat-treated steel can be controlled by specifying a tempering temperature. In Table A-1, the condition of heat-treated alloys is described in a manner like OQT 400. This means that the steel was heat-treated by quenching in oil and then tempered at 400°F. Similarly, WQT 1300 means water-quenched and tempered at 1300°F. Listing the properties of heat-treated steels at tempering temperatures of 400°F and 1300°F shows the practical range of properties that the steel can have. By specifying a tempering temperature between 400°F and 1300°F, intermediate properties can be achieved.

Stainless steels get their name because of their corrosion resistance. The primary alloying element in stainless steels is chromium, being present at about 17 percent in most alloys. A minimum of 10 percent chromium is used, and it may range as high as 27 percent.

Although over 40 grades of stainless steel are available from steel producers, they are usually categorized into three series containing alloys with similar properties. The 200 and 300 series steels have high strength and good corrosion resistance. They can be used at temperatures up to about 1200°F with good retention of properties. Because of their structure, these steels are essentially nonmagnetic. Good ductility and toughness and good weldability make them useful in chemical processing equipment, architectural products, and food-related products. They are not hardenable by heat treatment, but they can be strengthened by cold working.

The AISI 400 series steels are used for automotive trim and for chemical processing equipment such as acid tanks. Certain alloys can be heat-treated so they can be used as knife blades, springs, ball bearings, and surgical instruments. These steels are magnetic. The properties of some stainless steels are listed in Table A-2.

High strength steel is a term used to describe a variety of alloys having yield strengths above 180 000 psi (1240 MPa), although this strength level is somewhat arbitrary. Examples of some of these steels are as follows:

Medium carbon alloy steels, AISI 3140, AISI 4130, AISI 4140, AISI 4340. Yield strength greater than 180 000 psi (1240 MPa) when quenched and tempered at about 800°F or lower. (See Table A-1.)

Precipitation hardening steels, 17-4 PH, PH 13-8. Hardened by holding at an elevated temperature (about 900°F–1000°F). The process is often called artificial aging. (See Table A-2.)

Maraging steels. Many proprietary grades can achieve yield strengths above 300 000 psi (2070 MPa) by a simple aging treatment.

High carbon steel strip and wire. An example is music wire when drawn to a small diameter, as shown in Figure 2-6. Music wire is a plain carbon steel, number 1085.

Structural steels are produced in the forms of sheet, plate, bars, tubing, and structural shapes such as I-beams, wide flange beams, channels, and angles. The American Society for Testing and Materials (ASTM) assigns a number designation to these steels which is the number of the standard which defines the required minimum properties. Table A-3 lists four frequently used grades of structural steels and their properties.

A very popular steel for structural applications is ASTM A36, a carbon steel used for many commercially available shapes, plates, and bars. It has a minimum yield point of 36 kips per square inch (ksi) (248 MPa), is weldable, and is used in bridges, buildings, and for general structural purposes.

Steels called ASTM A242, A440, and A441 are members of a class called high strength, low alloy steels. Produced as shapes, plates, and bars, they may be specified instead of A36 steel to allow the use of a smaller, lighter member. In sizes up to $\frac{3}{4}$ in. thick, they have a minimum yield point of 50 ksi (345 MPa). From $\frac{3}{4}$ in. to $1\frac{1}{2}$ in. thick, the minimum yield point of 46 ksi (317 MPa) is specified. Alloy A242 is for general structural purposes and is called a weathering steel since its corrosion resistance is about four times that of plain carbon steel. Alloy A441 is used primarily for welded construction, and alloy A440 is used in riveted and bolted construction. Cost, of course, must be considered before specifying these alloys.

Alloy A514 is a high strength alloy, heat-treated by quenching and tempering to a yield point of 100 ksi (690 MPa) minimum. Produced as plates, it is used in welded bridges and similar structures.

Construction tubing is either round, square, rectangular, or of other shapes, and is frequently made of ASTM A500 steel. It can be made in either welded or seamless forms and is available in several grades. The minimum yield point of 33 ksi (228 MPa), listed in Table A-3, is for round tubing, grade A. Other grades have yield points ranging up to 46 ksi (317 MPa).

Another general purpose structural steel is ASTM A572, available as shapes, plates, and bars, and in grades 42 to 65. The grade number refers to the minimum yield point of the grade in ksi, and can be 42, 45, 50, 55, 60, and 65.

In summary, steels come in many forms and have a wide range of strengths and other properties. Selection of a suitable steel is indeed an art, supported by a knowledge of the significant features of each alloy.

2-3 ALUMINUM

Alloys of aluminum are designed to achieve optimum properties for specific uses. Some are produced primarily as sheet and plate. Others are rolled into bars or wire. Standard structural shapes and special sections are often extruded. Several alloys are used in forging operations.

Aluminum uses a four-digit designation to define the several alloys available. The first digit indicates the alloy group according to the principal alloying element. The second digit denotes a modification of the basic alloy. The last two digits identify a specific alloy within the group. A brief description of the seven major series of aluminum alloys follows.

1000 series, 99.0 percent aluminum or greater. Used in chemical and electrical fields. Excellent corrosion resistance, workability, and thermal and electrical conductivity. Low mechanical properties.

2000 series, copper alloying element. Heat-treatable with high mechanical properties. Lower corrosion resistance than most other alloys. Used in aircraft skins and structures.

3000 series, manganese alloying element. Non-heat-treatable, but moderate strength can be obtained by cold working. Good corrosion resistance and workability. Used in chemical equipment, cooking utensils, residential siding, and storage tanks.

4000 series, silicon alloying element. Non-heat-treatable with a low melting point. Used as welding wire and brazing alloy. Alloy 4032 used as pistons.

5000 series, magnesium alloying element. Non-heat-treatable, but moderate strength can be obtained by cold working. Good corrosion resistance and weldability. Used in marine service, pressure vessels, auto trim, builder's hardware, welded structures, TV towers, and drilling rigs.

6000 series, silicon and magnesium alloying elements. Heat-treatable to moderate strength. Good corrosion resistance, formability, and weldability. Used as heavy-duty structures, truck and railroad equipment, pipe, furniture, architectural extrusions, machined parts, and forgings. Alloy 6061 is one of the most versatile available.

7000 series, zinc alloying element. Heat-treatable to very high strength. Relatively poor corrosion resistance and weldability. Used mainly for aircraft structural members. Alloy 7075 is among the highest strength alloys available. It is produced in most rolled, drawn, and extruded forms and is also used in forgings.

Aluminum Temper Designations. Since the mechanical properties of virtually all aluminum alloys are very sensitive to cold working or heat treatment, suffixes are applied to the four-digit alloy designations to describe the temper. The most frequently used temper designations are described as follows:

O *temper*. Fully annealed to obtain the lowest strength. Annealing makes most alloys easier to form by bending or drawing. Parts formed in the annealed condition are frequently heat-treated later to improve properties.

H *temper*, strain-hardened. Used to improve the properties of non-heat-treatable alloys such as those in the 1000, 3000, and 5000 series. The H is always followed by a two- or three-digit number to designate a specific degree of strain hardening or special processing. The second digit following the H ranges from 0 to 8 and indicates a successively greater degree of strain hardening, resulting in higher strength. Table A-4 lists the properties of several aluminum alloys. Referring to alloy 3003 in that table shows that the yield strength is increased from 12 000 psi (83 MPa) to 24 000 psi (165 MPa) as the temper is changed from H12 to H18.

T *temper*, heat-treated. Used to improve strength and achieve a stable condition. The T is always followed by one or more digits indicating a particular heat treatment. For wrought products such as sheet, plate, extrusions, bars, and drawn tubes, the most frequently used designations are T4 and T6. The T6 treatment produces higher strength but generally reduces workability. In Table A-4, several heat-treatable alloys are listed in the O, T4, and T6 tempers to illustrate the change in properties.

Aluminum is also sensitive to the manner in which it is produced. Therefore, Table A-4 lists the kind of products for which the stated properties apply.

2-4 COPPER, BRASS, AND BRONZE

The name *copper* is properly used to denote virtually the pure metal having 99 percent or more copper. Its uses are primarily as electric conductors, switch parts, and motor parts which carry electric current. Copper and its alloys have good corrosion resistance, are readily fabricated, and have an attractive appearance. The main alloys of copper are the brasses, bronzes, and beryllium copper. Each has its special properties and applications.

Beryllium copper has very high strength and good electrical conductivity. Its uses include switch parts, fuse clips, electric connectors, bellows, Bourdon tubing for pressure gauges, and springs.

Brasses are alloys of copper and zinc. They have good corrosion resistance, workability, and a pleasing appearance, which leads to applications in automobile radiators, lamp bases, heat-exchanger tubes, marine hardware, ammunition cases, and home furnishings. Adding lead to brass improves its machinability, which makes it attractive to use for screw machine parts.

The major families of bronzes include phosphor bronze, aluminum bronze, and silicon bronze. Their high strength and corrosion resistance make them useful in marine applications, screws, bolts, gears, pressure vessels, springs, bushings, and bearings.

The strength of copper and its alloys is dependent on the hardness which is achieved by cold working. Successively higher strengths would result from the tempers designated annealed, quarter hard, half hard, three-quarters hard, hard, extra hard, spring, and extra spring tempers. The strengths of four copper alloys in the hard temper are listed in Table A-2.

2-5   ZINC, MAGNESIUM, AND TITANIUM

Zinc has moderate strength and toughness and excellent corrosion resistance. It is used in wrought forms such as rolled sheet and foil and drawn rod or wire. Dry-cell battery cans, builder's hardware, and plates for photo-engraving are some of the major applications.

Many zinc parts are made by die casting because the melting point is less then 800°F (427°C), much lower than other die-casting metals. The as-cast finish is suitable for many applications, such as business machine parts, pump bodies, motor housings, and frames for light-duty machines. Where a decorative appearance is required, electroplating with nickel and chromium is easily done. Such familiar parts as radio grilles, lamp housings, horn rings, and body moldings are made in this manner.

Table A-2 lists the properties of particular rolled and cast zinc alloys.

Magnesium is the lightest metal commonly used in load-carrying parts. Its density of only 0.066 lb/in.^{3} (1830 kg/m^3) is only about one-fourth that of steel and zinc, one-fifth that of copper, and two-thirds that of aluminum. It has moderate strength and lends itself well to applications where the final fabricated weight of the part or structure should be light. Ladders, hand trucks, conveyor parts, portable power tools, and lawn-mower housings use magnesium. In the automotive industry, body parts, blower wheels, pump bodies, and brackets are often made of magnesium. In aircraft, its lightness makes this metal attractive for floors, frames, fuselage skins, and wheels. The stiffness (modulus of elasticity) of magnesium is low, which is an advantage in parts where impact energy must be absorbed. Also, its lightness results in low weight designs when compared with other metals on an equivalent rigidity basis.

Titanium has very high strength, and its density is only about half that of steel. Although aluminum has a lower density, titanium is superior to both aluminum and most steels on a strength-to-weight basis. It retains a high percentage of its strength at elevated temperatures and can be used up to about 1000°F (538°C). Most applications of titanium are in the aerospace industry in engine parts, fuselage parts and skins, ducts, spacecraft structures, and pressure vessels. Because of its corrosion resistance and high temperature strength, the chemical industries use titanium in heat exchangers and as a lining for processing equipment. High cost is a major factor to be considered.

2-6   CAST IRON

The attractive properties of cast iron include low cost, good wear resistance, good machinability, and its ability to be cast into complex shapes. There are basically three varieties of cast iron having somewhat different properties: *gray iron, ductile iron*, and *malleable iron*.

Gray iron is used in automotive engine blocks, machinery bases, brake drums, and large gears. It is usually specified by giving a grade number corresponding to the minimum ultimate tensile strength. For example, Grade 20 gray cast iron has a minimum ultimate strength of 20 000 psi (138 MPa); Grade 60 has $s_u = 60$ 000 psi (414 MPa), etc. The usual grades available are from 20 to 60. Gray iron is somewhat brittle, so that yield strength is not usually reported as a property. An outstanding feature of gray iron is that its compressive strength is very high, about three to five times as high as the tensile strength. This should be taken into account in design, especially when a part is subjected to bending stresses, as will be discussed in Chapter 9.

Because of variations in the rate of cooling after the molten cast iron is poured into a mold, the actual strength of a particular section of a casting is dependent on the thickness of the section. Figure 2-8 illustrates this for Grade 40 gray iron. The range of in-place strength may range from as high as 52 000 psi (359 MPa) to as low as 27 000 psi (186 MPa).

Ductile iron differs from gray iron in that it does exhibit yielding, has a greater percent elongation, and has generally higher tensile strength. Grades of ductile iron are designated by a three-number system such as 80-55-6. The first number indicates the minimum ultimate tensile strength in ksi; the second is the yield strength in ksi; and the third is the percent elongation. Thus, Grade 80-55-6 has an ultimate strength of 80 000 psi, a yield strength of 55 000 psi, and a percent elongation of 6 percent. Uses for ductile iron include crankshafts and heavily loaded gears.

Malleable iron is used in automotive and truck parts, construction machinery, and electrical equipment. It does exhibit yielding, has tensile strengths comparable to ductile iron, and has ultimate compressive strengths which are somewhat higher than ductile iron. Generally a five-digit number is used to designate malleable iron grades. For example, Grade 40010 has a yield strength of 40 000 psi (276 MPa) and a percent elongation of 10 percent.

Table A-5 lists the mechanical properties of several grades of gray iron, ductile iron, and malleable iron.

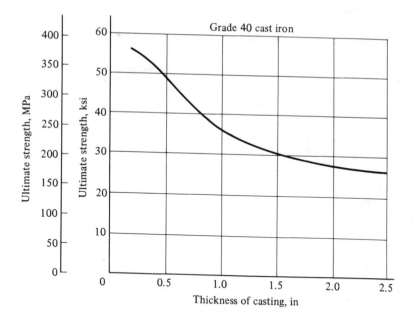

Figure 2-8

2-7 NONMETALS IN ENGINEERING DESIGN

Wood and concrete are widely used in construction. Plastics are found in virtually all fields of design, including consumer products, industrial equipment, automobiles, and architectural products. To the designer, the properties of strength and stiffness are of primary importance with nonmetals, as they are with metals. Because of the structural differences in the nonmetals, their behavior is quite different from the metals.

Wood, concrete, and many plastics have structures that are *anisotropic*. This means that the mechanical properties of the material are different depending on the direction of the loading. Also, because of natural chemical changes, the properties vary with time and often with climatic conditions. The designer must be aware of these factors.

2-8 WOOD

Since wood is a natural material, its structure is dependent on the way it grows and not on manipulation by man, as is the case in metals. The long, slender, cylindrical shape of trees results in an internal structure composed of longitudinal cells. As the tree grows, successive rings are added outside the older wood. Thus the inner core, called heartwood, has different properties than the sapwood, near the outer surface.

The species of the wood also affects its properties, as different kinds of trees produce harder or softer, stronger or weaker wood. Even in the same species variability occurs because of different growing conditions, such as differences in soil and amount of sun and rain.

The cellular structure of the wood gives it the grain which is so evident when sawn into boards and timber. The strength of the wood is dependent on whether it is loaded perpendicular to or parallel to the grain. Also, going across the grain, the strength is different in a radial direction than in a tangential direction with respect to the original cylindrical tree stem from which it was cut.

Another important variable affecting the strength of wood is moisture content. Changes in relative humidity can vary the amount of water absorbed by the cells of the wood.

Most construction lumber is stress-graded by standard rules adopted by the U.S. Forest Products Laboratory. Table A-6 lists allowable stresses for several species and grades of lumber. These allowable stresses account for variability due to natural imperfections.

2-9 CONCRETE

The components of concrete are cement and an aggregate. The addition of water and the thorough mixing of the components tend to produce a uniform structure with cement coating all the aggregate particles. After curing, the mass is securely bonded together. Some of the variables involved in determining the final strength of the concrete are the type of cement used, the type and size of aggregate, and the amount of water added.

A higher quantity of cement in concrete yields a higher strength. Decreasing the quantity of water relative to the amount of cement increases the strength of the concrete. Of course sufficient water must be added to cause the cement to coat the aggregate and to allow the concrete to be poured and worked before excessive curing takes place. The density of the concrete, affected by the aggregate, is also a factor. A mixture of sand, gravel, and broken stone is usually used for construction grade concrete.

Concrete is graded according to its compressive strength, which varies from about 2000 psi (14 MPa) to 7000 psi (48 MPa). The tensile strength of concrete is extremely low, and it is common practice to assume that it is zero. Of course, reinforcing concrete with steel bars allows its use in beams and wide slabs since the steel resists the tensile loads.

Concrete must be cured to develop its rated strength. It should be kept moist for at least 7 days, at which time it has about 75 percent of its rated compressive strength. Although its strength continues to increase for years, the strength after 28 days is often used to determine its rated strength.

2-10 PLASTICS

Plastics are being used in an increasing array of products. Although some applications are decorative, the use of plastics in sophisticated load-carrying parts accounts for a large part of the production. Plastics are composed of long chain-like molecules called polymers. They are synthetic organic materials which can be formulated and processed in literally thousands of ways.

One classification which can be made is between *thermoplastic* materials and *thermosetting* materials. Thermoplastics can be softened repeatedly by heating with no change in properties or chemical composition. Conversely, after initial curing of thermosetting plastics, they cannot be resoftened. A chemical change occurs during curing with heat and pressure.

Some examples of thermoplastics include ABS, acetals, acrylics, cellulose acetate, TFE fluorocarbons, nylon, polyethylene, polypropylene, polystyrene, and vinyls. Thermosetting plastics include phenolics, epoxies, polyesters, silicones, urethanes, alkyds, allylics, and aminos. *bakelite, melamines*

TABLE 2-3 Applications of Plastic Materials

Application	Desired Properties	Suitable Plastics
Housings, containers, ducts	High impact strength, stiffness, low cost, formability, environmental resistance, dimensional stability	ABS, polystyrene, polypropylene, polyethylene, cellulose acetate, acrylics
Low friction— bearings, slides	Low coefficient of friction; resistance to abrasion, heat, corrosion	TFE fluorocarbons, nylon, acetals
High strength components, gears, cams, rollers	High tensile and impact strength, stability at high temperatures, machinable	Nylon, phenolics, TFE-filled acetals
Chemical and thermal equipment	Chemical and thermal resistance, good strength, low moisture absorption	Fluorocarbons, polypropylene, polyethylene, epoxies, polyesters, phenolics
Electrostructural parts	Electrical resistance, heat resistance, high impact strength, dimensional stability, stiffness	Allylics, alkyds, aminos, epoxies, phenolics, polyesters, silicones
Light-transmission components	Good light transmission in transparent and translucent colors, formability, shatter resistance	Acrylics, polystyrene, cellulose acetate, vinyls

A particular plastic is often selected for a combination of properties such as light weight, flexibility, color, strength, chemical resistance, low friction, or transparency. Since the available products are so numerous, only a brief table of properties of plastics is included in the Appendix as Table A-7. Table 2-3 lists the primary plastic materials used for six different types of applications. An extensive comparative study of the design properties of plastics can be found in References 1 and 2.

REFERENCES

1. *Materials Selector, Materials Engineering Magazine*, Reinhold Publishing Company, Stamford, Conn., Vol. 80, No. 4.
2. *Materials Reference Issue, Machine Design Magazine*, Penton Publishing Company, Cleveland, Vol. 47, No. 6.
3. Baumeister, Theodore, ed., *Mark's Standard Handbook for Mechanical Engineers*, 7th ed., 1967.
4. *Aluminum Standards and Data*, 5th ed., The Aluminum Association, New York, 1976.

PROBLEMS

2-1   Name four kinds of metals commonly used for load-carrying members.
2-2   Name 11 factors which should be considered when selecting a material for a product.
2-3   Define ultimate tensile strength.
2-4   Define yield point.
2-5   Define yield strength.
2-6   When is yield strength used in place of yield point?
2-7   Define stiffness.
2-8   What material property is a measure of its stiffness?
2-9   State Hooke's law.
2-10   What material property is a measure of its ductility?
2-11   How is a material classified as to whether it is ductile or brittle?
2-12   Name five types of steels.
2-13   What does the designation AISI 4130 for a steel mean?
2-14   What are the ultimate strength, yield strength, and percent elongation of AISI 1050 hot-rolled steel? Is it a ductile or a brittle material?
2-15   Which has a greater ductility: AISI 1040 hot-rolled steel or AISI 1095 hot-rolled steel?
2-16   What does the designation AISI 1137 OQT 400 mean?
2-17   If the required yield strength of a steel is 150 ksi, could AISI 1141 be used? Why?
2-18   What is the modulus of elasticity for AISI 1137 steel?

2-19   A rectangular bar of steel is 1.0 in. by 4.0 in. by 14.5 in. How much does it weigh in pounds?

2-20   A circular bar is 50 mm in diameter and 250 mm long. How much does it weigh in newtons?

2-21   If a force of 400 N is applied to a bar of titanium and an identical bar of magnesium, which would stretch more?

2-22   Name four types of structural steels and list the yield point for each.

2-23   What does the aluminum alloy designation 6061-T6 mean?

2-24   List the ultimate strength, yield strength, modulus of elasticity, and density for an extruded form of 6061-O, 6061-T4, and 6061-T6 aluminum.

2-25   List five uses for bronze.

2-26   List three desirable characteristics of titanium as compared with aluminum or steel.

2-27   Name three varieties of cast iron.

2-28   Which type of cast iron is usually considered to be brittle?

2-29   What are the ultimate strengths in tension and in compression for ASTM A48 Grade 40 cast iron?

2-30   How does a ductile iron differ from gray iron?

2-31   List the allowable stresses in tension, compression, and shear for construction grade Douglas fir.

2-32   What is the normal range of compressive strengths for concrete?

2-33   Describe the difference between thermoplastic and thermosetting materials.

2-34   Name three suitable plastics for use as gears or cams in mechanical devices.

3

DIRECT TENSILE
AND
COMPRESSIVE STRESS

3-1 MEMBERS LOADED IN TENSION

Direct tension refers to that kind of loading in which the applied force pulls directly in line with the axis of the loaded member. An example of a member in tension is the tie rod for a C-frame type metal-forming press shown in Figure 3-1. The rod is fastened between the open ends of the press and then tightened so that there is a high pulling force in the rod. This holds the press frame and minimizes the tendency for the frame to open up as large forces are developed by the ram during the forming operation.

The direct tensile force (pulling force) in the rod tends to stretch it. The stress developed in the material from which the rod is made can be computed by application of the basic definition of stress as discussed in Chapter 1. Since the load is applied along the axis of the rod, all parts of the rod are subjected to some force. Also, at any cross section of the rod the force will be uniformly distributed over the entire area. Therefore, for a member loaded in tension along its axis,

$$\text{Stress} = \frac{\text{applied force}}{\text{area of cross section}}$$

In this book the symbol P will be used normally to indicate the applied force,

34

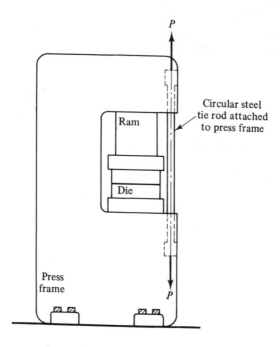

Figure 3-1

A will designate area, and s will designate tensile or compressive stress. Then, the general formula for stress due to a direct tensile load is

$$s = \frac{P}{A} \qquad (3\text{-}1)$$

In order for Equation (3-1) to be valid, the following conditions must be met.

1. The loaded member must be straight.
2. The loaded member must have a uniform cross section over the length under consideration.
3. The material from which the member is made must be homogeneous.
4. The load must be applied along the centroidal axis of the member so there is no tendency to bend it.

It is important to recognize that the concept of stress refers to the internal resistance provided by a *unit area*, that is, an infinitely small area. Stress is considered to act at a point and may, in general, vary from point to point in a particular body. Equation (3-1) indicates that, for a member subjected to direct axial tension or compression, the stress is uniform across the entire area if the four conditions are met. In many practical applications the minor variations which could occur in the local stress levels are accounted for by carefully selecting the allowable stress, as discussed later.

Example Problem 3-1

The tie rods for the press shown in Figure 3-1 are made of steel, and the maximum load on each rod during operation of the press is 40 000 lb. If the rods are circular and have a diameter of 2.00 in., compute the tensile stress in the rod.

Solution

Equation (3-1) is used to compute the stress. Note that all four conditions listed above are met for the rods.

$$s = \frac{P}{A}$$

$$P = 40\ 000\ \text{lb}$$

$$A = \frac{\pi D^2}{4} = \frac{\pi (2.0\ \text{in.})^2}{4} = 3.14\ \text{in.}^2$$

Then

$$s = \frac{40\ 000\ \text{lb}}{3.14\ \text{in.}^2} = 12\ 700\ \text{lb/in.}^2$$

Thus the tie rods of the press are each subjected to a tensile stress of 12 700 psi. Is that a safe stress? To answer this question, consideration must be given to what it means to have a safe stress. In this book, a member will be considered to be safe if the maximum calculated stress on it does not exceed a certain *design stress* for the material under the particular type of loading on the member. Design stress is described in Section 3-3.

Example Problem 3-2

A scientific experiment requires the use of a pendulum like that shown in Figure 3-2. The design criteria call for a very fine wire to support a spherical ball having a mass of 10.0 kg. Develop an equation which relates the stress in the wire to its diameter and the load due to the ball. Then compute the stress in the wire for two wire sizes: 0.50 mm and 1.00 mm.

Solution

Calling the load due to the ball P, the stress in the wire is

$$s = \frac{P}{A}$$

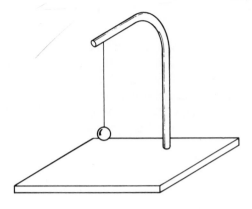

Figure 3-2

But

$$A = \frac{\pi D^2}{4}$$

Then

$$s = \frac{P}{\pi D^2/4} = \frac{4P}{\pi D^2} = \frac{1.27P}{D^2}$$

The force P is equal to the weight of the 10.0-kg ball.

$$P = m \cdot g \quad (g = 9.81 \text{ m/s}^2)$$
$$P = 10.0 \text{ kg} \cdot 9.81 \text{ m/s}^2 = 98.1 \text{ kg} \cdot \text{m/s}^2 = 98.1 \text{ N}$$

For $D = 0.50$ mm,

$$s = \frac{1.27P}{D^2} = \frac{(1.27)(98.1 \text{ N})}{(0.50 \text{ mm})^2} = 498 \text{ N/mm}^2$$

Since there are 10^3 mm in 1 m,

$$s = 498 \frac{\text{N}}{\text{mm}^2} \cdot \frac{(10^3 \text{ mm})^2}{\text{m}^2} = 498 \times 10^6 \text{N/m}^2 = 498 \text{ MPa}$$

Thus, a stress of 498 MPa would be developed in a wire 0.5 mm in diameter carrying a 10.0-kg ball.

Similarly, for the 1.00-mm wire,

$$s = \frac{1.27P}{D^2} = \frac{(1.27)(98.1 \text{ N})}{(1.00 \text{ mm})^2} \cdot \frac{(10^3 \text{ mm})^2}{\text{m}^2} = 125 \times 10^6 \text{N/m}^2$$
$$= 125 \text{ MPa}$$

Notice that doubling the wire diameter from 0.5 mm to 1.0 mm causes a decrease in the stress in the wire by a factor of 4. This could be significant when a material for the wire is selected.

3-2 MEMBERS LOADED IN COMPRESSION

Direct compression refers to that kind of loading in which the applied force pushes directly in line with the axis of the loaded member, tending to shorten it. An example of a member loaded in compression is shown in Figure 3-3. A large machine is positioned so that its total weight is equally divided among its four feet, which are resting on poured concrete foundation piers. Each pier must resist one-fourth of the load.

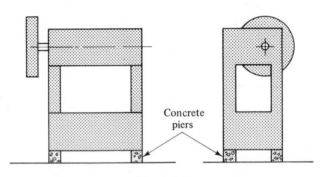

Concrete
piers

Figure 3-3

The stress in the concrete piers can be computed by application of the definition of stress used earlier. Assuming that the load acts directly along the axis of each pier, all sections of the pier are subjected to the same force. At any cross section of a pier, the force will be uniformly distributed over the entire area. Therefore, for a member loaded in compression along its axis,

$$\text{Stress} = \frac{\text{applied load}}{\text{area of cross section}}$$

or

$$s = \frac{P}{A}$$

This is the identical relationship used for tension loading. The conditions listed for tension loading also apply to compression with one important addition:

Under compression loading, the member must be short so there is no tendency to buckle.

Long compression members are called columns, and their design must consider the possibility of buckling as well as simple stress. Columns are discussed in Chapter 14, and in that chapter the method by which compression members are distinguished as being long or short is described.

38

Example Problem 3-3

For the concrete piers shown in Figure 3-3, determine what size each should be if the machine weighs 180 000 lb. Use a square cross section for the piers, and do not allow the stress in the concrete to exceed 2000 psi. Assume that the piers are short.

Solution

Each pier will carry one-fourth of the total load.

$$P = \frac{180\ 000\ \text{lb}}{4} = 45\ 000\ \text{lb}$$

The equation for computing stress is

$$s = \frac{P}{A}$$

Solving for the area A,

$$A = \frac{P}{s}$$

Using the maximum stress of 2000 psi,

$$A = \frac{P}{s} = \frac{45\ 000\ \text{lb}}{2000\ \text{lb/in.}^2} = 22.5\ \text{in.}^2$$

Let each side of the square pier have a length b. Then

$$b^2 = A$$

and

$$b = \sqrt{A} = \sqrt{22.5\ \text{in.}^2} = 4.74\ \text{in.}$$

If, for convenience, the piers were made an even 5.0 in. on each side, the stress would be

$$s = \frac{P}{A} = \frac{P}{b^2} = \frac{45\ 000\ \text{lb}}{(5.0\ \text{in.})^2} = 1800\ \text{lb/in.}^2$$

3-3   DESIGN STRESS

Design stress is that level of stress which may be developed in a material while ensuring that the loaded member is safe. In order to compute design stress, two factors must be specified: the *design factor N* and the *property of the material on which the design will be based*. Usually, for metals, the design stress is based on either the yield strength s_y or the ultimate strength s_u of the material.

The *design factor N* is a number by which the reported strength of a material is divided to obtain the *design stress* s_d. The following equations can be used to compute the design stress for a certain value of N.

$$s_d = \frac{s_y}{N} \text{ based on yield strength} \qquad (3\text{-}2)$$

or

$$s_d = \frac{s_u}{N} \text{ based on ultimate strength} \qquad (3\text{-}3)$$

The value of the design factor is normally determined by the designer, using judgment and experience. In some cases, codes, standards, or company policy may specify design factors or design stresses to be used. When the designer must determine the design factor, his or her judgment must be based on an understanding of how parts may fail and the factors which affect the design factor. Sections 3-4 and 3-5 give additional information about the design factor and about the choice of methods for computing design stresses.

Other references may use the term *factor of safety* in place of *design factor*. Also, *allowable stress* or *working stress* may be used in place of *design stress*. The choice of terms for use in this book is made to emphasize the role of the designer in specifying the design stress.

3-4 DESIGN FACTOR

Many different aspects of the design problem are involved in the specification of the design factor. In some cases, the precise conditions of service are not known. The designer must then make conservative estimates of the conditions, that is, estimates which would cause the resulting design to be on the safe side when all possible variations are considered. The final choice of a design factor depends on the following ten conditions.

Material Strength Basis. Most designs using metals are based on either yield strength or ultimate strength or both, as stated previously. This is because most theories of metal failure show a strong relationship between the stress at failure and these material properties. Also, these properties will almost always be reported for materials used in engineering design. The value of the design factor will be different depending on which material strength is used as the basis for design, as will be shown later.

Type of Material. A primary consideration with regard to the type of material is its ductility. The failure modes for brittle materials are quite different than those for ductile materials. Since brittle materials such as cast iron do not exhibit yielding, designs are always based on ultimate strength. Generally a metal is considered to be brittle if its percent elongation in a 2-in. gauge length is less than 5 percent. Except for highly hardened alloys, virtually all steels are ductile. Except for castings, aluminum is ductile. Other material factors which can affect the strength of a part are its uniformity and the confidence in the stated properties.

Manner of Loading. Three main types of loading can be identified. A *dead load* is one which is applied to a part slowly and gradually and which remains applied, or at least is applied and removed only infrequently during the design life of the part. *Repeated loads* are those which are applied and removed several thousand times during the design life of the part. Under repeated loading a part fails by the mechanism of fatigue at a stress level much lower than that which would cause failure under a dead load. This calls for the use of a higher design factor for repeated loads than for dead loads. Parts subject to *impact or shock* require the use of a large design factor for two reasons. First, a suddenly applied load causes stresses in the part which are several times higher than those which would be computed by standard formulas. Second, under impact loading the material in the part is usually required to absorb energy from the impacting body. The certainty with which the designer knows the magnitude of the expected loads also must be considered when specifying the design factor.

Possible Misuse of the Part. In most cases the designer has no control over actual conditions of use of the product he or she designs. Legally, it is the responsibility of the designer to consider any reasonably foreseeable use or *misuse* of the product and to insure the safety of the product. The possibility of an accidental overload on any part of a product must be considered.

Complexity of Stress Analysis. As the manner of loading or the geometry of a structure or a part becomes more complex, the designer is less able to perform a precise analysis of the stress condition. Thus the confidence in the results of stress analysis computations has an effect on the choice of a design factor.

Environment. Materials behave differently in different environmental conditions. Consideration should be given to the effects of temperature, humidity, radiation, weather, sunlight, and corrosive atmospheres on the material during the design life of the part.

Size Effect. Most metals exhibit different strengths as the cross-sectional area of a part varies. Most material property data were obtained using standard specimens about $\frac{1}{2}$ in. in diameter. Parts with larger sections usually have lower strengths. Parts of smaller size, for example drawn wire, have significantly higher strengths. An example of the size effect is shown in the following table for 1045 steel heat-treated by quenching in water and tempered at 1000°F.

Specimen Size, in.	Yield Strength s_y	
	psi	MPa
$\frac{1}{2}$	110 000	760
2	70 000	480
4	59 000	410

Quality Control. The more careful and comprehensive a quality control program is, the better a designer knows how the product will actually appear in service. With poor quality control, a larger design factor should be used.

Hazard Presented by a Failure. The designer must consider the consequences of a failure to a particular part. Would a catastrophic collapse occur? Would people be placed in danger? Would other equipment be damaged? Such considerations may justify the use of a higher than normal design factor.

Cost. Compromises must usually be made in design in the interest of limiting cost to a reasonable value under market conditions. Of course, where danger to life or property exists, compromises should not be made which would seriously affect the ultimate safety of the product or structure.

Experience in design and knowledge about the above conditions must be applied to determine a design factor. Table 3-1 includes guidelines which will be used in this book for selecting design factors. These should be considered to be average values. Special conditions or uncertainty about conditions may justify the use of other values.

TABLE 3-1

Manner of Loading	Design Factor N		
	Ductile Metals		Brittle Metals
	Yield Strength Basis	Ultimate Strength Basis	Ultimate Strength Basis
Dead load	2	4	6
Repeated load	4	8	10
Impact or shock	6	12	15

The design factor is used to determine design stress as shown in Equations (3-2) and (3-3).

If the stress in a part is already known and one wishes to choose a suitable material for a particular application, then the computed stress is considered to be the design stress. The required yield or ultimate strength is then found from

$$s_y = N \cdot s_d \qquad (N \text{ based on yield strength})$$

or

$$s_u = N \cdot s_d \qquad (N \text{ based on ultimate strength})$$

3-5 METHODS OF COMPUTING DESIGN STRESS

As mentioned in Section 3-3, an important factor to be considered when computing the design stress is the manner in which a part may fail when subjected to loads. This section will discuss failure modes relevant to parts subjected to tensile and compressive loads. Other kinds of loading are discussed in later chapters.

The failure modes and the consequent methods of computing design stresses can be classified according to the type of material and the manner of loading. Ductile materials, having more than 5 percent elongation, exhibit somewhat different modes of failure than do brittle materials. Dead loads, repeated loads, and shock loads produce different modes of failure.

Ductile Materials under Dead Loads. Ductile materials will undergo large plastic deformations when the stress reaches the yield strength of the material. Under most conditions of use, this would render the part unfit for its intended use. Therefore, for ductile materials subjected to dead loads, the design stress is usually based on yield strength. That is,

$$s_d = \frac{s_y}{N}$$

As indicated in Table 3-1, a design factor of $N = 2$ would be a reasonable choice under average conditions.

Ductile Materials under Repeated Loads. Under repeated loads, ductile materials fail by a mechanism called *fatigue*. The level of stress at which fatigue occurs is lower than the yield strength. By testing materials under repeated loads, the stress at which failure will occur can be measured. The terms *fatigue strength* or *endurance strength* are used to denote this stress level. However, fatigue-strength values are often not available. Also, factors such as surface finish, the exact pattern of loading, and the size of a part have a marked effect on the actual fatigue strength. To overcome these difficulties, it is often convenient to use a high value for the design factor when computing the design stress for a part subjected to repeated loads. It is also recommended that the ultimate strength be used as the basis for the design stress because tests show that there is a good correlation between fatigue strength and the ultimate strength. Therefore, for ductile materials subjected to repeated loads, the design stress can be computed from

$$s_d = \frac{s_u}{N}$$

A design factor of $N = 8$ would be reasonable under average conditions. Also, stress concentrations, which are discussed in Section 3-6, must be accounted for since fatigue failures often originate at points of stress concentrations.

Ductile Materials under Impact or Shock Loading. The failure modes for parts subjected to impact or shock loading are quite complex. They depend on the ability of the material to absorb energy and on the flexibility of the part. Because of the general inability of designers to perform precise analyses of stresses under shock loading, large design factors are recommended. In this book we will use

$$s_d = \frac{s_u}{N}$$

with $N = 12$ for ductile materials subjected to impact or shock loads.

Brittle Materials. Since brittle materials do not exhibit yielding, it is recommended that the design stress be based on ultimate strength. That is,

$$s_d = \frac{s_u}{N}$$

with $N = 6$ for dead loads, $N = 10$ for repeated loads, and $N = 15$ for impact or shock loads.

Structural Steel. According to the AISC specifications, the allowable stress in tension for structural steel is computed from

$$s_d = 0.60s_y$$

or

$$s_d = 0.50s_u$$

whichever is lower. In bending (discussed in Chapter 9),

$$s_d = 0.66s_y \qquad \text{(for compact beam cross sections)}$$

The allowable shear stress in beam webs (Chapter 10) is

$$s_{sd} = 0.40s_y$$

Special cases of columns and connections are discussed in later chapters.

Example Problem 3-4

As an illustration of the use of a design factor to determine the design stress, consider the application described in Example Problem 3-2. In that problem it was shown that for two wire sizes the following stresses would be developed in the pendulum wire carrying a 10.0-kg ball.

Wire Diameter, mm	Stress in Wire, MPa
0.50	498
1.00	125

Compute the required strength of the material for each wire size and specify suitable materials for the wires.

Solution

For this application it would be desirable to use a fine wire made of a ductile material. The load on the wire can be considered a dead load since the pendulum will swing very slowly. In the laboratory situation it is very unlikely that any impact or shock loading will be applied and there would be no adverse environmental conditions. Quality control should be good since only one pendulum is to be made. No great danger is presented to life or property. Under these conditions, the use of the design factor listed in Table 3-1 for a dead load on ductile metals should be satisfactory. Then we will use $N = 2$ based on yield strength. For the pendulum wire having a diameter of 0.50 mm, the material of the wire must have the following property:

Yield strength:

$$s_y = N \cdot s_d = 2 \cdot 498 \text{ MPa} = 996 \text{ MPa}$$

Similarly, for the 1.00-mm diameter wire

Yield strength:

$$s_y = N \cdot s_d = 2 \cdot 125 \text{ MPa} = 250 \text{ MPa}$$

Referring to Table A-1, we can see that many steel materials have the required strength for the 1.00-mm diameter wire. For example, AISI 1040, cold-drawn, has $s_y = 490$ MPa. Also, a standard steel wire can be chosen, such as one of those listed in the graph in Figure 2-6. Either the ASTM A228 or ASTM A229 would meet the stress requirements for both the 0.50-mm and the 1.00-mm wire.

Another possible material would be drawn aluminum, alloy 7075-T6, for the 1.00-mm wire. From Table A-4, the alloy has a yield strength of 455 MPa.

Example Problem 3-5

A designer working on the C-frame press shown in Figure 3-1 is considering using AISI 4150 steel, quenched in oil and tempered at 1300°F for the tie rods. Evaluate this material to determine if it has adequate strength under the conditions described in Example Problem 3-1. Use a rod diameter of 2.00 in.

Solution

It was computed in Example Problem 3-1 that under a 40 000-lb load, the 2.00-in. diameter rod would be subjected to a stress of 12 700 psi. Therefore, in evaluating the suitability of a material, this will be considered to be the design stress.

A suitable design factor must now be specified. During the normal operation of a metal-forming press, a mild shock loading occurs as the ram strikes the workpiece. The load is repeated for each cycle of the press, and there would be several thousand cycles of operation during the expected life of a press. Referring to Table 3-1, we can conclude that the design factor should be at least 12 based on ultimate strength. From Table A-1 it can be found that AISI 4150 OQT 1300 has an ultimate strength $s_u = 128\ 000$ psi. Since the design stress and the material strength are known, the resulting design factor can be computed.

Based on ultimate strength,

$$s_u = N \cdot s_d$$

or

$$N = \frac{s_u}{s_d} = \frac{128\ 000\ \text{psi}}{12\ 700\ \text{psi}} = 10.1$$

Therefore, it appears that the proposed material does not have sufficient ultimate strength for this application. A logical next step would be to consider the same alloy, AISI 4150, but with a lower tempering temperature, which would improve strength properties, as discussed in Chapter 2. Tempering at 1100°F would increase the ultimate strength to approximately 160 000 psi. The resulting design factor would be

$$N = \frac{s_u}{s_d} = \frac{160\ 000\ \text{psi}}{12\ 700\ \text{psi}} = 12.6 \qquad \text{(based on ultimate strength)}$$

This is an acceptable value for the design factor.

3-6 STRESS CONCENTRATIONS

In defining the method for computing stress due to a direct tensile or compressive load on a member, it was emphasized that the member must have a uniform cross section in order for the equation $s = P/A$ to be valid. The reason for this restriction is that wherever a change in the geometry of a loaded member occurs, the actual stress developed is higher than would be predicted by the standard equation. This phenomenon is called *stress concentration* because detailed studies reveal that localized high stresses appear to concentrate around sections where geometry changes occur.

An example of a member where a stress concentration occurs is shown in Figure 3-4. A circular bar is to be used as the cylinder rod in an automated packaging system. As the rod pulls down under the influence of the hydraulic cylinder, two pads attached to the rod compress a bulk granular chemical into containers. A maximum force of 2700 lb is exerted on the rod. The end of the rod has two circumferential grooves cut into it in order to use rings to secure the pads in position on the rod. Therefore, the rod is subjected to a tensile stress. A larger drawing of the end of the rod and a detailed drawing of the grooves are shown in parts (b) and (c) of the figure.

The stress in the rod at the location of the grooves is much higher than

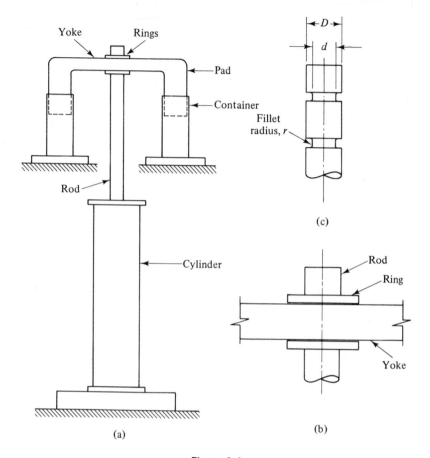

Figure 3-4

that which would be calculated by the axial stress formula, even when the smallest area at the bottom of the groove is used. In order to account for the higher stress, a modified stress formula is used whenever a change in geometry occurs. This formula is

$$s = \frac{K_t P}{A} \tag{3-4}$$

where K_t is the *stress concentration factor*. The value of K_t represents the factor by which the actual stress is higher than the nominal stress computed by the standard formulas. In the present example it is related to direct tensile stress. However, stress concentration factors are used for other kinds of stresses, such as bending and torsion, which will be discussed in later chapters.

The actual geometry has a large effect on the magnitude of K_t. In the case of the grooved rod, the ratio of the major rod diameter to the smaller diameter at the root of the groove is important. Also, the sharpness of the radius at the bottom of the groove affects K_t. Data for stress concentration factors are usually presented in graphical form such as Figure A-1. Thus, for any geometry, the value of K_t can be found.

It is most important to determine the basis on which the nominal stress is to be calculated. Normally the smaller dimensions at a section where a change in geometry occurs are used in calculating the nominal stress. For example, the area at the root of the grooves in the rod in Figure 3-4 would be used. However, this is not universally true. In all cases where the method of computing the nominal stress is not stated, it should be assumed that the smaller section is used, in order to be on the conservative side.

Example Problem 3-6

The rod shown in Figure 3-4 is subjected to a tensile force of 2700 lb. The detailed dimensions of the rod and grooves are:

$$D = 0.50 \text{ in.}$$
$$d = 0.44 \text{ in.}$$
$$r = 0.04 \text{ in.}$$

Compute the maximum tensile stress in the rod.

Solution

Equation (3-4) will be used with $P = 2700$ lb. The area at the root of the grooves is

$$A = \frac{\pi d^2}{4} = \frac{\pi (0.44 \text{ in.})^2}{4} = 0.152 \text{ in.}^2$$

In order to determine K_t, the ratios of D/d and r/d are needed.

$$\frac{D}{d} = \frac{0.50}{0.44} = 1.136$$

$$\frac{r}{d} = \frac{0.04}{0.44} = 0.09$$

Using Figure A-1, $K_t = 2.25$. Then,

$$s = \frac{K_t P}{A}$$

$$s = \frac{(2.25)(2700 \text{ lb})}{0.152 \text{ in.}^2} = 39\ 950 \text{ lb/in.}^2$$

Notice that neglecting the stress concentration and using $s = P/A$ would have produced a calculated stress of only 17 800 lb/in.². Stress concentrations can be very critical in design.

Other situations where stress concentrations occur are described in the Appendix, along with the graphs for determining the value of K_t. It is expected that all problem solutions will consider the effect of stress concentrations where they exist.

3-7 DESIGN EXAMPLE: DIRECT TENSION AND COMPRESSION

In a manufacturing plant, an overhead conveyor system is being designed to transport heavy castings through a paint spray booth. Figure 3-5 shows a drawing of the casting, made of ductile cast iron.

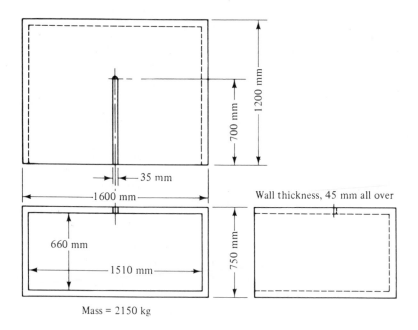

Figure 3-5

The objective of this project is to design the carrier which connects the traveling fixture of the conveyor to the casting. At the pickup point, the top of the casting will be 1.80 m below the conveyor fixture. The carrier must be capable of supporting the weight of the casting safely. The basic plan is to utilize the slot in the top surface of the casting for attachment. The completed design must specify the entire geometry of the carrier and the material from which the parts are to be made.

Possible Solution

Step 1.   Define the design requirements.

1. An attachment means must be designed which will engage the slot in the casting easily and be capable of supporting the casting in a safe and stable manner.
2. An attachment means must be designed which will secure the carrier to the conveyor fixture.
3. The overall length of the carrier must be such that the slot in the casting is engaged at a point 1.80 m below the conveyor.
4. The load on the carrier is the weight of the casting. To determine the weight, the volume of the cast iron in the casting will first be calculated. The dimensions are taken from Figure 3-5, converting them to metres for convenience by using the factor 1.0 m = 1000 mm.

$$\text{Bottom:} \quad V_1 = 1.60 \text{ m} \cdot 1.20 \text{ m} \cdot 0.045 \text{ m}$$
$$= 0.086 \text{ m}^3$$
$$\text{Top:} \quad V_2 = V_1 - \text{volume of slot}$$
$$V_2 = 0.086 \text{ m}^3 - 0.70 \text{ m} \cdot 0.035 \text{ m} \cdot 0.045 \text{ m}$$
$$= 0.085 \text{ m}^3$$
$$\text{2 sides:} \quad V_3 = 2 \cdot 0.66 \text{ m} \cdot 1.20 \text{ m} \cdot 0.045 \text{ m}$$
$$= 0.071 \text{ m}^3$$
$$\text{Back:} \quad V_4 = 1.51 \text{ m} \cdot 0.66 \text{ m} \cdot 0.045 \text{ m}$$
$$= 0.045 \text{ m}^3$$
$$\text{Total volume} = V_1 + V_2 + V_3 + V_4 = 0.287 \text{ m}^3$$

Then the mass of the casting is equal to its volume multiplied by the density of the cast iron. In Table A-5 it can be found that the density varies from 6920 kg/m³ to 7480 kg/m³. Using the larger number in order to be conservative, we have

$$\text{Mass} = 7480 \text{ kg/m}^3 \cdot 0.287 \text{ m}^3 = 2147 \text{ kg}$$

To find the weight, the mass must be multiplied by *g*.

$$\text{Weight} = \text{mass} \cdot g = 2147 \text{ kg} \cdot 9.81 \text{ m/s}^2 = 21\ 062 \text{ kg} \cdot \text{m/s}^2$$

Calling the weight *P*, we will use $P = 21\ 100$ N.

5. The temperature in which the carrier will be used is the normal variation for a plant in a temperate zone, say from 10°C to 40°C. This range should not affect the properties of the materials significantly.
6. The environment will include normal conditions of humidity, with no excessively corrosive atmosphere.

7. The loading will normally be steady, although some mild shock may occur during engagement.
8. The cost of downtime and damaged castings which would result from a failed carrier makes a high design factor justified. Assuming a ductile material will be used, a design factor of 4 based on yield strength is specified.
9. The design should provide for some adjustment of the position of the attachment to the casting. Connections must provide for easy installation of the carriers.
10. It is estimated that 120 carriers will be made. Therefore, standard hardware and fabrication methods should be used.

Step 2.   Proposed design.

Note that only one design is proposed here. It is usually advisable to conceptualize a number of possible designs from which one optimum design can be selected.

Figure 3-6 shows a sketch of the proposed design. It is composed of a

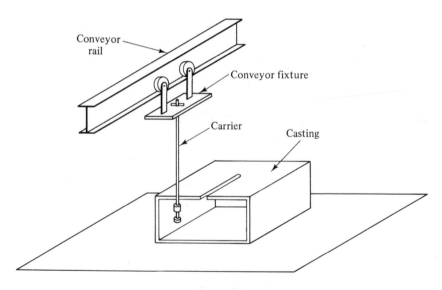

Figure 3-6

machined steel, lower attachment device for engaging the slot in the casting. The long part is made of bar steel, threaded on the lower end. The upper end passes through a hole in the conveyor fixture and is secured by pressing a round pin through a hole drilled in the bar.

Step 3.   Detailed design and strength analysis.

The following items must be specified to complete the design.

1. Geometry of the lower attachment device and its material.
2. Diameter and thread specifications for the long bar.
3. Diameter of the pin at the upper end of the bar.
4. Material for the long bar.

Lower Attachment Device. A proposed geometry is shown in Figure 3-7. The 25-mm diameter was selected to facilitate engagement with the 35-mm

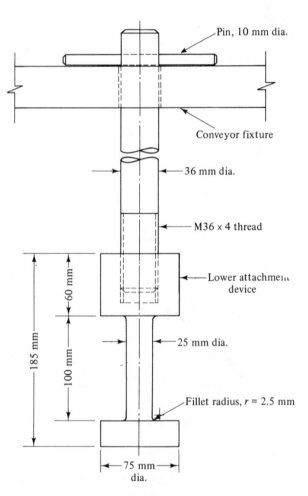

Figure 3-7

slot in the casting. The lower 75-mm diameter should provide a stable support surface on which the casting can rest. The upper 75-mm diameter allows the drilling and tapping of the threaded hole for fastening to the long bar. The entire piece will be machined from a steel bar, 75 mm in diameter and 185 mm long.

A strength analysis is required to determine what material should be used for the lower attachment device. It appears obvious that the 25-mm diameter section is the most highly stressed part under the direct tensile load. Furthermore, the sections where the diameters change are more critical because of the stress concentrations presented by the change in geometry.

In order to decrease the effect of the geometry change, rounded fillets should be used as shown in Figure 3-7. The graph of stress concentration factors for a stepped round tension bar with a fillet is found in the Appendix, Figure A-2. It can be seen that the value of K_t depends on the ratio of the larger diameter to the smaller diameter (D/d) and on the ratio of the fillet radius to the smaller diameter (r/d). For our design,

$$\frac{D}{d} = \frac{75 \text{ mm}}{25 \text{ mm}} = 3.0$$

A fillet radius must be specified. In this application, a small radius would be desirable so that the fillet would not interfere with the slot in the casting. However, it is recommended that the r/d ratio be no smaller than about 0.05, if possible, in order to keep K_t to a reasonable value. Let's specify $r/d = 0.10$. Then from the graph of K_t versus r/d at a value of $D/d = 3.0$, $K_t = 2.1$. Then the fillet radius will be

$$r = 0.10 \cdot d = 0.10 \cdot 25 \text{ mm} = 2.5 \text{ mm}$$

The stress in the vicinity of the fillet can now be calculated from Equation (3-4).

$$s = K_t \cdot \frac{P}{A}$$

$$A = \frac{\pi d^2}{4} = \frac{\pi (25 \text{ mm})^2}{4} = 491 \text{ mm}^2$$

$$s = \frac{(2.1)(21\ 100 \text{ N})}{491 \text{ mm}^2} = 90.3 \text{ N/mm}^2 = 90.3 \text{ MPa}$$

A material must now be specified. Using the design factor specified previously, we can calculate the required yield strength. The computed stress of 90.3 MPa will be used as the design stress s_d.

Yield strength:

$$s_y = N \cdot s_d = 4 \cdot 90.3 \text{ MPa} = 361.2 \text{ MPa}$$

Referring to Table A-1, we find that AISI 1137 steel has sufficient strength.

The part should be heat-treated by quenching in oil and tempered at a temperature of 1300°F.

This completes the design of the lower attachment device, except for the internal threads in the body. The strength analysis of the long bar must precede the specification of the threads.

Long Bar. The maximum stress in the long, round bar will occur at either the threaded end or at the section near the top where the pin is inserted through the bar. Notice that at both of these sections, the cross-sectional area is smaller than in the rest of the bar. Also, there are stress concentrations due to the change in the geometry of the bar. In order to accommodate the reduced diameter at the lower end where the threads are to be cut, a basic bar diameter greater than the 25 mm used for the attachment device should be chosen. A standard metric screw thread is available with a nominal outside diameter of 36 mm. Let's make a trial design using a 36-mm diameter bar.

The screw thread has the designation M36 with the standard coarse thread having a pitch of 4 mm. The maximum stress in the threaded portion of the bar will occur at the root of the threads. The tensile load is 21 100 N, as used previously. The cross-sectional area at the root, computed using the minor diameter of the thread (found in the Appendix, Table A-10), is

$$A = \frac{\pi D_r^2}{4} = \frac{\pi (30.521 \text{ mm})^2}{4} = 731.6 \text{ mm}^2$$

Using a stress concentration factor of 3.0 for the thread root, we find the stress in the threaded bar to be

$$s = \frac{K_t \cdot P}{A} = \frac{(3.0)(21\ 100 \text{ N})}{731.6 \text{ mm}^2} = 86.5 \text{ N/mm}^2 = 86.5 \text{ MPa}$$

Looking now at the upper end of the rod, we note that the highest stress will occur at the section where the pin is inserted. Assume that the pin will be 10 mm in diameter.* Figure 3-8 shows the cross section through the rod at the pin hole. The net area is

$$A = \frac{\pi (36 \text{ mm})^2}{4} - (36 \text{ mm})(10 \text{ mm}) = 658 \text{ mm}^2$$

This calculation assumes that the area removed by the pin hole is rectangular, which is very nearly true.

The stress concentration factor for this case can be found from Figure A-3 to be a function of the ratio of the pin diameter a to the outside diameter of the bar d.

$$\frac{a}{d} = \frac{10 \text{ mm}}{36 \text{ mm}} = 0.28$$

---

*Of course, the stress in the pin would also have to be checked. The pin is subjected to shear stress, which is covered in Chapter 5.

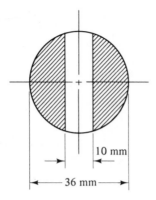

Figure 3-8

Then $K_t = 2.65$. Finally, the stress is

$$s = \frac{K_t \cdot P}{A} = \frac{(2.65)(21\ 100\ N)}{658\ mm^2} = 85.0\ N/mm^2 = 85.0\ MPa$$

Therefore, the most highly stressed part of the long bar is at its lower end, and the stress of 86.5 MPa is used to select a suitable material. In this case, since the stress in the long bar is so close to the 90.3 MPa stress calculated for the lower attachment device, it would be desirable to specify the same material, AISI 1137 OQT 1300, for the bar also. It should be noted that 1137 is a free-machining alloy which should facilitate the fabrication of both parts.

In summary, a carrier has been designed incorporating a lower attachment device and a long bar as shown in Figure 3-7. All parts will be fabricated from AISI 1137 OQT 1300, which will provide a design factor of a least 4 based on yield strength. Thus the carrier will be safe.

PROBLEMS

3-1   Compute the stress in a round bar subjected to a direct tensile force of 3200 N if the diameter of the bar is 10 mm.

3-2   Compute the stress in a rectangular bar having cross-sectional dimensions of 10 mm by 30 mm if a direct tensile force of 20 kN is applied.

3-3   A link in a mechanism for an automated packaging machine is subjected to a tensile force of 860 lb. If the link is square, 0.40 in. on a side, compute the stress in the link.

3-4   A circular rod, $\frac{3}{8}$ in. in diameter supports a heater assembly weighing 1850 lb. Compute the stress in the rod.

3-5   A tension member in a wood truss is subjected to 5200 lb of force. Would a Douglas fir member, $1\frac{1}{2}$ in. by $3\frac{1}{2}$ in., construction grade, be satisfactory for use in the truss? (See Table A-6.)

3-6 A guy wire for an antenna tower is to be aluminum, having an allowable stress of 12 000 psi. If the expected maximum load on the wire is 6400 lb, determine the required diameter of the wire.

3-7 A hopper having a mass of 1150 kg is designed to hold a load of bulk salt having a mass of 6350 kg. The hopper is to be suspended by four rectangular straps, each carrying one-fourth of the load. Steel plate with a thickness of 8.0 mm is to be used to make the straps. What should be the width in order to limit the stress to 70 MPa?

3-8 A shelf is being designed to hold crates having a total mass of 1840 kg. Two support rods like that shown in Figure 3-9 will hold the shelf. Assume that the center of gravity of the crates is at the middle of the shelf. Specify the required diameter of the circular rods to limit the stress to 110 MPa.

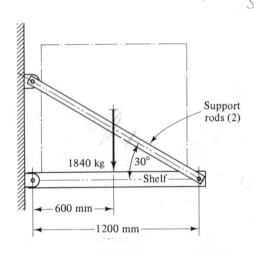

Figure 3-9

3-9 A concrete column base is circular, 8 in. in diameter, and carries a direct compressive load of 70 000 lb. Compute the compressive stress in the concrete.

3-10 Three short, square, wood blocks, $3\frac{1}{2}$ in. on a side, support a machine weighing 29 500 lb. Compute the compressive stress in the blocks.

3-11 A short link in a mechanism carries an axial compressive load of 3500 N. If it has a square cross section, 8.0 mm on a side, compute the stress in the link.

3-12 A pier to support a column is to be made of concrete, with an allowable stress of 27 MPa. If the expected load on the circular pier is 1.50 MN, compute the required diameter of the pier.

3-13 An aluminum ring has an outside diameter of 12.0 mm and an inside diameter of 10 mm. If the ring is short and is made of 2014-T6, compute the force required to produce ultimate compressive failure in the ring. Assume that s_u is the same in both tension and compression.

3-14 A wood block is a cube having dimensions 40 mm on a side. Compute the allowable compressive force which could be applied to the block if it is western hemlock, construction grade. Consider the load to be applied either parallel to or perpendicular to the grain.

3-15 For the round bar in Problem 3-1, specify a suitable aluminum alloy if the load is a repeated tensile load.

3-16 Specify a suitable steel alloy for the rectangular bar in Problem 3-2 if the load is expected to be applied with heavy shock.

3-17 In Problem 3-7, a steel plate with an allowable stress of 70 MPa was used to make some straps. If the straps are to be subjected to a repeated tensile load, specify a suitable steel alloy.

3-18 For the shelf support rods described in Problem 3-8, specify a suitable steel alloy if a design factor of 3 based on yield strength is desired.

3-19 A rectangular bar, 0.50 in. thick by 1.75 in. wide, is used as a tension link in a truss. If the bar is made of A36 structural steel, what load can be applied to the link?

3-20 A round bar of A242 steel is to be used as a tension member to stiffen a structural steel frame. If a maximum load of 4000 lb is expected, what diameter is required?

3-21 A portion of a casting made of ASTM A48, Grade 20 gray cast iron has the shape shown in Figure 3-10 and is subjected to compressive force in line with the centroidal axis of the section. If the member is short and carries a load of 52 000 lb, compute the stress in the section and the design factor.

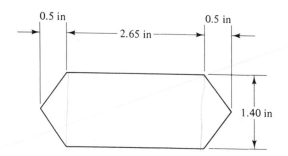

Figure 3-10

3-22 A tension member in a truss is a steel angle with legs of 2 in. and a thickness of $\frac{1}{4}$ in. Its cross-sectional area is 0.938 in². (See Table A-15.) If the angle is A36 structural steel, how much tensile load could be applied according to AISC specifications?

3-23 A valve stem in an automotive engine is subjected to an axial tensile load of 900 N due to the valve spring, as shown in Figure 3-11. Compute the maximum stress in the stem at the place where the spring force acts against the shoulder.

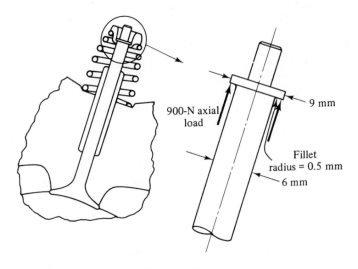

Figure 3-11

3-24 A round shaft has two grooves in which rings are placed to retain a gear in position, as shown in Figure 3-12. If the shaft is subjected to an axial tensile force of 36 kN, compute the maximum tensile stress in the shaft.

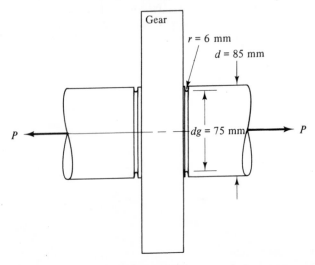

Figure 3-12

3-25 *Design problem.* Figure 3-13 shows a linkage used to activate a block brake by hand. The operator pulls on lever 1 with a force F_1. Tension link 2 pulls on lever 3. Lever 3 then applies the braking force F_3 to the rotating disk. Link 2 is to be designed connecting points B and C. Assume that the maximum force F_1 applied to the handle will be 880 N. Specify the configuration of link 2, its cross-sectional dimensions, its material, and the manner of con-

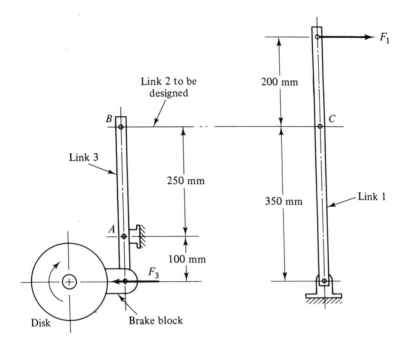

Figure 3-13

nection to links 1 and 3. Design all parts subjected to tensile or compressive loads and identify the parts which cannot yet be designed because a particular kind of stress analysis has not been covered.

3-26   *Design problem.* A portion of a steel roof truss is shown in Figure 3-14, along

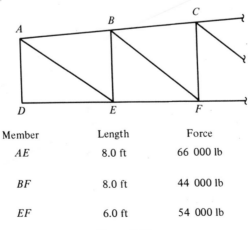

Member	Length	Force
AE	8.0 ft	66 000 lb
BF	8.0 ft	44 000 lb
EF	6.0 ft	54 000 lb

Figure 3-14

with the expected loads in three of its tension members. Select a configuration, a material, and the dimensions of each member. Discuss how the member could be attached to the other members of the truss. *Note:* If it is desired to use standard structural shapes, see the Appendix for dimensions and areas. Consider the loads to be dead loads.

4

DEFORMATION
AND
THERMAL STRESSES

4-1 ELASTIC DEFORMATION IN TENSION
AND COMPRESSION MEMBERS

Deformation refers to some change in the dimensions of a load-carrying member. Being able to compute the magnitude of deformation is important in the design of precision mechanisms, machine tools, building structures, and machine structures.

An example of where deformation is important is shown in Figure 4-1. The C-frame press shown is the same as that in Chapter 3. The tie rods are subjected to tension when in operation. Since they contribute to the rigidity of the press, the amount that they deform under load is something the designer needs to be able to determine.

In order to develop the relationship from which deformation can be computed for members subjected to axial tension or compression, some concepts from Chapter 2 must be reviewed. *Strain* is defined as the ratio of the total deformation to the original length of a member. Using the symbols ϵ for strain, δ for total deformation, and L for length, the formula for strain becomes

$$\epsilon = \frac{\delta}{L} \tag{4-1}$$

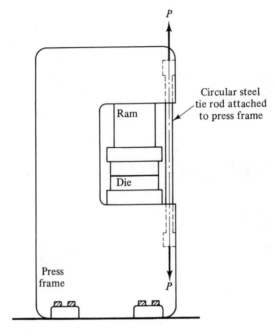

Figure 4-1

The stiffness of a material is a function of its modulus of elasticity E, defined as

$$E = \frac{\text{stress}}{\text{strain}} = \frac{s}{\epsilon} \qquad (4\text{-}2)$$

Solving for strain gives

$$\epsilon = \frac{s}{E} \qquad (4\text{-}3)$$

Now Equations (4-1) and (4-3) can be equated to each other.

$$\frac{\delta}{L} = \frac{s}{E} \qquad (4\text{-}4)$$

Solving for deformation gives

$$\delta = \frac{sL}{E} \qquad (4\text{-}5)$$

Since this formula applies to members which are subjected to either direct tensile or compressive forces, the direct-stress formula can be used to compute the stress s. That is, $s = P/A$, where P is the applied load and A is the cross-sectional area of the member. Substituting this into Equation (4-5) gives

$$\delta = \frac{sL}{E} = \frac{PL}{AE} \qquad (4\text{-}6)$$

Equation (4-6) can be used to compute the total deformation of any load-carrying member, provided that it meets the conditions defined for direct tensile and compressive stress. That is, the member must be straight and have a constant cross section; the material must be homogeneous; the load must be directly axial; and the stress must be below the proportional limit of the material.

Example Problem 4-1

The tie rods in the press in Figure 4-1 are made of steel and nave a diameter of 2.00 in. The length of each rod before installation is 68.5 in. If each rod is subjected to an axial tensile load of 40 000 lb, compute the deformation of each rod.

Solution

Equation (4-6) will be used to compute the deformation, in this case an elongation, or stretching, of the rod. The conditions on Equation (4-6) as described above are met if the stress is below the proportional limit.

$$s = \frac{P}{A}$$

$$P = 40\ 000\ \text{lb}$$

$$A = \frac{\pi D^2}{4} = \frac{\pi (2.0\ \text{in.})^2}{4} = 3.14\ \text{in.}^2$$

Then

$$s = \frac{40\ 000\ \text{lb}}{3.14\ \text{in.}^2} = 12\ 700\ \text{psi.}$$

This level of stress is well below the proportional limit of any steel.
Now from Equation (4-6),

$$\delta = \frac{PL}{AE}$$

We know $P = 40\ 000$ lb, $L = 68.5$ in., and $A = 3.14$ in.^{2}. From the notes for Table A-1, it is found that $E = 30 \times 10^6$ psi is approximately correct for any carbon or alloy steel. Then

$$\delta = \frac{PL}{AE} = \frac{(40\ 000\ \text{lb})(68.5\ \text{in.})}{(3.14\ \text{in.}^2)(30 \times 10^6\ \text{lb/in.}^2)} = 0.029\ \text{in.}$$

Example Problem 4-2

A large pendulum is composed of a 10.0-kg ball suspended by an aluminum wire having a diameter of 1.00 mm and a length of 6.30 m. The aluminum is the alloy 7075-T6. Compute the elongation of the wire due to the weight of the 10-kg ball.

Solution

In order to use Equation (4-6) to compute the elongation, the stress in the wire must be below the proportional limit for the 7075-T6 aluminum alloy.

$$s = \frac{P}{A}$$

The force P is equal to the weight of the 10.0-kg ball.

$$P = m \cdot g = 10.0 \text{ kg} \cdot 9.81 \text{ m/s}^2 = 98.1 \text{ N}$$

$$A = \frac{\pi D^2}{4} = \frac{\pi (1.00 \text{ mm})^2}{4} = 0.785 \text{ mm}^2$$

Then

$$s = \frac{P}{A} = \frac{98.1 \text{ N}}{0.785 \text{ mm}^2} = 125 \text{ N/mm}^2 = 125 \text{ MPa}$$

The proportional limit for 7075-T6 aluminum alloy would be close to its yield strengh. From Table A-4, the yield strength is 455 MPa. Therefore, the stress is well below the proportional limit.

In this case it is most convenient to use Equation (4-5) to compute the deformation since stress is already known.

$$\delta = \frac{sL}{E}$$

Also from Table A-4, $E = 72$ GPa for 7075-T6 aluminum.
Then

$$\delta = \frac{sL}{E} = \frac{(125 \text{ MPa})(6.30 \text{ m})}{72 \text{ GPa}} = \frac{(125 \times 10^6 \text{ Pa})(6.30 \text{ m})}{(72 \times 10^9 \text{ Pa})}$$

$$\delta = 10.9 \times 10^{-3} \text{ m} = 10.9 \text{ mm}$$

Example Problem 4-3

A tension link in a machine must have a length of 610 mm and will be subjected to a maximum axial load of 3000 N. It has been proposed that the link be made of steel and that it have a square cross section. Determine the required dimensions of the link if the elongation under load must not exceed 0.05 mm.

Solution

The objective of the solution is to determine the dimensions of the square link. Thus the area is unknown. Then in Equation (4-6),

$$\delta = \frac{PL}{AE}$$

we can solve for A.

$$A = \frac{PL}{E\delta}$$

we let $P = 3000$ N, $L = 610$ mm, and $\delta = 0.05$ mm. In order to keep consistent units, it is convenient to express E in the units of N/m². Table A-1 gives $E = 207$ GPa. Then we will use $E = 207 \times 10^9$ N/m². Now

$$A = \frac{PL}{E\delta} = \frac{(3000 \text{ N})(610 \text{ mm})}{(207 \times 10^9 \text{ N/m}^2)(0.05 \text{ mm})} = 176.8 \times 10^{-6} \text{ m}^2$$

Converting to mm²,

$$A = 176.8 \times 10^{-6} \text{ m}^2 \times \frac{(10^3 \text{ mm})^2}{\text{m}^2} = 176.8 \text{ mm}^2$$

If we call each side of the square cross section of the link d,

$$A = d^2$$

and

$$d = \sqrt{A} = \sqrt{176.8 \text{ mm}^2} = 13.3 \text{ mm}$$

Therefore, the link must have cross-sectional dimensions of at least 13.3 mm by 13.3 mm.

Example Problem 4-4

Figure 4-2 shows a steel pipe being used as a bracket to support equipment through cables attached as shown. The forces are $F_1 = 8000$ lb and $F_2 = 2500$ lb. Select the smallest standard Schedule 40 steel pipe which will limit the stress to no more than 18 000 psi. Then for

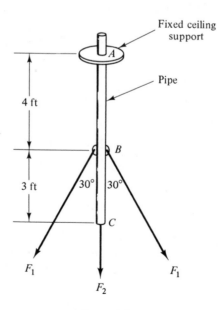

Figure 4-2

the selected pipe, determine the total downward deflection of point
C at the bottom of the pipe as the loads are applied.

Solution

The maximum load in the pipe will occur between points A and B,
where the pipe is subjected to the sum of F_2 and the vertical components of both F_1 forces.

That is,

$$P_{A\text{-}B} = F_2 + 2F_1 \cos 30°$$

$$= 2500 \text{ lb} + 2(8000 \text{ lb}) \cos 30°$$

$$P_{A\text{-}B} = 16\ 400 \text{ lb}$$

Then, letting $s = 18\ 000$ psi, the required cross-sectional area of the
metal in the pipe is

$$A = \frac{P}{s} = \frac{16\ 400 \text{ lb}}{18\ 000 \text{ lb/in.}^2} = 0.911 \text{ in.}^2$$

From Table A-20, listing the properties of steel pipe, the standard
size with the next larger cross-sectional area is the 2-in. Schedule 40
pipe with $A = 1.075$ in.^{2}.

In computing the deflection of point C, it is necessary to observe
that the force in the pipe is 16 400 lb between A and B but only
2500 lb between B and C. Therefore, the deformations of the two
segments must be computed separately.

$$\delta_{A\text{-}B} = \left(\frac{PL}{AE}\right)_{A\text{-}B} = \frac{(16\ 400 \text{ lb})(48 \text{ in.})}{(1.075 \text{ in.}^2)(30 \times 10^6 \text{ lb/in.}^2)} = 0.024 \text{ in.}$$

$$\delta_{B\text{-}C} = \left(\frac{PL}{AE}\right)_{B\text{-}C} = \frac{(2500 \text{ lb})(36 \text{ in.})}{(1.075 \text{ in.}^2)(30 \times 10^6 \text{ lb/in.}^2)} = 0.003 \text{ in.}$$

Then

$$\delta_C = \delta_{A\text{-}B} + \delta_{B\text{-}C} = 0.027 \text{ in.}$$

Point C moves downward 0.027 in. as the loads are applied to the
2-in. Schedule 40 pipe.

4-2 DEFORMATION DUE TO TEMPERATURE CHANGES

A machine or a structure could undergo deformation or be subjected to
stress by changes in temperature in addition to the application of loads.
Bridge members and other structural components see temperatures as low as
$-30°$F($-34°$C) to as high as $110°$F($43°$C) in some areas. Vehicles and machinery operating outside experience similar temperature variations. Frequently,

a machine part will start at room temperature and then become quite hot as the machine operates. Examples are parts of engines, furnaces, metal-cutting machines, rolling mills, plastics molding and extrusion equipment, food-processing equipment, air compressors, hydraulic and pneumatic devices, and high speed automation equipment.

As a metal part is heated, it tends to expand. If the expansion is unrestrained, the dimensions of the part will grow but no stress will be developed in the metal. However, in some cases the part is restrained, preventing the change in dimensions. Under such circumstances, stresses will occur.

Different materials expand at different rates when subjected to temperature changes. The property of a material which indicates its tendency to change dimensions with a change in temperature is its coefficient of thermal expansion. The greek letter α denotes this coefficient. It is a measure of the change in length of a material per unit length for a one-degree change in temperature. The units, then, for α in the English system would be

$$\text{inches/(inch} \cdot °F) \quad \text{or} \quad 1/°F \quad \text{or} \quad °F^{-1}$$

In SI units, α would be

$$\text{m/(m} \cdot °C) \quad \text{or} \quad \text{mm/(mm} \cdot °C) \quad \text{or} \quad 1/°C \quad \text{or} \quad °C^{-1}$$

For use in computations, the last form of each unit type is most convenient. However, the first form will help you remember the physical meaning of the term.

It follows from the definition of the coefficient of thermal expansion that the change in length δ of a member can be computed from the equation

$$\delta = \alpha \cdot L \cdot (\Delta t) \tag{4-7}$$

where L = original length of the member
Δt = change in temperature

Table 4-1 gives values for the coefficient of thermal expansion for several common materials. The actual values vary somewhat with temperature. However, the numbers in Table 4-1 are approximately average values over the range of temperature from 0°C (32°F) to 100°C (212°F).

Example Problem 4-5

A rod made from AISI 1040 steel is used as a link in a steering mechanism of a large truck. If its nominal length is 56 in, compute its change in length as the temperature changes from −30°F to 110°F.

Solution

Using Equation (4-7), we find the change in length to be

$$\delta = \alpha \cdot L \cdot (\Delta t)$$

From Table 4-1, we find $\alpha = 6.3 \times 10^{-6}$ °F^{-1}. The term Δt is the change in temperature; thus

$$\Delta t = 110°\text{F} - (-30°\text{F}) = 140°\text{F}$$

Then

$$\delta = (6.3 \times 10^{-6}°\text{F}^{-1})\,(56 \text{ in.})(140°\text{F}) = 0.049 \text{ in.}$$

TABLE 4-1   Coefficients of Thermal Expansion

Material	$\alpha \times 10^6$	
	°F^{-1}	°C^{-1}
Steel, AISI No.		
1020	6.5	11.7
1040	6.3	11.3
1050	6.1	11.0
4140	6.2	11.2
Structural steel	6.5	11.7
Gray cast iron	6.0	10.8
Stainless steel		
AISI 301	9.4	16.9
AISI 430	5.8	10.4
AISI 501	6.2	11.2
Aluminum alloys		
2014	12.8	23.0
6061	13.0	23.4
7075	12.9	23.2
Brass, CDA 260	11.1	20.0
Bronze, CDA 220	10.2	18.4
Copper, CDA 145	9.9	17.8
Magnesium, ASTM AZ63A-T6	14.0	25.2
Titanium 6Al-4V	5.3	9.5
Plate Glass	5.0	9.0
Wood (pine)	3.0	5.4
Concrete	6.0	10.8

Example Problem 4-6

A pushrod in the valve mechanism of an automotive engine has a nominal length of 203 mm. If the rod is made of AISI 4140 steel, compute the elongation due to a temperature change from $-20°$C to $140°$C.

Solution

Again using Equation (4-7), we have

$$\delta = \alpha \cdot L \cdot (\Delta t)$$

From Table 4-1, we find $\alpha = 11.2 \times 10^{-6}°$C^{-1}. The change in temperature is

$$\Delta t = 140°C - (-20°C) = 160°C$$

Then

$$\delta = (11.2 \times 10^{-6}°C^{-1})(203 \text{ mm})(160°C) = 0.364 \text{ mm}$$

It would be important to accommodate this expansion in the design of the valve mechanism.

Example Problem 4-7

An aluminum frame of 6061 alloy for a window is 4.350 m long and holds a piece of plate glass 4.347 m long when the temperature is 35°C. At what temperature would the aluminum and glass be the same length?

Solution

The temperature would have to decrease in order for the aluminum and glass to reach the same length, since aluminum contracts at a greater rate than glass. As the temperature decreases, the change in temperature Δt would be the same for both the aluminum and the glass. After the temperature change, the length of the aluminum would be

$$L_{al} = 4.350 \text{ m} - \alpha_{al} \cdot L_{al} \cdot \Delta t$$

The length of the glass would be

$$L_g = 4.347 \text{ m} - \alpha_g \cdot L_g \cdot \Delta t$$

But

$$L_{al} = L_g$$

Then

$$4.350 \text{ m} - \alpha_{al} \cdot L_{al} \cdot \Delta t = 4.347 \text{ m} - \alpha_g \cdot L_g \cdot \Delta t$$

Solving for Δt gives

$$\Delta t = \frac{4.350 \text{ m} - 4.347 \text{ m}}{\alpha_{al} \cdot L_{al} - \alpha_g \cdot L_g}$$

Actually, the new values of L_{al} and L_g are not known. However, they will be very close to their original lengths, and little error will result if they are both taken to be about 4.340 m. Then from Table 4-1 we find $\alpha_{al} = 23.4 \times 10^{-6}°C^{-1}$ and $\alpha_g = 9.0 \times 10^{-6}°C^{-1}$.
Then

$$\Delta t = \frac{4.350 \text{ m} - 4.347 \text{ m}}{(23.4 \times 10^{-6}°C^{-1})(4.340 \text{ m}) - (9.0 \times 10^{-6}°C^{-1})(4.340 \text{ m})}$$

$$\Delta t = \frac{0.003}{(0.000102) - (0.000039)} °C = 48°C$$

If the temperature decreases by 48°C from the original temperature of 35°C, the resulting temperature would be −13°C. Since this is well within the possible ambient temperature for a building, a dangerous condition could be created by this window. The window frame and glass would contract without stress until a temperature of −13°C was reached. If the temperature continued to decrease, the frame would contract faster than the glass and would generate stress in the glass. Of course, if the stress is great enough, the glass would fracture, possibly causing injury. The window should be reworked so there is a larger difference in size between the glass and the aluminum frame.

Stresses due to thermal expansion and contraction when members are restrained is discussed in the next section.

4-3 THERMAL STRESSES

In the preceding section, parts which were subjected to changes in temperature were unrestrained, so that they could grow or contract freely. If the parts were held in such a way that deformation was resisted, then stresses would be developed.

Consider a steel structural member in a furnace which is heated while the members to which it is attached are kept at a lower temperature. Assuming the ideal case, the supports would be considered rigid and immovable. Thus, all expansion of the steel member would be prevented.

If the steel part were allowed to expand, it would elongate by an amount $\delta = \alpha \cdot L \cdot (\Delta t)$. But since it is restrained, this represents the apparent total strain in the steel. Then the unit strain would be

$$\epsilon = \frac{\delta}{L} = \frac{\alpha \cdot L \cdot (\Delta t)}{L} = \alpha(\Delta t) \tag{4-8}$$

The resulting stress in the part can be found from

$$s = E\epsilon$$

or

$$s = E\alpha(\Delta t) \tag{4-9}$$

Example Problem 4-8

A steel structural member in a furnace is made from AISI 1020 steel and undergoes an increase in temperature of 95°F while being held rigid at its ends. Compute the resulting stress in the steel.

Solution

Using Equation (4-9),

$$s = E\alpha(\Delta t)$$

From Table 4-1 we find $\alpha = 6.5 \times 10^{-6}\,^\circ\text{F}^{-1}$. For steel,

$$E = 30 \times 10^6 \text{ psi}$$

Then

$$s = (30 \times 10^6 \text{ psi})(6.5 \times 10^{-6}\,^\circ\text{F}^{-1})(95\,^\circ\text{F}) = 18\ 500 \text{ psi}$$

Example Problem 4-9

An aluminum rod of alloy 2014-T6 in a machine is held at its ends while being cooled from 95°C. At what temperature would the tensile stress in the rod equal half of the yield strength of the aluminum if it is originally at zero stress?

Solution

In Equation (4-9), we can solve for the change in temperature Δt.

$$s = E\alpha(\Delta t)$$

$$\Delta t = \frac{s}{E\alpha}$$

From Table A-4, for aluminum alloy 2014-T6, $s_y = 379$ MPa and $E = 73$ GPa. From Table 4-1, $\alpha = 23.0 \times 10^{-6}\,^\circ\text{C}^{-1}$.
Then

$$s = \frac{s_y}{2} = \frac{379 \text{ MPa}}{2} = 190 \text{ MPa}$$

and

$$\Delta t = \frac{s}{E\alpha} = \frac{190 \text{ MPa}}{(73 \text{ GPa})(23.0 \times 10^{-6}\,^\circ\text{C}^{-1})}$$

$$\Delta t = \frac{190 \times 10^6 \text{ Pa}}{(73 \times 10^9 \text{ Pa})(23.0 \times 10^{-6}\,^\circ\text{C}^{-1})} = 113\,^\circ\text{C}$$

Since the rod had zero stress when its temperature was 95°C, the temperature at which the stress would be 190 MPa would be

$$t = 95\,^\circ\text{C} - 113\,^\circ\text{C} = -18\,^\circ\text{C}$$

4-4 MEMBERS MADE OF MORE THAN ONE MATERIAL

When two or more materials in a load-carrying member share the load, a special analysis is required to determine what portion of the load each material takes. Consideration of the elastic properties of the materials is required.

Figure 4-3 shows a steel pipe filled with concrete and used to support part of a large structure. The load is distributed evenly across the top of the support. It is desired to determine the stress in both the steel and the concrete.

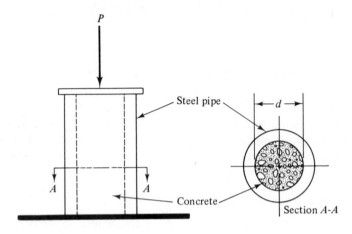

Figure 4-3

Two concepts must be understood in deriving the solution to this problem.

1. The total load P is shared by the steel and the concrete such that $P = P_s + P_c$.
2. Under the compressive load P, the composite support deforms and the two materials deform in equal amounts. That is, $\delta_s = \delta_c$.

Now since the steel and the concrete were originally the same length,

$$\frac{\delta_s}{L} = \frac{\delta_c}{L}$$

But

$$\frac{\delta_s}{L} = \epsilon_s \quad \text{and} \quad \frac{\delta_c}{L} = \epsilon_c$$

Also,

$$\epsilon_s = \frac{s_s}{E_s} \quad \text{and} \quad \epsilon_c = \frac{s_c}{E_c}$$

Then,

$$\frac{s_s}{E_s} = \frac{s_c}{E_c}$$

Solving for s_s,

$$s_s = \frac{s_c E_s}{E_c} \tag{4-10}$$

This equation gives the relationship between the two stresses.

Now considering the loads,

$$P_s + P_c = P$$

Since both materials are subjected to axial stress,

$$P_s = s_s A_s \quad \text{and} \quad P_c = s_c A_c$$

where A_s and A_c are the areas of the steel and concrete, respectively.

Then

$$s_s A_s + s_c A_c = P \qquad (4\text{-}11)$$

Substituting Equation (4-10) into Equation (4-11) gives

$$\frac{A_s s_c E_s}{E_c} + s_c A_c = P$$

Now, solving for s_c,

$$s_c = \frac{PE_c}{A_s E_s + A_c E_c} \qquad (4\text{-}12)$$

Equations (4-10) and (4-12) can now be used to compute the stresses in the steel and the concrete. Of course, these equations can be used for any other composite member of two materials by substituting the appropriate values for area and modulus of elasticity.

Example Problem 4-10

For the support shown in Figure 4-3, the pipe is a standard 6-in Schedule 40 steel pipe completely filled with concrete. If the load P is 155 000 lb, compute the stress in the concrete and the steel. For the concrete use $E = 2 \times 10^6$ psi, and for steel use $E = 30 \times 10^6$ psi.

Solution

Equations (4-10) and (4-12) can be used. The areas of the steel pipe and the concrete are required. From Table A-20, the cross-sectional area of metal in a 6-in. Schedule 40 pipe is 5.581 in.². Its inside diameter d is 6.065 in. Then the area of the concrete is

$$A_c = \frac{\pi d^2}{4} = \frac{\pi (6.065 \text{ in.})^2}{4} = 28.89 \text{ in.}^2$$

Then in Equation (4-12),

$$s_c = \frac{(155\ 000 \text{ lb})(2 \times 10^6 \text{ psi})}{(5.584 \text{ in.}^2)(30 \times 10^6 \text{ psi}) + (28.89 \text{ in.}^2)(2 \times 10^6 \text{ psi})}$$

$$s_c = 1376 \text{ psi}$$

Using Equation (4-10),

$$s_s = s_c \frac{E_s}{E_c} = \frac{(1376 \text{ psi})(30 \times 10^6 \text{ psi})}{2 \times 10^6 \text{ psi}} = 20\ 600 \text{ psi}$$

PROBLEMS

4-1   A post of western hemlock, construction grade, is $3\frac{1}{2}$ in. square and 6.0 ft long. How much would it be shortened when it is loaded in compression up to its allowable load parallel to the grain?

4-2   Determine the elongation of a strip of plastic, 0.75 mm thick by 12 mm wide by 375 mm long, if it is subjected to a load of 90 N and is made of (a) glass-reinforced TFE or (b) melamine. (See Table A-7 in the Appendix.)

4-3 A hollow aluminum cylinder made of 2014-T5 has an outside diameter of 2.50 in. and a wall thickness of 0.085 in. Its length is 14.5 in. What axial compressive force would cause the cylinder to shorten by 0.005 in.? What is the resulting stress in the aluminum?

4-4 A square bar 0.25 in. on a side is found in a stock bin, and it is not known if it is aluminum or magnesium. Discuss two ways in which you could determine what it is.

4-5 A tensile member is being designed for a car. It must withstand a repeated load of 3500 N and not elongate more than 0.12 mm in its 630-mm length. Use a design factor of 8 based on ultimate strength, and compute the required diameter of a round rod to satisfy these requirements using (a) AISI 1020 hot-rolled steel, (b) AISI 4140 OQT 400 steel, and (c) aluminum alloy 6061-T6. Compare the mass of the three options.

4-6 A steel bolt has a diameter of 12.0 mm in the unthreaded portion. Determine the elongation in a length of 220 mm if a force of 17.0 kN is applied.

4-7 In an aircraft structure, a rod is designed to be 1.25 m long and have a square cross section 8.0 mm on a side. Determine the amount of elongation which would occur if it is made of (a) titanium 6Al-4V and (b) AISI 501 OQT 1000 stainless steel. The load is 5000 N.

4-8 A tension member in a welded steel truss is 13.0 ft long and subjected to a force of 35,000 lb. Choose an equal leg angle made of A36 steel which will limit the stress to 21 600 psi. Then compute the elongation in the angle due the force. Use $E = 29.0 \times 10^6$ psi for structural steel.

4-9 A link in a mechanism is subjected alternately to a tensile load of 450 lb and a compressive load of 50 lb. Compute the elongation and compression of the link if it is a rectangular steel bar $\frac{1}{4}$ in. by $\frac{1}{8}$ in. in cross section and 8.40 in. long.

4-10 A concrete slab in a highway is 80 ft long. Determine the change in length of the slab if the temperature changes from $-30°F$ to $+110°F$.

4-11 A steel rail for a railroad siding is 12.0 m long. Determine the change in length of the rail if the temperature changes from $-34°C$ to $+43°C$.

4-12 Determine the stress which would result in the rail described in Problem 4-11 if it were completely restrained from expanding.

4-13 The pushrods which actuate the valves on a six-cylinder engine are AISI 1040 steel and are 625 mm long and 8.0 mm in diameter. Calculate the change in length of the rods if their temperature varies from $-40°C$ to $+116°C$ and the expansion is unrestrained.

4-14 If the pushrods described in Problem 4-13 were installed with zero clearance with other parts of the valve mechanism at 25°C, compute the following:
(a) The clearance between parts at $-40°C$.
(b) The stress in the rod due to a temperature rise to 116°C.
Assume that mating parts are rigid.

4-15 A bridge deck is made as one continuous concrete slab 140 ft long. Determine the required width of expansion joints at the ends of the bridge if no stress is to be developed when the temperature varies from $-30°F$ to $+110°F$.

4-16 For the bridge deck in Problem 4-15, what stress would be developed if only

1.0-in. expansion joints were used and the supports were rigid? For concrete use $E = 2 \times 10^6$ psi.

4-17 For the bridge deck in Problem 4-15, assume that the deck is to be just in contact with its support at the temperature of 110°F. If the deck is to be installed when the temperature is 60°F, what should the gap be between the deck and its supports? $\Delta L_T = \alpha L_b \Delta T$

4-18 A ring of AISI 301 stainless steel is to be placed on a shaft having a temperature of 20°C and a diameter of 55.200 mm. The inside diameter of the ring is 55.100 mm. To what temperature must the ring be heated to make it 55.300 mm in diameter and thus allow it to be slipped onto the shaft?

4-19 When the ring of Problem 4-18 is placed on the shaft and then cooled back to 20°C, what tensile stress will be developed in the ring?

4-20 A heat exchanger is made by arranging several brass (CDA 260) tubes inside a stainless steel (AISI 430) shell. Initially, when the temperature is 10°C, the tubes are 4.20 m long and the shell is 4.50 m long. Determine how much each will elongate when heated to 85°C.

4-21 In Alaska, a 40-ft section of steel pipe may see a variation in temperature from −50°F when it is at ambient temperature to +140°F when it is carrying heated oil. Compute the change in the length of the pipe under these conditions.

4-22 A short post is made by welding steel plates into a square, as shown in Figure 4-4, and then filling the area inside with concrete. Compute the stress in the steel and in the concrete if $b = 150$ mm, $t = 10$ mm, $E_s = 207$ GPa, $E_c = 13.8$ GPa, and the post carries an axial load of 1.40 MN.

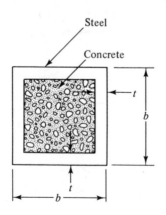

Steel

Concrete

t

b

t

b

Figure 4-4

4-23 A short post is made by welding steel plates into a square, as shown in Figure 4-4, and then filling the area inside with concrete. The steel has an allowable stress of 21 600 psi, and the concrete has an allowable stress of 2000 psi. If $b = 6.00$ in. and $t = \frac{3}{8}$ in., compute the allowable axial load on the post. Use $E_s = 30 \times 10^6$ psi and $E_c = 2 \times 10^6$ psi.

4-24   A short post is being designed to support an axial compressive load of 500 000 lb. It is to be made by welding $\frac{1}{2}$-in. thick plates of A36 steel into a square and filling the area inside with concrete, as shown in Figure 4-4. It is required to determine the dimension of the side of the post *b* in order to limit the stress in the steel to no more than 21 600 psi and in the concrete to no more than 2000 psi. Use $E_c = 2 \times 10^6$ psi and $E_s = 30 \times 10^6$ psi.

4-25   Two disks are connected by four rods as shown in Figure 4-5. All rods are 6.0-mm in diameter and have the same length. Two rods are steel ($E = 207$ GPa), and two are aluminum ($E = 69$ GPa). Compute the stress in each rod when an axial force of 11.3 kN is applied to the disks.

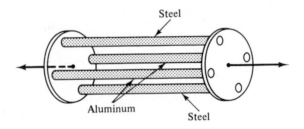

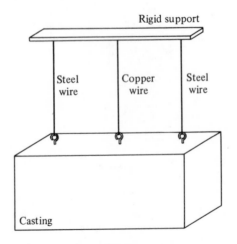

Figure 4-5

4-26   An array of three wires is used to suspend a casting having a mass of 2265 kg in such a way that the wires are symmetrically loaded (see Figure 4-6). The outer two wires are AISI 430 stainless steel, cold-worked. The middle wire is hard beryllium copper, CDA 172. All three wires have the same diameter and length. Determine the required diameter of the wires if none is to be stressed beyond one-half of its yield strength.

Figure 4-6

5

DIRECT SHEAR

5-1 MEMBERS LOADED IN DIRECT SHEAR

Direct shear refers to a kind of loading in which the force on a part acts perpendicular to the cross section of the part. This action tends to cut the part at the section resisting the shear force. The resulting effect on the material from which the part is made is quite different from the pulling and stretching caused by a tensile force, and from the crushing caused by a compressive force.

Figure 5-1 illustrates several cases where direct shear is one of the kinds of stress produced. The clevis joint in part (a) is used in many applications where the ability to rotate one part relative to another is required when transferring a load across the joint. It is composed of a link with a hole in it, a pin, and a pair of ears. The pin passes through the ears and the link, fastening them together while allowing rotation around the pin. When a pushing or pulling force is applied to the link, that force is transferred through the pin to the ears across the two sections between the link and the ears. These sections see direct shear stress.

Part (b) of Figure 5-1 shows a boat propeller attached to its drive shaft in such a way that the turning couple must be transferred to the propeller through the drive pin. Then as the shaft rotates, it tends to cut the pin at the top and

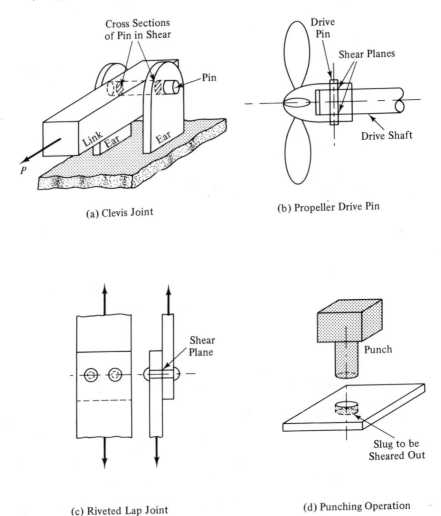

(a) Clevis Joint

(b) Propeller Drive Pin

(c) Riveted Lap Joint

(d) Punching Operation

Figure 5-1

bottom of the shaft between the shaft and the propeller housing. Thus these two sections of the pin are subjected to direct shear stress.

Many fasteners are subjected to direct shear stresses. The lap joint in part (c) of Figure 5-1 connects the two plates with two rivets. If the plates receive a tensile pulling force F, there is a tendency to cut the rivets at the section where the lower plate contacts the upper plate. A direct shear stress is produced at that section in the rivets.

When a hole must be punched in a piece of metal, the action of punching shears the slug of metal from the hole, as shown in part (d) of Figure 5-1.

In this case, the area of metal which is sheared is the surface of the cylindrical slug, shown shaded in the figure.

In computing direct shear stress, it is necessary that the correct shear area be recognized. In doing this, you should ask: What *area* or *areas* of the part will tend to be *cut* as a force is applied? You must then determine what *force* is applied to that area.

5-2   DIRECT SHEAR STRESS

In computing the direct shear stress on parts such as those shown in Figure 5-1, it is assumed that the shear force on a section is uniformly distributed across the entire area resisting the shear force. Thus the direct shear stress can be found from

$$\text{Shear stress} = \frac{\text{applied force}}{\text{shear area}}$$

In symbol form this equation can be written

$$s_s = \frac{P}{A_s} \tag{5-1}$$

Equation (5-1) is valid only for *direct shear*. Other kinds of loading produce shear stresses also, but the analysis is different. Torsional shear will be discussed in Chapter 6. Shear stresses in beams are presented in Chapter 10. There are also cases in which direct shear stresses are combined with other kinds of stresses, as described in Chapter 11.

Example Problem 5-1

In a clevis joint like that shown in part (a) of Figure 5-1, a force P of 4000 lb is applied to the link. Calculate the magnitude of the shear stress in the pin of the joint. The pin has a diameter of $\frac{1}{2}$ in.

Solution

Since the pin is subjected to direct shear, Equation (5-1) can be used. The applied force is $P = 4000$ lb. To determine the shear area, notice that *two* cross sections of the pin resist the force. Then the shear area is

$$A_s = \frac{2 \cdot \pi D^2}{4}$$

Where D is the pin diameter. Then

$$A_s = \frac{2 \cdot \pi \cdot (0.5 \text{ in.})^2}{4} = 0.393 \text{ in.}^2$$

The shear stress can now be calculated.

$$s_s = \frac{P}{A_s} = 4000 \text{ lb}/0.393 \text{ in.}^2 = 10\ 200 \text{ lb/in.}^2$$

Example Problem 5-2

A punching operation is being planned in which a 20-mm diameter hole is to be punched in a piece of aluminum 6 mm thick. If the ultimate shear strength of the aluminum is 165 MPa, compute the force which must be applied to the punch to complete the operation.

Solution

From Equation (5-1),

$$s_s = \frac{P}{A_s}$$

we can solve for the applied force P.

$$P = s_s A_s$$

Let $s_s = 165$ MPa. Recall that one MPa = one N/mm^2. The shear area is the surface area of the cylindrical plug removed from the plate, as illustrated in Figure 5-1(d).

$$A_s = \pi D \cdot t$$

where D = hole diameter t = plate thickness, then

$$A_s = \pi(20 \text{ mm})(6 \text{ mm}) = 377 \text{ mm}^2$$

Now the required applied force is

$$P = s_s A_s = 165 \frac{N}{mm^2} \cdot 377 \text{ mm}^2 = 62.2 \text{ kN}$$

That is, 62.2 kN of force is required to punch the hole.

5-3 SINGLE SHEAR AND DOUBLE SHEAR

A set of terms used frequently to describe the manner of loading of a part, particularly a pin or rivet joint, is *single shear* and *double shear*. The terms refer to an observation of how many cross sections of the part resist the applied force. This concept was applied in the example problems presented in the preceding section. In Example Problem 5-1, it was observed that the applied force of 4000 lb was resisted by *two* cross sections of the pin. Thus the pin was in *double shear*.

If the pin joint had been designed as shown in Figure 5-2, it can be seen that the applied force is resisted by only *one* cross section of the pin. This is *single shear*. If the force was the same and the pin diameter was the same, then

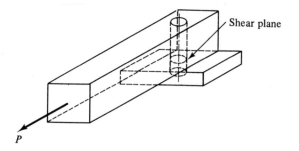

Figure 5-2 Pin Joint in Single Shear

the shear stress would be doubled since the shear area is only half of what it was in Example Problem 5-1. The stress on the pin in Figure 5-2 would be

$$s_s = \frac{P}{A_s}$$

$$A_s = \frac{\pi D^2}{4} = \frac{\pi (0.5 \text{ in.})^2}{4} = 0.196 \text{ in.}^2.$$

$$s_s = \frac{4000 \text{ lb}}{0.196 \text{ in.}^2} = 20\ 400 \text{ lb/in.}^2$$

This is twice the value found for the pin in double shear.

Another frequently encountered example of single and double shear is in the design of lap joints. Figure 5-3 shows two joints made with rivets. In each case, the force F on the joint is resisted by one rivet. Thus the applied force for each rivet is F. In part (a) of Figure 5-3 the entire force F must be transferred across one cross section of the rivet. However, in part (b) the

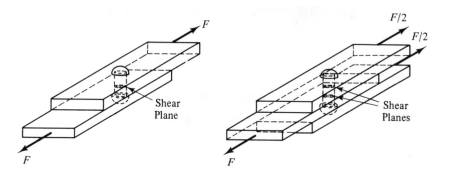

(a) Rivet in Single Shear   (b) Rivet in Double Shear

Figure 5-3

force F is transferred across two cross sections. Part (a) illustrates single shear, and part (b) illustrates double shear. Because the shear area for double shear is twice that for single shear, the resulting shear stress will be half as much for double shear as for single shear.

5-4 DESIGN OF MEMBERS IN DIRECT SHEAR

As in all design problems, the objective of the design of shear members is to determine the required geometry of the member in order to limit the stress to a safe value when the member is subjected to expected loads. This can be done by modifying the basic direct shear stress equation, Equation (5-1),

$$s_s = \frac{P}{A_s}$$

Let s_s equal a *design shear stress* s_{sd}. Then the required shear area to limit the stress to s_{sd} is

$$A_s = \frac{P}{s_{sd}} \tag{5-2}$$

Knowing A_s would allow the dimensions of the member to be specified.

The specification of the design shear stress s_{sd} depends somewhat on the application. The design stress can be based on the yield strength in shear s_{ys} or on the ultimate strength in shear s_{us}. The method of computing the design shear stress is similar to the method of computing direct tensile and compressive stresses. That is,

$$s_{sd} = \frac{s_{ys}}{N} \text{ based on yield strength in shear} \tag{5-3}$$

or

$$s_{sd} = \frac{s_{us}}{N} \text{ based on ultimate strength in shear} \tag{5-4}$$

One difficulty with this method is that the values of s_{ys} and s_{us} are frequently not reported with material properties data. For many ductile metals, particularly steel, estimates can be made as follows:

$$s_{ys} = 0.5s_y \tag{5-5}$$

$$s_{us} = 0.75s_u \tag{5-6}$$

where s_y and s_u are the yield strength and ultimate strength, respectively, determined from the standard tensile test. These data are normally reported. Combining Equations (5-3) and (5-4) with Equations (5-5) and (5-6), we can compute the design shear stress from

$$s_{sd} = \frac{0.5s_y}{N} \tag{5-7}$$

or

$$S_{sd} = \frac{0.75s_u}{N} \qquad (5\text{-}8)$$

The values to be used for the design factor N can be taken from Table 3-1, with the same discussion applying. Normally, for a dead load on a ductile metal, Equation (5-7) would be used with $N = 2$. For repeated loads on a ductile metal, Equation (5-8) would be used with $N = 8$. For impact or heavy shock, use Equation (5-8) with $N = 12$. For brittle materials, use Equation (5-8) only, with $N = 6$ for a dead load, $N = 10$ for repeated loads, and $N = 15$ (or more) for impact and heavy shock.

Example Problem 5-3

Design a round pin to be used as the drive pin for a boat propeller, as illustrated in Figure 5-1(b). The normal torque developed by the 3-in. diameter drive shaft is 1575 in.·lb. Consider the torque (turning couple) to be transferred to the propeller through the drive pin in such a way that equal forces are exerted at the top and bottom of the pin. Assume that the forces will be steady so that they can be treated as a dead load.

Solution

The basic geometry of the drive pin will be cylindrical. Therefore, the objective is to determine the required diameter of the pin. Also, the material from which the pin is to be made must be specified.

  In selecting a material, an inexpensive, ductile metal having a reasonably high strength is desirable. Although many alloys could be used, let us complete the design using AISI 1020 cold-drawn steel. From Table A-1, $s_y = 51\ 000$ psi.

  The design stress can be computed using Equation (5-7) since the load is being treated as a dead load and the selected material is ductile. Also, we will use $N = 2$. Then the design stress is

$$S_{sd} = \frac{0.5s_y}{N} = \frac{(0.5)(51\ 000\ \text{psi})}{2} = 12\ 750\ \text{psi}$$

  It is now necessary to determine the magnitude of the forces exerted on the pin. Figure 5-4 shows the pin passing through the hub of the propeller and the drive shaft. The two forces labeled F form a couple which must equal the applied torque of 1575 in.·lb. That is,

$$F \cdot D = 1575\ \text{in.·lb}$$

Solving for F with $D = 3.0$ in. gives

$$F = \frac{1575\ \text{in.·lb}}{3.0\ \text{in.}} = 525\ \text{lb}$$

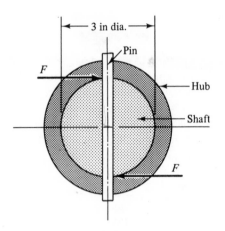

Figure 5-4

This force acts at both the top and the bottom. Therefore, the stress at either section would be

$$S_s = \frac{F}{A_p}$$

where A_p is the cross-sectional area of the pin. Now, letting $s_s = s_{sd}$ and solving for A_p gives

$$A_p = \frac{F}{s_{sd}} = \frac{525 \text{ lb}}{12\ 750 \text{ psi}} = 0.041 \text{ in.}^2$$

Since $A_p = \pi d^2/4$,

$$d = \sqrt{\frac{4A_p}{\pi}} = \sqrt{\frac{(4)(0.041 \text{ in.}^2)}{\pi}} = 0.228 \text{ in.}$$

Thus the minimum required pin diameter d would be 0.228 in. To use a more convenient size, a diameter of $\frac{1}{4}$ in. would be specified. It should be noted that the design process used here did not take into account the probability that a higher torque or a shock loading would be subjected to the pin. In reality it is very likely that this will occur and that the pin we have designed will be sheared. In a device such as this, shearing the inexpensive pin would be preferred to the alternative of damaging an expensive propeller or the transmission of the motor.

PROBLEMS

5-1 A clevis joint like that shown in Figure 5-1(a) is subjected to a force of 16.5 kN. Determine the shear stress in the 12.0-mm diameter pin.

5-2   Determine the required diameter of a pin for a clevis joint like that shown in Figure 5-1(a) if it is subjected to a force of 12.3 kN. Use AISI 1040 cold-drawn steel and a design factor of 8 based on the ultimate strength in shear.

5-3   A support for a beam is made as shown in Figure 5-5. Determine the required thickness of the projecting ledge *a* if the maximum shear stress is to be 6000 psi. The load on the support is 21 000 lb.

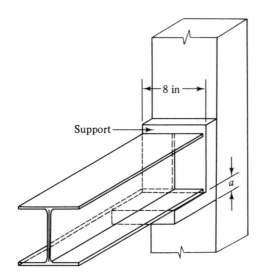

Figure 5-5

5-4   A section of pipe is supported by a saddle-like structure which, in turn, is supported on two steel pins, as illustrated in Figure 5-6. If the load on the

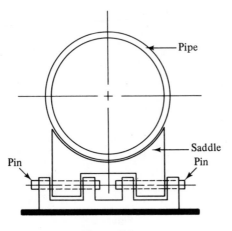

Figure 5-6

saddle is 42 000 lb, determine the required diameter of the pins. Use AISI 1050 cold-drawn steel.

5-5   The hanger designed in Chapter 3 and shown in Figures 3-6 and 3-7 used a cylindrical pin to secure the upper end of the rod to the carrier. It was assumed that the pin would be 10.0 mm in diameter. The load on the hanger was 21.1 kN. Specify a material for the pin which will allow the use of a 10.0-mm diameter pin with the load considered to be impact.

5-6   In a pair of pliers, the hinge pin is subjected to direct shear, as indicated in Figure 5-7. If the pin has a diameter of 3.0 mm and the force exerted at the handle, F_h, is 55 N, compute the stress in the pin.

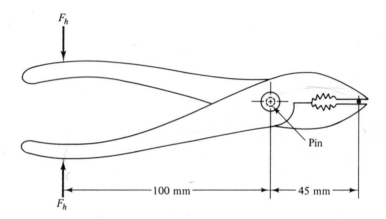

Figure 5-7

5-7   The lower control arm on an automotive suspension system is connected to the frame by a round steel pin, 16 mm in diameter. Two sides of the arm

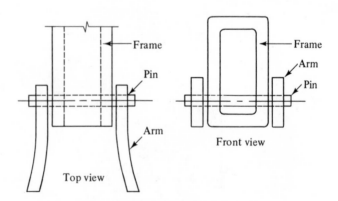

Figure 5-8

transfer loads from the frame to the arm, as sketched in Figure 5-8. How much shear force could the pin withstand if it is made of AISI 1050 cold-drawn steel and a design factor of 6 based on the yield strength in shear is desired?

5-8 A centrifuge is used to separate liquids according to their densities using centrifugal force. Figure 5-9 illustrates one arm of a centrifuge having a bucket at its end to hold the liquid. In operation, the bucket and the liquid have a mass of 0.40 kg. The centrifugal force has the magnitude in newtons of

$$F = 0.010 \ 97 \cdot m \cdot R \cdot n^2$$

where m = rotating mass of bucket and liquid (kilograms)
$\quad\quad R$ = radius to center of mass (metres)
$\quad\quad n$ = rotational speed (revolutions per minute)

The centrifugal force places the pin holding the bucket in direct shear. Compute the stress in the pin due to a rotational speed of 3000 rpm.

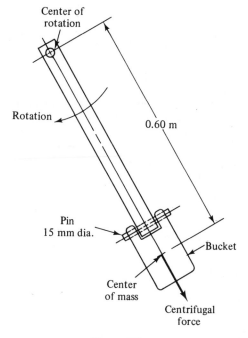

Figure 5-9

5-9 For the pin in the centrifuge described in Problem 5-8, specify a suitable steel for the pin, considering the load to be repeated.

5-10 A notch is made in a piece of wood, as shown in Figure 5-10, in order to support an external load F of 1800 lb. Compute the shear stress in the wood. Is the notch safe? (See Table A-6.)

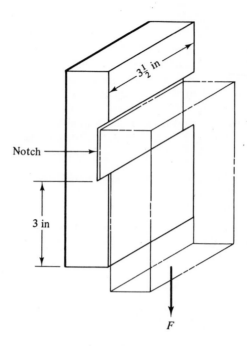

Figure 5-10

5-11 A circular punch is used to punch a 20.0-mm diameter hole in a sheet of steel having a thickness of 8.0 mm. If the ultimate shear strength of the steel is 500 MPa, compute the force required to punch the hole.

5-12 Determine the force required to punch a slug the shape shown in Figure 5-11 from an aluminum sheet 5.0 mm thick. The ultimate shear strength of the aluminum is 88 MPa.

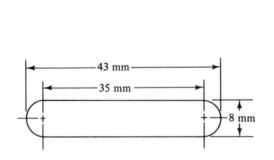

Figure 5-11

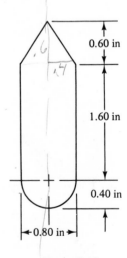

Figure 5-12

5-13 Compute the force required to punch a slug the shape shown in Figure 5-12 from a sheet of AISI 1020 hot-rolled steel having a thickness of 0.194 in. Use $s_{su} = 0.75\ s_u$.

5-14 Compute the force required to shear a straight edge of a sheet of AISI 1040 cold-drawn steel having a thickness of 0.105 in. Use $s_{su} = 0.75\ s_u$. The length of the edge is 7.50 in.

5-15 The key in Figure 5-13 has the dimensions $b = 10$ mm, $h = 8$ mm, and $L = 22$ mm. Determine the shear stress in the key when 95 N·m of torque is transferred from the 35 mm-diameter shaft to the hub.

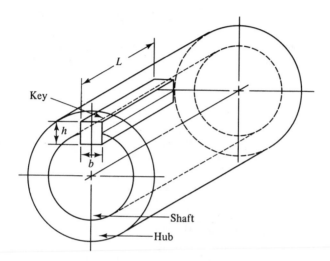

Figure 5-13

5-16 A key is used to connect a hub of a gear to a shaft, as shown in Figure 5-13. It has a rectangular cross section with $b = \frac{1}{2}$ in. and $h = \frac{3}{8}$ in. The length is 2.25 in. Compute the shear stress in the key when it transmits 8000 lb·in. of torque from the 2.0-in. diameter shaft to the hub.

5-17 Determine the required length of the key in Figure 5-13 between the gear hub and shaft if it must transmit a torque of 565 N·m. The key is made of AISI 1040 cold-drawn steel and is rectangular with $b = 16$ mm and $h = 10$ mm. The shaft is 60 mm in diameter. Use a design factor of 4 based on the yield strength of key material in shear.

5-18 A square key is used to connect a shaft to the hub of a gear, as shown in Figure 5-13. The shaft has a diameter of 1.500 in. and the key is $\frac{3}{8}$ in. square. The key is made of AISI 1020 cold-drawn steel. Determine the required length of the key to provide a design factor of 4 based on the yield strength in shear when the key transmits 7200 lb·in. of torque.

5-19 The base of a large crane boom is attached to the frame through a pin as shown in Figure 5-14. Determine the required diameter of the pin if a repeated load of 110 000 lb is applied to the boom. Use AISI 4140 OQT 1300 steel.

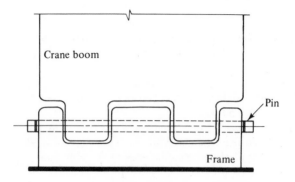

Crane boom

Pin

Frame

Figure 5-14

5-20 A set of two tubes is connected in the manner shown in Figure 5-15. Under a compressive load of 20 000 lb, the load is transferred from the upper tube

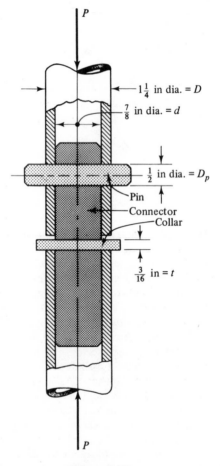

P

$1\frac{1}{4}$ in dia. $= D$

$\frac{7}{8}$ in dia. $= d$

$\frac{1}{2}$ in dia. $= D_p$

Pin

Connector

Collar

$\frac{3}{16}$ in $= t$

P

Figure 5-15

through the pin to the connector, then through the collar to the lower tube. Compute the shear stress in the pin and in the collar.

5-21 A small, hydraulic crane like that shown in Figure 5-16 carries an 800-lb load. Determine the shear stress that occurs in the pin at B, which is in double shear. The pin diameter is $\frac{3}{8}$ in.

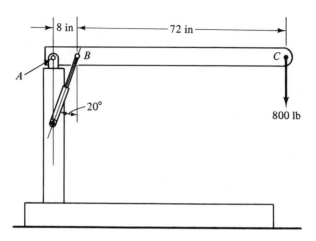

Figure 5-16

5-22 The frame for a 50-ton hydraulic press is made as shown in Figure 5-17. Determine the required diameter of the pins which connect the channel beams to the uprights. Assume that the force is centered on the beams. Use AISI 1050 cold-drawn steel for the pins, and assume the load to be a dead load.

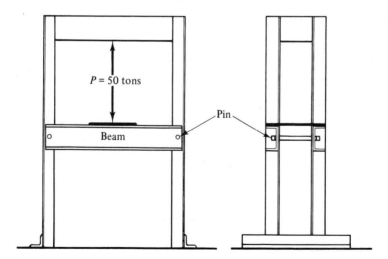

Figure 5-17

5-23 A ratchet device on a jack stand for a truck has a tooth configuration as shown in Figure 5-18. For a load of 88 kN compute the shear stress at the base of the tooth.

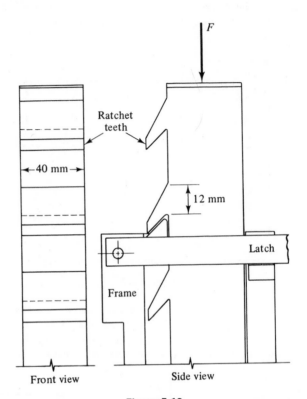

Figure 5-18

6

TORSION

6-1 MEMBERS LOADED IN TORSION

Torsion refers to the loading of a member which would tend to twist it. A twisting moment or couple must be applied to cause the member to twist. Figure 6-1 shows several examples of members subjected to torsion. Part (a) shows a socket wrench with an extension shaft being used to tighten a bolt. The *torque*, or twisting moment, applied to both the bolt and the socket extension is the product of the applied force and the distance from the line of action of the force to the axis of the bolt. That is,

$$\text{Torque} = T = Fd \qquad (6\text{-}1)$$

Torque is expressed in the units of force times distance, which is $N \cdot m$ in the SI metric system and $lb \cdot in.$ or $lb \cdot ft$ in the English system.

Part (b) of Figure 6-1 shows a drive system for a boat, consisting of an engine, transmission, drive shaft, and propeller. Power developed by the engine flows to the propeller, where it drives the boat forward. In this example, there are several members subjected to torsion. The drive shaft is the most obvious. When transmitting power, a torque is developed in the shaft, thus rotating the propeller. The power transmitted is related to the torque T and rotational speed n by

$$\text{Power} = P = Tn \qquad (6\text{-}2)$$

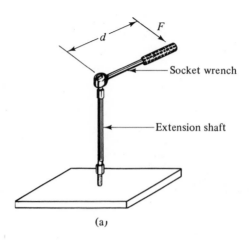

(a)

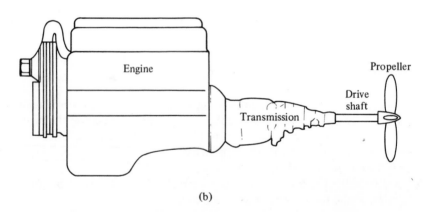

(b)

Figure 6-1

This is a very useful relationship, since if any two values, P, n, or T, are known, the third can be computed. In using Equation (6-2), careful attention must be paid to the units for all terms, illustrated as follows:

SI Metric System units:

$$\text{Power} = \text{watts(W)} = \text{J/s} = \text{N} \cdot \text{m/s}$$

$$\text{Torque} = \text{N} \cdot \text{m}$$

$$\text{Rotational speed} = \text{rad/s}$$

Note that J is the symbol for joule, the unit for energy $(1.0\,\text{J} = 1\,\text{N} \cdot \text{m})$. Some confusion can result from these units, in that the radian unit would appear to remain when torque is multiplied by rotational speed to get power. Two approaches can be used to overcome this difficulty. The most straightforward approach would be to consider the unit for torque to be N·m/rad instead of simply N·m. This is dimensionally correct and serves to identify torque as a phenomenon which causes rotation. Using this form, power would then be computed directly from

$$\text{Power} = Tn = (\text{N·m/rad}) \cdot (\text{rad/s}) = \text{N·m/s} = \text{W}$$

The second approach to dealing with these units would be to consider the unit of radian *to be no unit at all* when it is encountered in an equation. Thus, with torque in N·m and rotational speed in rad/s, power would be

$$\text{Power} = Tn = (\text{N·m})(\text{rad/s}) = \text{N·m/s} = \text{W}$$

English Gravitational Unit System:

$$\text{Torque} = \text{lb·in.}$$

$$\text{Rotational speed} = \text{rev/min (rpm)}$$

$$\text{Power} = \text{hp}$$

where 1 hp = 550 ft·lb/sec. In this case, conversion factors must be used as follows:

$$\text{Power} = T(\text{lb·in.}) \cdot n\left(\frac{\text{rev}}{\text{min}}\right) \cdot \left(\frac{1 \text{ min}}{60 \text{ sec}}\right) \cdot \frac{2\pi \text{ rad}}{\text{rev}} \cdot \frac{1 \text{ ft}}{12 \text{ in.}} \cdot \frac{1 \text{ hp}}{550 \text{ ft·lb/sec}}$$

or

$$\boxed{\text{Power} = \frac{Tn}{63\ 000}} \quad English \tag{6-3}$$

Example Problem 6-1

For the wrench in Figure 6-1(a), compute the amount of torque applied to the socket if a force of 50 N is exerted at a point 200 mm out from the axis of the socket.

Solution

Using Equation (6-1),

$$T = Fd = (50 \text{ N})(200 \text{ mm}) = 10\ 000 \text{ N·mm} \times \frac{1 \text{ m}}{1000 \text{ mm}}$$

$$T = 10 \text{ N·m}$$

Example Problem 6-2

A drive system for a boat like that shown in Figure 6-1(b) transmits 95 kW of power through the drive shaft, which rotates at 54 rad/s. Compute the torque in the drive shaft.

Solution

Using Equation (6-2) and solving for T,

$$P = Tn$$

$$T = \frac{P}{n} = \frac{95 \text{ kW}}{(54 \text{ rad/s})}$$

But $1 \text{ kW} = 10^3 \text{ N} \cdot \text{m/s}$. Then

$$T = 95(10^3)\frac{\text{N} \cdot \text{m}}{\text{s}} \cdot \frac{1}{54 \text{ rad/s}} = 1.76(10^3)\text{N} \cdot \text{m/rad}$$

Expressing this in the conventional units,

$$T = 1.76 \text{ kN} \cdot \text{m}$$

Example Problem 6-3

Compute the power being transmitted by a shaft if it is developing a torque of 15 000 lb·in. and rotating at 525 rpm.

Solution

Using Equation (6-3), the power in horsepower is

$$P = \frac{Tn}{63\ 000}$$

when T is in lb·in. and n is in rpm.

$$P = \frac{(15\ 000)(525)}{63\ 000} = 125 \text{ hp}$$

6-2   TORSIONAL SHEAR STRESS IN CIRCULAR MEMBERS

When a member is subjected to an externally applied torque, an internal resisting torque must be developed in the material from which the member is made. The internal resisting torque is the result of stresses developed in the material. For torsional loading, *shear stresses* are developed.

The relationship used to compute the magnitude of the maximum shear stress due to torsion in a circular member is

$$S_{s_{max}} = \frac{Tc}{J} \quad \text{or} \quad \frac{16\ T}{\pi D^3} \tag{6-4}$$

where $T =$ applied torque
$c =$ radius of the circular member
$J =$ polar moment of inertia of the cross section of the member

For a solid circular member, having a diameter D,

$$J = \frac{\pi D^4}{32} \qquad (6-5)$$

The derivation of Equation (6-4) is now shown in order that you will better understand its proper application. Figure 6-2 illustrates what happens to a circular bar when subjected to a torque T. Considering two cross sections M and N, at different places on the bar, section N would be rotated through an angle θ relative to section M. The fibers of the material would undergo a strain which would be maximum at the outside surface of the bar and varying linearly with radial position to zero at the center of the bar. Since, for elastic materials obeying Hooke's law, stress is proportional to

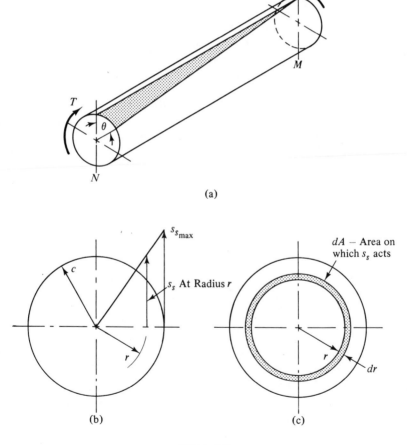

(a)

(b)

(c)

Figure 6-2

strain, the maximum stress would also occur at the outside of the bar. Part (b) of Figure 6-2 shows the cross section N of the bar. Calling the maximum shear stress $s_{s_{max}}$, the stress at any radial distance r can be expressed as

$$s_s = s_{s_{max}} \cdot \frac{r}{c} \qquad (6\text{-}6)$$

It should be noted that the shear stress s_s acts uniformly on a small ring-shaped area, dA, of the shaft, as illustrated in part (c) of Figure 6-2. Now since force equals stress times area, the force on the area dA is

$$dF = s_s \, dA = s_{s_{max}} \frac{r}{c} \, dA$$

The next step is to consider that the torque dT developed by this force is the product of dF and the radial distance to dA. Then

$$dT = dF \cdot r = s_{s_{max}} \frac{r}{c} r \, dA$$

This equation is the internal resisting torque developed on the small area dA. The total torque on the entire area would be the sum of all the individual torques on all areas of the cross section. The process of summing is accomplished by the mathematical technique of integration, illustrated as follows:

$$T = \int_A dT = \int_A s_{s_{max}} \frac{r}{c} r \, dA$$

In the process of integration, constant terms such as $s_{s_{max}}$ and c can be brought outside the integral sign, leaving

$$T = \frac{s_{s_{max}}}{c} \int_A r^2 \, dA \qquad (6\text{-}7)$$

In mechanics, the term $\int r^2 \, dA$ is given the name *polar moment of inertia* and is identified by the symbol J. Equation (6-7) can then be written

$$T = s_{s_{max}} \frac{J}{c} \qquad (6\text{-}8)$$

or

$$s_{s_{max}} = \frac{Tc}{J}$$

as stated in Equation (6-4).

Two different kinds of members can be analyzed for torsional shear stress using Equation (6-4) or Equation (6-8). They are solid circular bars and hollow circular tubes. The difference between the two is in the evaluation of J.

Refer to Figure 6-3, showing a solid circular cross section. To evaluate J from

$$J = \int r^2 \, dA$$

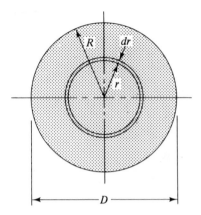

Figure 6-3

it must be seen that dA is the area of a small ring located at a distance r from the center of the section and having a thickness dr. Then

$$dA = 2\pi r \, dr$$

and

$$J = \int_0^R r^2 \, dA = \int_0^R 2\pi r^3 \, dr = \frac{2\pi R^4}{4}$$

It is usually more convenient to use diameter rather than radius. Then since $R = D/2$,

$$J = \frac{2\pi (D/2)^4}{4} = \frac{\pi D^4}{32}$$

This is the same as Equation (6-5), introduced earlier.

A similar process can be used for finding J for a hollow tube. Refer to Figure 6-4.

$$J = \int r^2 dA$$

Again, $dA = 2\pi r \, dr$. But for the hollow tube, r varies only from R_i to R_o. Then

$$J = \int_{R_i}^{R_o} r^2 (2\pi r) \, dr = 2\pi \int_{R_i}^{R_o} r^3 \, dr = \frac{2\pi (R_o^4 - R_i^4)}{4}$$

$$J = \frac{\pi}{2} (R_o^4 - R_i^4)$$

Substituting $R_o = D_o/2$ and $R_i = D_i/2$ gives

$$J = \frac{\pi}{32} (D_o^4 - D_i^4) \tag{6-9}$$

This is the equation for the polar moment of inertia for a hollow circular tube.

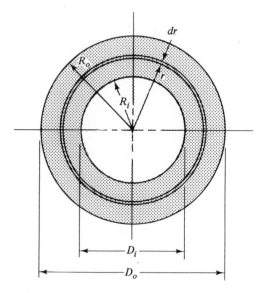

Figure 6-4

Summary of Relationships for Torsional Shear Stresses in Circular Members.

Maximum shear stress: *solid shaft*

$$S_{s_{max}} = \frac{Tc}{J} \quad \cap$$

(6-4)

occurs at the outer surface of the bar or tube where c is the radius of the bar or tube.

Shear stress at any radial position r:

$$S_s = S_{s_{max}} \frac{r}{c} = \frac{Tr}{J}$$

(6-10)

Polar moment of inertia for solid round bars:

$$J = \frac{\pi D^4}{32}$$

(6-5)

Polar moment of inertia for hollow tubes:

hollow shaft

$$J = \frac{\pi}{32}(D_o^4 - D_i^4)$$

(6-9)

Example Problem 6-4

For the socket wrench extension shown in Figure 6-1(a), compute the maximum torsional shear stress in the central portion, where the diameter is 9.5 mm. The applied torque is 10 N·m.

hollow shaft

$$S_{s \, max} = \frac{16 \cdot T \cdot D_o}{\pi (D_o^4 - D_i^4)}$$

Solution

Using Equations (6-4) and (6-5),

$$s_{s_{max}} = \frac{Tc}{J}$$

where $T = 10 \text{ N} \cdot \text{m}$

$$c = \frac{D}{2} = \frac{9.5 \text{ mm}}{2} = 4.75 \text{ mm}$$

$$J = \frac{\pi D^4}{32} = \frac{\pi (9.5 \text{ mm})^4}{32} = 800 \text{ mm}^4$$

Then

$$s_{s_{max}} = \frac{(10 \text{ N} \cdot \text{m})(4.75 \text{ mm})}{800 \text{ mm}^4} \times \frac{10^3 \text{ mm}}{\text{m}} = 59.4 \text{ N/mm}^2$$

$$s_{s_{max}} = 59.4 \text{ MPa}$$

Example Problem 6-5

For the propeller drive shaft of Figure 6-1(b), compute the torsional shear stress when it is transmitting a torque of 1.76 kN·m. The shaft is a hollow tube having an outside diameter of 60 mm and an inside diameter of 40 mm. Find the stress at both the outer and inner surfaces.

Solution

Using Equations (6-4), (6-9), and (6-10),

$$s_{s_{max}} = \frac{Tc}{J} \qquad \text{at the outer surface}$$

$$J = \frac{\pi}{32}(D_o^4 - D_i^4) = \frac{\pi}{32}(60^4 - 40^4) \text{ mm}^4 = 1.02 \times 10^6 \text{ mm}^4$$

$$c = \frac{D_o}{2} = \frac{60 \text{ mm}}{2} = 30 \text{ mm}$$

$$T = 1.76 \text{ kN} \cdot \text{m} = 1.76 \times 10^3 \text{ N} \cdot \text{m}$$

Then

$$s_{s_{max}} = \frac{Tc}{J} = \frac{(1.76 \times 10^3 \text{ N} \cdot \text{m})(30 \text{ mm})}{1.02 \times 10^6 \text{ mm}^4} \times \frac{10^3 \text{ mm}}{\text{m}}$$

$$s_{s_{max}} = 51.8 \text{ N/mm}^2 = 51.8 \text{ MPa}$$

At the inner surface,

$$s_s = s_{s_{max}} \frac{r}{c}$$

and

$$r = \frac{D_t}{2} = \frac{40 \text{ mm}}{2} = 20 \text{ mm}$$

Then

$$s_s = 51.8 \text{ MPa} \times \frac{20 \text{ mm}}{30 \text{ mm}} = 34.5 \text{ MPa}$$

Example Problem 6-6

Calculate the torsional shear stress which would be developed in a solid circular shaft having a diameter of 1.25 in. if it is transmitting 125 hp while rotating at 525 rpm.

Solution

The first steps in a problem involving power transmission by shafts is to determine the torque developed in the shaft. Since this problem is stated with English units, Equation (6-3) can be used.

$$\text{Power} = P = \frac{Tn}{63\ 000}$$

Solving for the torque T,

$$T = \frac{63\ 000P}{n}$$

Recall that this equation will give the value of torque directly in lb·in. when P is in horsepower and n is in rpm. Then

$$T = \frac{63\ 000(125)}{525} = 15\ 000 \text{ lb·in.}$$

Now the torsional shear stress can be computed using Equation (6-4).

$$s_{s\text{max}} = \frac{Tc}{J}$$

where $c = $ radius of shaft $= \dfrac{D}{2}$

$$c = \frac{1.25 \text{ in.}}{2} = 0.625 \text{ in.}$$

$$J = \frac{\pi D^4}{32} = \frac{\pi (1.25 \text{ in.})^4}{32} = 0.240 \text{ in.}^4$$

Then

$$s_{s\text{max}} = \frac{Tc}{J} = \frac{(15\ 000 \text{ lb·in.})(0.625 \text{ in.})}{0.240 \text{ in.}^4}$$

$$s_{s\text{max}} = 39\ 100 \text{ psi}$$

6-3 DESIGN OF CIRCULAR MEMBERS UNDER TORSION

In a design problem, the loading on a member is known, and it is required to determine the geometry of the member to ensure that it will carry the loads safely. Material selection and the determination of design stresses are integral parts of the design process. *The techniques developed in this section are for circular members only, subjected only to torsion.* Of course, both solid and hollow circular members are covered. Torsion in noncircular members is covered in a later section of this chapter. The combination of torsion with bending and axial loads is presented in Chapter 11.

The basic torsional shear stress equation, Equation (6-4), was expressed as

$$s_{s\,max} = \frac{Tc}{J} \qquad (6\text{-}4)$$

In design, we can substitute a certain design stress s_{sd} for $s_{s\,max}$. As in the case of members subjected to direct shear stress and made of ductile materials, the design stress is related to the yield strength of the material in shear. That is,

$$s_{sd} = \frac{s_{ys}}{N} \leftarrow pg\ 42$$

where N is the design factor chosen by the designer based on the manner of loading. Table 3-1 can be used as a guide to determine the value of N. Where the data for s_{ys} are not available, the value can be estimated as $s_y/2$. This will give a reasonable, and usually conservative, estimate for ductile metals, especially steel.

The torque T would be known in a design problem. Then, in Equation (6-4), only c and J are left to be determined. Notice that both c and J are properties of the geometry of the member which is being designed. For solid circular members (shafts), the geometry is completely defined by the diameter. It has been shown that

$$c = \frac{D}{2}$$

and

$$J = \frac{\pi D^4}{32}$$

It is now convenient to note that if the quotient J/c is formed, a simple expression involving D is obtained.

$$\frac{J}{c} = \frac{\pi D^4}{32} \cdot \frac{1}{D/2} = \frac{\pi D^3}{16} \qquad (6\text{-}11)$$

103

In the study of strength of materials, the term J/c is given the name *polar section modulus*, and the symbol Z_p is used to denote it.

Substituting Z_p for J/c in Equation (6-4) gives

$$s_{s_{max}} = \frac{T}{Z_p} \tag{6-12}$$

To use this equation in design, we can let $s_{s_{max}} = s_{sd}$ and then solve for Z_p.

$$Z_p = \frac{T}{s_{sd}} \tag{6-13}$$

Equation (6-13) will give the required value of the polar section modulus of a circular shaft to limit the torsional shear stress to s_{sd} when subjected to a torque T. Then Equation (6-11) can be used to find the required diameter of a solid circular shaft. If a hollow shaft is to be designed,

$$Z_p = \frac{J}{c} = \frac{\pi}{32}(D_o^4 - D_i^4) \cdot \frac{1}{D_o/2}$$
$$Z_p = \frac{\pi}{16} \frac{D_o^4 - D_i^4}{D_o} \tag{6-14}$$

In this case, one of the diameters *or* the relationship between the two diameters would have to be specified in order to solve for the complete geometry of the hollow shaft.

Example Problem 6-7

The final drive to a conveyor which feeds coal into a railroad car is a shaft loaded in pure torsion and carrying 800 N·m of torque. Determine the required diameter of the shaft if it is to have a solid circular cross section. Specify a suitable steel for the shaft.

Solution

In selecting a material, it should be noted that many alloys could be used. Because of the rough loading likely from a coal conveyor, a high ductility is desirable. Also, machinability would be important, as some machining would most likely be required on the shaft. Alloy AISI 1137 OQT 1300, from Table A-1, fits these requirements. The 1100 series steels generally have good machinability, and the 28 percent elongation indicates good ductility. The yield strength is 414 MPa.

A design factor of $N = 6$ would be reasonable based on yield strength since the shaft would likely be subjected to some shock or impact loading. Then

$$s_{sd} = \frac{s_y}{2N} = \frac{414 \text{ MPa}}{2(6)} = 34.5 \text{ MPa}$$

Now, from Equation (6-13), the required polar section modulus of the shaft can be found.

$$Z_p = \frac{T}{S_{sd}} = \frac{800 \text{ N} \cdot \text{m}}{34.5 \text{ MPa}}$$

Restating the units in other forms gives

$$Z_p = \frac{T}{S_{sd}} = \frac{800 \text{ N} \cdot \text{m}}{34.5 \text{ N/mm}^2} \cdot \frac{10^3 \text{ mm}}{\text{m}} = 23.2 \times 10^3 \text{ mm}^3$$

Now the diameter can be found using Equation (6-11).

$$Z_p = \frac{J}{c} = \frac{\pi D^3}{16}$$

Then

$$D = \sqrt[3]{\frac{16Z_p}{\pi}} = \sqrt[3]{\frac{16(23.2 \times 10^3) \text{ mm}^3}{\pi}}$$

$$D = 49.1 \text{ mm}$$

A preferred nominal size of 50 mm should be specified.

Example Problem 6-8

An alternative design for the shaft described in Example Problem 6-7 would be to use a hollow tube for the shaft. Assume that a tube having an outside diameter of 60 mm is available in the same material as specified for the solid shaft (AISI 1137 OQT 1300). Compute what maximum inside diameter the tube can have which would result in the same stress in the steel as the 50-mm solid shaft.

32.4

Solution

It was shown in Example Problem 6-7 that the 50-mm solid shaft would encounter a torsional shear stress less than 34.5 MPa. The actual stress in the shaft caused by the 800-N·m applied torque would be

$$S_{s_{max}} = \frac{T}{Z_p}$$

For the 50-mm diameter shaft,

$$Z_p = \frac{\pi D^3}{16} = \frac{\pi (50)^3 \text{ mm}^3}{16} = 24.5 \times 10^3 \text{ mm}^3$$

Then

$$S_{s_{max}} = \frac{800 \text{ N} \cdot \text{m}}{24.5 \times 10^3 \text{ mm}^3} \times \frac{10^3 \text{ mm}}{1 \text{ m}} = 32.6 \text{ N/mm}^2 = 32.6 \text{ MPa}$$

Since torsional shear stress is inversely proportional to the polar section modulus, it is necessary that the hollow tube have the same value for Z_p as does the solid shaft if it is to experience the same stress. For the hollow tube,

$$Z_p = \frac{\pi}{16} \frac{D_o^4 - D_i^4}{D_o}$$

The outside diameter D_o is known to be 60 mm. We can then solve for the required inside diameter D_i.

$$D_i = \left[D_o^4 - \frac{16 Z_p D_o}{\pi} \right]^{1/4}$$

Now, for $D_o = 60$ mm and $Z_p = 24.5 \times 10^3$ mm³,

$$D_i = \left[(60)^4 - \frac{(16)(24.5 \times 10^3)(60)}{\pi} \right]^{1/4} \text{mm}$$

$$D_i = 48.4 \text{ mm}$$

This is the maximum allowable inside diameter, since to use a larger value for D_i would result in a thinner tube and higher stress. Figure 6-5 shows a comparison of the two designs for the shaft, drawn full scale.

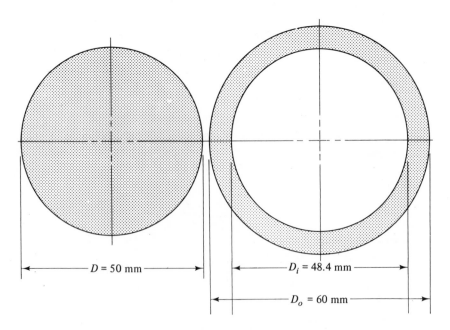

$D = 50$ mm $D_i = 48.4$ mm

$D_o = 60$ mm

Figure 6-5

6-4 COMPARISON OF SOLID AND HOLLOW CIRCULAR MEMBERS

In many design situations, economy of material usage is a major criterion of performance for a product. In aerospace applications, every reduction in the mass of the aircraft or space vehicle allows increased payload. Automobiles achieve higher fuel economy when they are lighter. Also, since raw materials are purchased on a price per unit mass basis, a lighter part generally costs less.

Providing economy of material usage for load-carrying members requires that all the material in the member be stressed to a level approaching the safe design stress. Then every portion is carrying its share of the load.

Example Problems 6-7 and 6-8 can be used to illustrate this point. Recall that both designs shown in Figure 6-5 result in the same maximum torsional shear stress in the steel shaft. The hollow shaft is slightly larger in outside diameter, but it is the *volume* of metal which determines the mass of the shaft. Consider a length of shaft 1.0 m long. For the solid shaft, the volume is the cross-sectional area times the length.

$$V_S = AL = \frac{\pi D^2}{4} L$$

$$V_S = \frac{\pi (50 \text{ mm})^2}{4} \cdot 1.0 \text{ m} \cdot \frac{1 \text{ m}^2}{(10^3 \text{ mm})^2} = 1.96 \times 10^{-3} \text{ m}^3$$

The mass is the volume times the density. Table A-1 gives the density of steel to be 7680 kg/m³. Then the mass of the solid shaft is

$$M_S = 1.96 \times 10^{-3} \text{ m}^3 \cdot 7680 \text{ kg/m}^3 = 15.1 \text{ kg}$$

Now for the hollow shaft the volume is

$$V_H = AL = \frac{\pi}{4}(D_o^2 - D_i^2)(L)$$

$$V_H = \frac{\pi}{4}(60^2 - 48.4^2)\text{mm}^2(1.0 \text{ m}) \cdot \frac{1 \text{ m}^2}{(10^3 \text{ mm})^2}$$

$$V_H = 0.988 \times 10^{-3} \text{ m}^3$$

The mass of the hollow shaft is

$$M_H = 0.988 \times 10^{-3} \text{ m}^3 \cdot 7680 \text{ kg/m}^3 = 7.58 \text{ kg}$$

Thus it can be seen that the hollow shaft has almost exactly *one-half the mass* of the solid shaft, even though both are subjected to the same stress level for a given applied torque. Why?

The reason for the hollow shaft being lighter is that a greater portion of its material is being stressed to a higher level than in the solid shaft. Figure 6-6 shows a sketch of the stress distribution in the solid shaft. The maximum stress, 32.6 MPa, occurs at the outer surface. The stress then varies linearly with the radius for other points within the shaft to *zero* at the center. From this it can be seen that the material near the middle of the shaft is not being used efficiently.

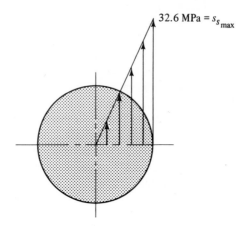

32.6 MPa = $s_{s_{max}}$

Figure 6-6

Contrast this with the sketch of the hollow shaft in Figure 6-7. Again the stress at the outer surface is the maximum, 32.6 MPa. The stress at the inner surface of the hollow shaft can be found from Equation (6-10).

$$s_s = s_{s_{max}} \frac{r}{c}$$

At the inner surface, $r = D_i/2 = 48.4$ mm$/2 = 24.2$ mm. Also, $c = D_o/2 = 60$ mm$/2 = 30$ mm. Then

$$s_s = 32.6 \text{ MPa} \frac{24.2}{30} = 26.3 \text{ MPa}$$

The stress at points between the inner and outer surfaces varies linearly with the radius to each point. Thus it can be seen that all of the material in the hollow shaft shown in Figure 6-7 is being stressed to a fairly high but safe level. This illustrates why the hollow section requires less material.

Of course the specific data used in the above illustration cannot be generalized to all problems. However, it can be said that for torsional loading of circular members, a hollow section can be designed which is lighter than a solid section while subjecting the material to the same maximum torsional shear stress.

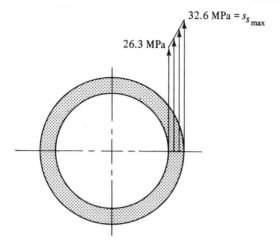

Figure 6-7

6-5   STRESS CONCENTRATIONS
IN TORSIONALLY LOADED MEMBERS

Changes in the cross section of a member loaded in torsion cause the local stress near the changes to be higher than would be predicted by using the torsional shear stress formula. The actual level of stress in such cases is determined experimentally. Then a *stress concentration factor* is determined which allows the stress in similar designs to be computed from the relationship

$$s_s = K_t s_{s_{nom}}$$ (6-15)

The term $s_{s_{nom}}$ is the nominal stress due to torsion which would be developed in the parts if the stress concentration were not present. Thus the standard torsional shear stress formulas, Equation (6-4) and Equation (6-12), can be used to compute the nominal stress. The value of K_t is a factor by which the actual stress is greater than the nominal stress.

Three cases where stress concentrations occur and which are encountered frequently are illustrated in Figures 6-8, 6-9, and 6-10. A shaft carrying gears, pulleys, chain sprockets, or other power-transmitting devices may have a key seat cut into it. In Figure 6-8(a), a key transfers the torque from the motor shaft to the hub of the pulley. The key seat is usually cut with a circular milling cutter, resulting in the sled-runner type shown in Figure 6-8(b). For this, the value of K_t is taken as 1.6. If an end mill or profile milling cutter is used, the profile type key seat shown in Figure 6-8(c) is created. Because of the more abrupt end of this key seat, higher stresses are developed, and the value of K_t is about 2.0.

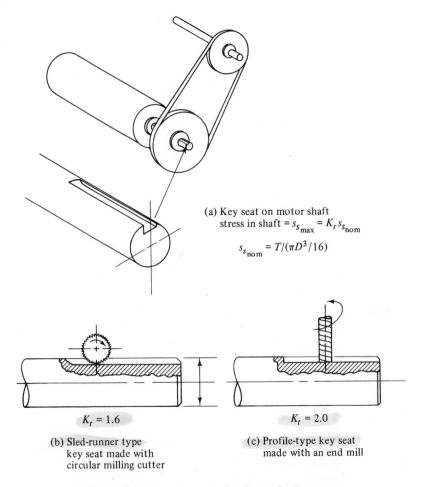

(a) Key seat on motor shaft
stress in shaft $= s_{s_{max}} = K_t s_{s_{nom}}$

$$s_{s_{nom}} = T/(\pi D^3/16)$$

$K_t = 1.6$

$K_t = 2.0$

(b) Sled-runner type
key seat made with
circular milling cutter

(c) Profile-type key seat
made with an end mill

Figure 6-8 Stress Concentration Factors for Keyseats

The axial location of parts on a shaft is often maintained by placing a circular ring in a groove cut in the shaft, as shown in Figure 6-9. The ring provides a shoulder against which to locate a gear, bearing, pulley, or chain sprocket. The groove in the shaft is a point of stress concentration. For this case, the value of K_t is dependent on the geometry of the groove and on the relative depth of the groove. The graph shown in Figure 6-9 can be used to determine K_t if the precise geometry is known. Frequently the radius at the bottom of the groove is quite small and not accurately known. For this case, using $K_t = 3.0$ would be reasonable. Notice that the nominal stress to be used in Equation (6-15) is to be computed using the root diameter of the groove.

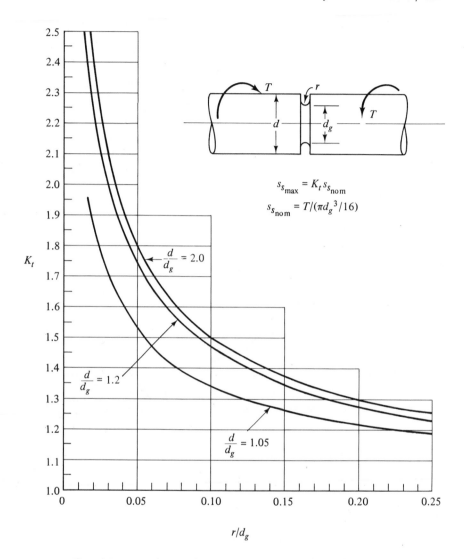

Figure 6-9   Stress Concentration Factor for Grooved Shaft in Torsion

Stepped shafts provide shoulders for locating parts, as shown in Figure 5-10. The stress concentration due to the change in diameter is dependent on the relative diameters at the shoulder and on the sharpness of the fillet provided at the base of the shoulder. To reduce the stress concentration, a large fillet radius should be provided if possible.

The use of stress concentration factors is illustrated in the following example problem.

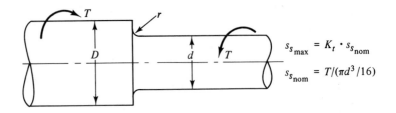

$$ss_{max} = K_t \cdot ss_{nom}$$

$$ss_{nom} = T/(\pi d^3/16)$$

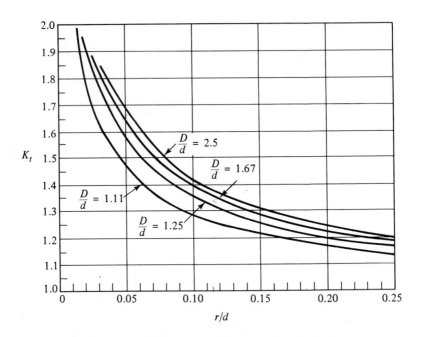

Figure 6-10 Stress Concentration Factor for Stepped Shaft in Torsion.

Example Problem 6-9

Figure 6-11 shows a portion of a shaft on which a pulley is mounted. The pulley rests against a shoulder at its left face. A ring in a groove locates the pulley on its right face. A profile key seat is used to transmit 20 N·m of torque from the shaft to the pulley. Compute the stress in the shaft at the sections 1, 2, 3, 4, and 5. Assume that the torque is the same at all sections. Then, using a design factor $N = 4$ based on yield strength, specify a suitable material for the shaft.

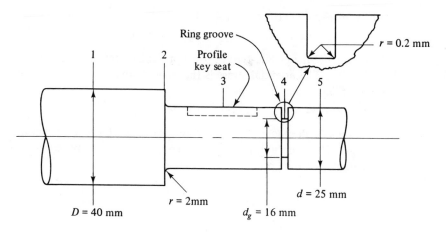

Figure 6-11

Solution

The stress at *section 1* is computed for the larger 40-mm diameter portion of the shaft with no stress concentration. Then

$$s_s = \frac{T}{Z_p}$$

$$Z_p = \frac{\pi D^3}{16} = \frac{\pi (40 \text{ mm})^3}{16} = 12\ 570 \text{ mm}^3$$

$$s_s = \frac{20 \text{ N} \cdot \text{m}}{12\ 570 \text{ mm}^3} \cdot \frac{10^3 \text{ mm}}{\text{m}} = 1.59 \frac{\text{N}}{\text{mm}^2} = 1.59 \text{ MPa}$$

At *section 2*, the stress concentration due to the shoulder fillet must be considered. See Figure 6-10.

$$s_s = K_t s_{s_{\text{nom}}}$$

$$s_{s_{\text{nom}}} = \frac{T}{\pi d^3 / 16} = \frac{20 \text{ N} \cdot \text{m}}{[\pi (25)^3 / 16] \text{ mm}^3} \cdot \frac{10^3 \text{ mm}}{\text{m}}$$

$$s_{s_{\text{nom}}} = 6.52 \text{ N/mm}^2 = 6.52 \text{ MPa}$$

The value of K_t depends on the ratios D/d and r/d.

$$\frac{D}{d} = \frac{40 \text{ mm}}{25 \text{ mm}} = 1.60$$

$$\frac{r}{d} = \frac{2 \text{ mm}}{25 \text{ mm}} = 0.08$$

Then from Figure 6-10, $K_t = 1.45$. Then

$$s_s = (1.45)(6.52 \text{ MPa}) = 9.45 \text{ MPa}$$

At *section 3*, the profile-type key seat presents a stress concentration factor of 2.0. The nominal stress is the same as that computed at the shoulder fillet. Then

$$s_s = K_t s_{s\text{nom}} = (2.0)(6.52 \text{ MPa}) = 13.04 \text{ MPa}$$

Section 4 is the location of the ring groove of the type shown in Figure 6-9. Here the nominal stress is computed on the basis of the root diameter of the groove.

$$s_{s\text{nom}} = \frac{T}{\pi d_g^3/16} = \frac{20 \text{ N·m}}{[\pi(16)^3/16] \text{ mm}^3} \cdot \frac{10^3 \text{ mm}}{\text{m}} = 24.9 \frac{\text{N}}{\text{mm}^2}$$

$$s_{s\text{nom}} = 24.9 \text{ MPa}$$

The value of K_t depends on d/d_g and r/d_g.

$$\frac{d}{d_g} = \frac{25 \text{ mm}}{19 \text{ mm}} = 1.56$$

$$\frac{r}{d_g} = \frac{0.2 \text{ mm}}{16 \text{ mm}} = 0.013$$

From Figure 6-9, $K_t = 2.5$. Then

$$s_s = K_t s_{s\text{nom}} = (2.5)(24.9 \text{ MPa}) = 62.3 \text{ MPa}$$

Section 5 is in the smaller portion of the shaft, where no stress concentration occurs. Then

$$s_s = \frac{T}{Z_p} = \frac{T}{\pi d^3/16}$$

Notice that this is identical to the nominal stress computed for sections 2 and 3. Then at section 5,

$$s_s = 6.52 \text{ MPa}$$

It can be seen that there is a large variation in the magnitude of the torsional shear stress at the various points around the pulley mounted on the shaft. The maximum computed stress occurs at section 4, where the ring groove is cut into the shaft. There, both the smallest diameter and the largest value of K_t occur. The selection of a material must be based on the stress at section 4. Letting that stress, 62.3 MPa, be the design stress, the required yield strength for the material is

$$s_y = s_{sd} \cdot 2N = 62.3 \text{ MPa } (2)(4) = 498 \text{ MPa}$$

From Table A-1, two suitable materials would be AISI 1050, cold-drawn, and AISI 4130 OQT 1300. Certainly other alloys and heat treatments could be used.

6-6 TWISTING—ELASTIC TORSIONAL DEFORMATION

Stiffness in addition to strength is an important design consideration for torsionally loaded members. The measure of torsional stiffness is the angle of twist of one part of a shaft relative to another part when a certain torque is applied.

In mechanical power transmission applications, excessive twisting of a shaft may cause vibration problems which would result in noise and improper synchronization of moving parts. One guideline for torsional stiffness is that the shaft should not twist more than 0.08 degrees per foot of length (0.0046 rad/m).

In structural design, load-carrying members are sometimes loaded in torsion as well as tension or bending. The rigidity of the structure then depends on the torsional stiffness of the components. Any load applied off from the axis of a member and transverse to the axis will produce torsion. This section will discuss twisting of circular members, both solid and hollow. Noncircular sections will be covered in a later section. It is very important to note that the behavior of an open-section shape such as a channel or angle is much different from that of a closed section such as a pipe or rectangular tube. In general, the open sections have very low torsional stiffness.

To aid in the development of the relationship for computing the angle of twist of a circular member, consider the shaft shown in Figure 6-12. One end of the shaft is held fixed while a torque T is applied to the other end. Under these conditions the shaft will twist between the two ends through an angle θ. The amount of twist depends on the applied torque T, the length of the shaft L, its polar moment of inertia J, and the stiffness of the shaft material in shear, called G, the shear modulus of elasticity. The governing equation for angle of twist is

$$\theta = \frac{TL}{JG} \tag{6-16}$$

Careful attention should be paid to units in this equation. For any set of consistent units it will appear that all units cancel, making the angle θ dimensionless. Actually the resulting unit for θ is *radians*. This same phenomenon was observed in Section 6-1 during the discussion of power. The use of Equation (6-16) is illustrated by the following example problems. Then the derivation of the formula is shown.

The material property involved in Equation (6-16) is G, the shear modulus of elasticity. It is a measure of how much deformation the material would undergo when subjected to a shear stress. Since torsion produces shear stress, G is the property which relates to the stiffness of the material. The units for G are force per unit area: N/m^2 or Pa in the SI units, and $lb/in.^2$ in the English

115

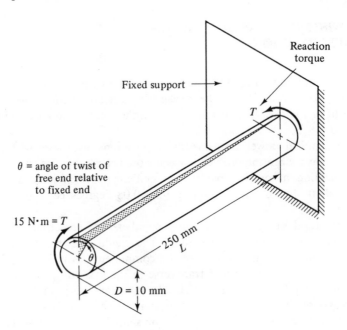

Figure 6-12

units. For common materials, G has the values given in Table 6-1. The shear modulus of elasticity G is related to shear stress in the same manner as the tensile modulus of elasticity E is related to tensile stress. That is, G is the ratio of shear stress to shear strain,

$$G = \frac{S_s}{\gamma} \qquad (6\text{-}17)$$

The term γ (Greek letter gamma) is the shearing strain measured by an angle in radians. Remember, in Chapter 2, E was defined as

$$E = \frac{s}{\epsilon}$$

where ϵ is the tensile strain occurring as a resulf of the tensile stress s.

TABLE 6-1

Material	Shear Modulus G	
	GPa	psi
Plain carbon and alloy steels	80	11.5×10^6
Stainless steel Type 304	69	10.0×10^6
Aluminum 6061-T6	26	3.75×10^6
Beryllium copper	48	7.0×10^6
Magnesium	17	2.4×10^6
Titanium alloy	43	6.2×10^6

Example Problem 6-10

Determine the angle of twist between two sections 250 mm apart in a steel rod having a diameter of 10 mm when a torque of 15 N·m is applied. Figure 6-12 shows a sketch of the arrangement.

Solution

Equation (6-16) can be used directly.

$$\theta = \frac{TL}{JG}$$

For steel, $G = 80$ GPa $= 80 \times 10^9$ N/m². The value of J is

$$J = \frac{\pi D^4}{32} = \frac{\pi (10 \text{ mm})^4}{32} = 982 \text{ mm}^4$$

Then

$$\theta = \frac{TL}{JG} = \frac{(15 \text{ N·m})(250 \text{ mm})}{(982 \text{ mm}^4)(80 \times 10^9 \text{ N/m}^2)} \times \frac{(10^3 \text{ mm})^3}{1 \text{ m}^3}$$

$$\theta = 0.048 \text{ rad}$$

Note that all units cancel. Expressing the angle in degrees,

$$\theta = 0.048 \text{ rad} \times \frac{180 \text{ deg}}{\pi \text{ rad}} = 2.73 \text{ deg}$$

Example Problem 6-11

Determine the required diameter of a round shaft made of aluminum alloy 6061-T6 if it is to twist not more than 0.08 degree in one foot of length when a torque of 75 lb·in. is applied.

Solution

Equation (6-16) must be solved for J since that factor is related to the unknown diameter of the shaft and all other factors are specified.

$$\theta = \frac{TL}{JG}$$

$$J = \frac{TL}{\theta G}$$

The angle of twist must be expressed in radians.

$$\theta = 0.08 \text{ deg} \times \frac{\pi \text{ rad}}{180 \text{ deg}} = 0.0014 \text{ rad}$$

From Table 6-1, $G = 3.75 \times 10^6$ psi. Then

$$J = \frac{TL}{\theta G} = \frac{(75 \text{ lb·in.})(12 \text{ in.})}{(0.0014)(3.75 \times 10^6 \text{ lb/in.}^2)} = 0.171 \text{ in.}^4$$

Now since $J = \pi D^4/32$,

$$D = \left(\frac{32J}{\pi}\right)^{1/4} = \left[\frac{(32)(0.171 \text{ in.}^4)}{\pi}\right]^{1/4}$$

$$D = 1.15 \text{ in.}$$

Equation (6-16) applies to any circular member, solid or hollow, which is subjected to a uniform level of torque throughout its length and which has a uniform cross section. If a member has varying torque or cross section, it can be divided into sections where these values are uniform for purposes of analysis. Then the twist for all sections can be summed algebraically to get the total angle of twist.

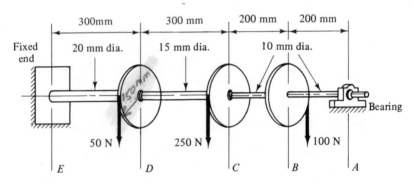

Figure 6-13

Example Problem 6-12

Figure 6-13 shows a steel rod to which three disks are attached. The rod is fixed against rotation at its left end, but free to rotate in a bearing at its right end. Each disk is 300 mm in diameter. Downward forces act at the outer surfaces of the disks so that torques are applied to the rod. Determine the angle of twist of the section A relative to the fixed section E.

Solution

Obviously the diameter of the rod and the magnitude of the applied torque vary along the length of the rod. However, for each segment AB, BC, CD, and DE the conditions are constant. Then Equation (6-16) will be applied separately to each segment.

The first thing to be done is to determine the torque applied to the shaft at the points B, C, and D due to the loads on the disks. Since torque is the product of force times distance, we have:

Torque on disk B, clockwise:

$$T_B = (100 \text{ N})(150 \text{ mm}) = 15\ 000 \text{ N} \cdot \text{mm} = 15 \text{ N} \cdot \text{m}$$

Torque on disk C, counterclockwise:

$$T_C = (250 \text{ N})(150 \text{ mm}) = 37\ 500 \text{ N} \cdot \text{mm} = 37.5 \text{ N} \cdot \text{m}$$

Torque on disk D, counterclockwise:

$$T_D = (50 \text{ N})(150 \text{ mm}) = 7500 \text{ N} \cdot \text{mm} = 7.5 \text{ N} \cdot \text{m}$$

Now, to compute the angle of twist in the rod, the level of internal torque *in each segment of the rod* must be determined. Here, the use of the free-body approach can be helpful. Start first at the right end of the rod from the bearing A up to *but not including B*. If the rod were cut at a section to the right of B, there would be no internal torque because no torque is applied to that segment. Therefore T_{AB}, the torque in the rod from A to B, is zero.

For segment BC, cut the shaft just to the right of C and observe that the internal torque in the rod must be 15 N·m to balance the applied torque at B. Then $T_{BC} = 15$ N·m.

For any section in the segment CD, the torque in the rod would be the difference between T_C and T_B. Then

$$T_{CD} = T_C - T_B = 37.5 \text{ N} \cdot \text{m} - 15 \text{ N} \cdot \text{m} = 22.5 \text{ N} \cdot \text{m}$$

Between D and E in the rod, the torque is the resultant of all the applied torques at D, C, and B.

$$T_{DE} = T_D + T_C - T_B = 7.5 \text{ N} \cdot \text{m} + 37.5 \text{ N} \cdot \text{m} - 15 \text{ N} \cdot \text{m}$$

$$T_{DE} = 30 \text{ N} \cdot \text{m}$$

The fixed support at E must be capable of providing a reaction torque of 30 N·m to maintain the rod in equilibrium.

In summary, the distribution of torque in the rod can be shown in graphical form as in Figure 6-14. Notice that the applied torques T_B, T_C, and T_D are the *changes* in torque which occur at B, C, and D but that they are not the magnitudes of the torque *in the rod* at those points.

Now the computation for angle of twist can be made.

Segment AB

$$\theta_{AB} = T_{AB} \frac{L}{JG}$$

Since $T_{AB} = 0$, $\theta_{AB} = 0$. There is no twisting of the rod between A and B.

Segment BC

$$\theta_{BC} = T_{BC} \frac{L}{JG}$$

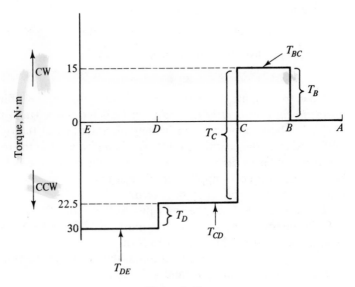

Figure 6-14

We know $T_{BC} = 15 \text{ N·m}$, $L = 200 \text{ mm}$, and $G = 80$ GPa for steel. For the 10-mm diameter rod,

$$J = \frac{\pi D^4}{32} = \frac{\pi (10 \text{ mm})^4}{32} = 982 \text{ mm}^4$$

Then

$$\theta_{BC} = \frac{(15 \text{ N·m})(200 \text{ mm})}{(982 \text{ mm}^4)(80 \times 10^9 \text{ N/m}^2)} \times \frac{(10^3)^3 \text{ mm}^3}{\text{m}^3} = 0.038 \text{ rad}$$

This means that section B is rotated 0.038 rad clockwise relative to section C, since θ_{BC} is the total angle of twist in the segment BC.

Segment CD

$$\theta_{CD} = T_{CD} \frac{L}{JG}$$

Here, $T_{CD} = 22.5 \text{ N·m}$, $L = 300 \text{ mm}$, and the rod diameter is 15 mm. Then

$$J = \frac{\pi D^4}{32} = \frac{\pi (15 \text{ mm})^4}{32} = 4970 \text{ mm}^4$$

$$\theta_{CD} = \frac{(22.5 \text{ N·m})(300 \text{ mm})}{(4970 \text{ mm}^4)(80 \times 10^9 \text{ N/m}^2)} \times \frac{(10^3)^3 \text{ mm}^3}{\text{m}^3} = 0.017 \text{ rad}$$

Section C is rotated 0.017 rad counterclockwise relative to Section D.

Segment DE

$$\theta_{DE} = T_{DE}\frac{L}{JG}$$

Here, $T_{DE} = 30$ N·m, $L = 300$ mm, and $D = 20$ mm. Then

$$J = \frac{\pi D^4}{32} = \frac{\pi(20 \text{ mm})^4}{32} = 15\ 700 \text{ mm}^4$$

$$\theta_{DE} = \frac{(30 \text{ N·m})(300 \text{ mm})}{(15\ 700 \text{ mm}^4)(80 \times 10^9 \text{ N/mm}^2)} \times \frac{(10^3)^3 \text{ mm}^3}{\text{m}^3} = 0.007 \text{ rad}$$

Section D is rotated 0.007 rad counterclockwise relative to E.

Total Angle of Twist from E to A

$$\theta_{AE} = \theta_{AB} + \theta_{BC} - \theta_{CD} - \theta_{DE}$$
$$\theta_{AE} = 0 + 0.038 - 0.017 - 0.007 = 0.014 \text{ rad}$$

The angle of twist at any section can be found by referring to the graph in Figure 6-15. Notice that the angle varies linearly with position within any segment.

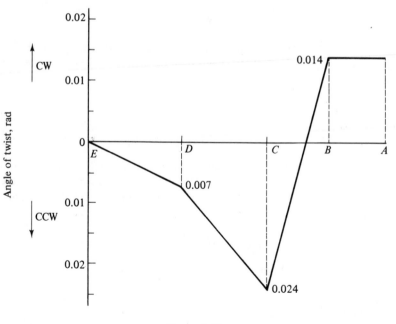

Figure 6-15

6-7 DERIVATION OF ANGLE-OF-TWIST FORMULA

The derivation of the angle-of-twist formula, Equation (6-16), depends on a few basic assumptions about the behavior of a circular member when subjected to torsion. Figure 6-16 illustrates this behavior. As the torque is applied, an element along the outer surface of the member, which was initially

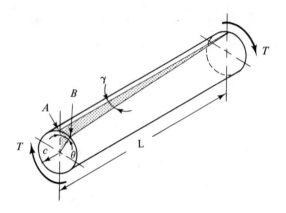

Figure 6-16

straight, rotates through an angle γ (gamma). Likewise, a radius of the member rotates through an angle θ. It can then be seen that arc length AB can be expressed as either

$$AB = \gamma L \qquad (6\text{-}18)$$

or

$$AB = \theta c \qquad (6\text{-}19)$$

where θ is the angle of twist of the member over a length L. The angle γ is a measure of the maximum shearing strain at the outer surface of the member. Recalling the definition of the shear modulus of elasticity from the preceding section, the shear strain can be related to the shear stress.

$$G = \frac{S_s}{\gamma}$$

At the outer surface, then,

$$S_{s_{max}} = G\gamma$$

From the torsional shear stress formula,

$$S_{s_{max}} = \frac{Tc}{J}$$

122

Then

$$Gy = \frac{Tc}{J} \qquad (6\text{-}20)$$

Equations (6-18) and (6-19) can be used to relate the angle of twist θ to γ.

$$\gamma = \frac{\theta c}{L} \qquad (6\text{-}21)$$

Substituting into Equation (6-20) gives

$$\frac{G\theta c}{L} = \frac{Tc}{J}$$

Now, solving for θ,

$$\theta = \frac{TL}{JG}$$

This is Equation (6-16), as used earlier.

6-8   POWER TRANSMISSION SHAFTING

Examples were given in Section 6-2 illustrating the computation of the torsional shear stress in simple power-transmitting shafts where the torque was constant along the entire length of interest. The direct application of the torsional shear stress formula is all that was required. In this section we will work with shafts carrying several power-transmitting elements. Here it will be necessary to determine the distribution of torque in the shaft before proceeding with the design of the shaft.

It must be understood that only the effects of torsion are considered in these examples. In reality, both torsion and bending exist in shafts, and their combined effects must be considered to insure a safe design. Combined stresses are presented in a later chapter.

Figure 6-17 shows an example where variations in the torque in the shafts occur. All power is delivered to shaft 1 from a motor through a set of gears at A. The gears at B deliver part of the power to an adjacent shaft to drive a conveyor. At C, a set of gears delivers power to a blower. In problems such as this, it is assumed that the system is in rotational equilibrium, with all elements rotating at a constant speed. Therefore all the input power must be balanced by the output power. Also, the driving torque at the input must be balanced by the sum of resisting torques at the outputs. Now, using the data in Figure 6-17, we can see that the portion of the shaft from C to B has to transmit only 7.5 kW of power required by the blower. The portion from A to B, however, must transmit 20 kW for the conveyor *plus* 7.5 kW for the blower, or a total of 27.5 kW. Also, the motor must then supply 27.5 kW to

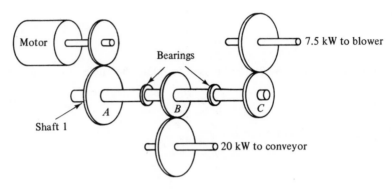

Figure 6-17

the shaft at *A*. The torque in the shaft can be computed from the power transmitted and the rotational speed of the shaft.

Example Problem 6-13

For shaft 1 shown in Figure 6-17, compute the torsional shear stress in all parts of the shaft if it rotates at 80 rad/s. The shaft has a uniform diameter of 40 mm throughout its length, and all gears are keyed to the shaft with sled-runner key seats.

Solution

First let's compute the torque in each segment of shaft 1. Equation (6-2) states that power = *Tn*. Then

metric

$$T = \frac{\text{power}}{n}$$

For segment *BC*, power = 7.5 kW = 7.5(10^3) N·m/s. Then

$$T_{BC} = \frac{7.5(10^3) \text{ N·m/s}}{80 \text{ rad/s}} = 93.8 \text{ N·m}$$

For segment *AB*, power = 27.5 kW = 27.5(10^3) N·m/s. Then

$$T_{AB} = \frac{27.5(10^3) \text{ N·m/s}}{80 \text{ rad/s}} = 344 \text{ N·m}$$

The variation in torque is graphed in Figure 6-18.

Stresses can be computed from the torsional shear stress formula modified to account for the stress concentration due to the key seats at sections *A*, *B*, and *C*.

$P = TN$

$$s_s = \frac{TK_t}{Z_p}$$

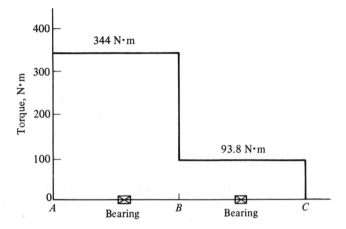

Figure 6-18

For the entire shaft,

$$Z_p = \frac{\pi D^3}{16} = \frac{\pi (40 \text{ mm})^3}{16} = 12.6(10^3)\text{mm}^3$$

At each key seat, $K_t = 1.6$ (Figure 6-8). Then at section A,

$$S_{S_A} = \frac{T_{AB} K_t}{Z_p} = \frac{(344 \text{ N} \cdot \text{m})(1.6)}{12.6(10^3) \text{ mm}^3} \times \frac{10^3 \text{ mm}}{\text{m}}$$

$$S_{S_A} = 43.7 \text{ N/mm}^2 = 43.7 \text{ MPa}$$

Notice that at section B the torque in the shaft changes from 344 N·m to the left of B to 93.8 N·m to the right of B. Of course the higher value must be used to compute the stress at B. Since this is the same as the data used to compute the stress at A, the stress at B is also 43.7 MPa.

Between sections A and B, the torque in the shaft is still 344 N·m. But there is no stress concentration since there are no changes in the shaft geometry. Therefore,

$$S_{S_{AB}} = \frac{T}{Z_p} = \frac{344 \text{ N} \cdot \text{m}}{12.6(10^3) \text{ mm}^3} \times \frac{10^3 \text{ mm}}{\text{m}} = 27.3 \text{ MPa}$$

At section C, the torque is 93.8 N·m, and the stress concentration due to the key seat exists. Then

$$S_{S_C} = \frac{T_{BC} K_t}{Z_p} = \frac{93.8 \text{ N} \cdot \text{m} (1.6)}{12.6(10^3) \text{ mm}^3} \times \frac{10^3 \text{ mm}}{\text{m}} = 11.9 \text{ MPa}$$

Between B and C, there is no stress concentration. Then

$$S_{S_{BC}} = \frac{T_{BC}}{Z_p} = \frac{93.8 \text{ N} \cdot \text{m}}{12.6(10^3) \text{ mm}^3} \times \frac{10^3 \text{ mm}}{\text{m}} = 7.4 \text{ MPa}$$

In summary, the shear stresses in shaft 1 are:

At section A:	43.7 MPa
Between A and B:	27.3 MPa
At section B:	43.7 MPa
Between B and C:	7.4 MPa
At section C:	11.9 MPa

Example Problem 6-14

In Figure 6-19, 25 hp is delivered from the motor to shaft 1 through gears at B. At A, 20 hp is delivered to a machine which grinds tree branches into small chips. At D, 5 hp is delivered to a bucket elevator

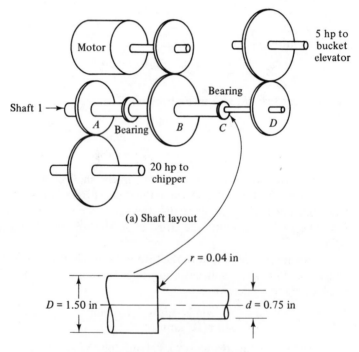

(a) Shaft layout

(b) Shaft detail at section C

Figure 6-19

which carries the chips to a storage hopper. The shaft from A to the bearing at C is 1.50 in. in diameter. At C, the diameter reduces to 0.75 in., with the geometry shown in part (b) of Figure 6-19. Gears at A, B, and C are keyed to shaft 1 with profile key seats. Shaft 1 rotates at 560 rpm. Compute the torsional shear stress at all parts of shaft 1.

Solution

The power and torque distributions in the shaft are computed first. From A to B, the shaft transmits 20 hp. From B to D, only 5 hp is transmitted. Then, using Equation (6-3),

english

$$T = \frac{(63\ 000)\ \text{power}}{n}$$

use largest T when figuring S_s max

$$T_{AB} = \frac{(63\ 000)(20)}{560} = 2250\ \text{lb} \cdot \text{in}.$$

$$T_{BD} = \frac{63\ 000(5)}{560} = 563\ \text{lb} \cdot \text{in}.$$

Stress at section A:

$$S_{s_A} = T_{AB}\frac{K_t}{Z_p}$$

For the profile key seat, $K_t = 2.0$ (Figure 6-8).

$$Z_p = \frac{\pi D^3}{16} = \frac{\pi(1.50\ \text{in}.)^3}{16} = 0.663\ \text{in}.^3$$

Then

$$S_{s_A} = \frac{2250\ \text{lb} \cdot \text{in}.\ (2.0)}{0.663\ \text{in}.^3} = 6790\ \text{psi}$$

Stress between A and B (no stress concentration):

$$S_{s_{AB}} = \frac{T_{AB}}{Z_p} = \frac{2250\ \text{lb} \cdot \text{in}.}{0.663\ \text{in}.^3} = 3390\ \text{psi}$$

Stress at B:

Here the torque in the shaft changes from 2250 lb·in. to the left of B to 563 lb·in. to the right of B. Using the higher value and $K_t = 2.0$ for the key seat, the stress is the same as at A.

$$S_{s_B} = 6790\ \text{psi}$$

Stress between B and C (no stress concentration):

$$S_{s_{BC}} = \frac{T_{BD}}{Z_p} = \frac{563\ \text{lb} \cdot \text{in}.}{0.663\ \text{in}.^3} = 849\ \text{psi}$$

Stress at C:

Here the stress concentration due to the reduction in diameter must be considered (see Figure 6-10).

$$\frac{D}{d} = \frac{1.50\ \text{in}.}{0.75\ \text{in}.} = 2.0$$

$$\frac{r}{d} = \frac{0.04\ \text{in}.}{0.75\ \text{in}.} = 0.053$$

Then $K_t = 1.6$ based on the nominal stress in the smaller diameter of the section. For the 0.75-in. diameter shaft,

$$Z_p = \frac{\pi d^3}{16} = \frac{\pi(0.75 \text{ in.})^3}{16} = 0.083 \text{ in.}^3$$

Then

$$S_{sc} = \frac{T_{BD}K_t}{Z_p} = \frac{(563 \text{ lb} \cdot \text{in.})(1.6)}{0.083 \text{ in.}^3} = 10\ 850 \text{ psi}$$

Stress between C and D (no stress concentration):

$$S_{sCD} = \frac{T_{BD}}{Z_p} = \frac{563 \text{ lb} \cdot \text{in.}}{0.083 \text{ in.}^3} = 6780 \text{ psi}$$

Stress at D:

For the profile key seat, $K_t = 2.0$. Then

$$S_{sD} = \frac{T_{BD}K_t}{Z_p} = \frac{563 \text{ lb} \cdot \text{in.} (2.0)}{0.083 \text{ in.}^3} = 13\ 600 \text{ psi}$$

Therefore, the critical point in shaft 1, where the torsional shear stress is maximum, is section D.

6-9 TORSION IN NONCIRCULAR SECTIONS

The behavior of noncircular sections when subjected to torsion is vastly different from that of circular sections, for which the discussions earlier in this chapter applied. There is a large variety of shapes which can be imagined, and the analysis of stiffness and strength is different for each. The development of the relationships involved will not be done here. Compilations of the pertinent formulas occur in References 1 and 2 listed at the end of this chapter.

Some generalizations can be made. Solid sections having the same cross-sectional area are stiffer when their shape more closely approaches a circle

Square is two times stiffer than rectangle even though both have the same area: $ab = h^2$

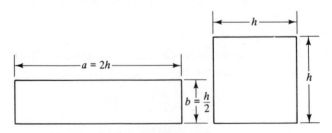

Figure 6-20

(see Figure 6-20). Conversely, a member made up of long, thin sections which do not form a closed, tube-like shape are very weak and flexible in torsion. Examples of flexible sections are common structural shapes such as wide flange beams, standard I-beams, channels, angles, and tees, as illustrated in Figure 6-21. Pipes, solid bars, and structural rectangular tubes have high rigidity, or stiffness (see Figure 6-22).

An interesting illustration of the lack of stiffness of open, thin sections is shown in Figure 6-23. The thin plate, the angle, and the channel have the

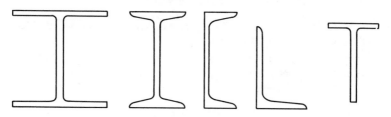

Figure 6-21   Torsionally Flexible Sections.

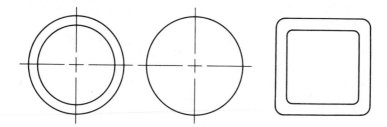

Figure 6-22   Torsionally Stiff Sections.

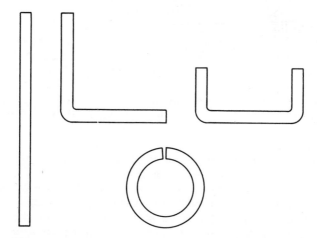

Figure 6-23   Sections Having nearly Equal (and low) Torsional Stiffness.

same thickness and cross-sectional area, and all have nearly the same torsional stiffness. Likewise, if the thin plate were formed into a circular shape, but with a slit remaining, its stiffness would remain low. However, closing the tube completely by welding or by drawing a seamless tube would produce a relatively stiff member. Understanding these comparisons is an aid to selecting a reasonable shape for members loaded in torsion.

REFERENCES

1. Roark, R. J., and W. C. Young, *Formulas for Stress and Strain*, 5th ed., McGraw-Hill Book Company, New York, 1975.
2. Seeley, F. B., and J. O. Smith, *Advanced Mechanics of Materials*, 2nd ed., John Wiley & Sons, Inc., New York, 1952.

PROBLEMS

6-1 Compute the torsional shear stress which would be produced in a solid circular shaft having a diameter of 20 mm when subjected to a torque of 280 N·m.

6-2 A hollow shaft has an outside diameter of 35 mm and an inside diameter of 25 mm. Compute the torsional shear stress in the shaft when it is subjected to a torque of 560 N·m.

6-3 Compute the torsional shear stress in a shaft having a diameter of 1.25 in. when carrying a torque of 1550 lb·in.

6-4 A steel tube is used as a shaft carrying 5500 lb·in. of torque. The outside diameter is 1.75 in., and the wall thickness is $\frac{1}{8}$ in. Compute the torsional shear stress at the outside and the inside surfaces of the tube.

6-5 A movie projector drive mechanism is driven by a 0.08-kW motor whose shaft rotates at 180 rad/s. Compute the torsional shear stress in its 3.0-mm diameter shaft.

6-6 The impeller of a fluid agitator rotates at 42 rad/s and requires 35 kW of power. Compute the torsional shear stress in the shaft which drives the impeller if it is hollow and has an outside diameter of 40 mm and an inside diameter of 25 mm.

6-7 A gear drive shaft for a milling machine transmits 7.5 hp at a speed of 240 rpm. Compute the torsional shear stress in the 0.860-in. diameter solid shaft.

6-8 The input shaft for the gear drive described in Problem 6-7 also transmits 7.5 hp, but rotates at 1140 rpm. Determine the required diameter of the input shaft to give it the same stress as the output shaft.

6-9 Determine the stress which would result in a $1\frac{1}{2}$-in. Schedule 40 steel pipe if a plumber applies a force of 80 lb at the end of a wrench handle 18 in. long.

6-10 A rotating sign makes 1 rev every 5 sec. In a high wind, a torque of 30 lb·ft is required to maintain rotation. Compute the power required to drive the sign. Also compute the stress in the final drive shaft if it is 0.60 in. in diameter.

Specify a suitable steel for the shaft to provide a design factor of 4 based on yield strength in shear.

6-11 A short, cylindrical bar is welded to a rigid plate at one end, and then a torque is applied at the other. If the bar has a diameter of 15 mm and is made of AISI 1020 cold-drawn steel, compute the torque which must be applied to it to subject it to a stress equal to its yield strength in shear. Use $s_{ys} = s_y/2$.

6-12 A propeller drive shaft on a ship is to transmit 2500 hp at 75 rpm. It is to be made of AISI 1040 WQT 1300 steel. Use a design factor of 6 based on the yield strength in shear. The shaft is to be hollow, with the inside diameter equal to 0.80 times the outside diameter. Determine the required diameter of the shaft.

6-13 If the propeller shaft of Problem 6-12 was to be solid instead of hollow, determine the required diameter. Then compute the ratio of the weight of the solid shaft to that of the hollow shaft.

6-14 A power screwdriver uses a shaft with a diameter of 5.0 mm. What torque can be applied to the screwdriver if the limiting stress due to torsion is 80 MPa?

6-15 An extension for a socket wrench similar to that shown in Figure 6-1(a) has a diameter of 6.0 mm and a length of 250 mm. Compute the stress and angle of twist in the extension when a torque of 5.5 N·m is applied. The extension is steel.

6-16 Compute the angle of twist in a steel shaft, 15 mm in diameter and 250 mm long, when a torque of 240 N·m is applied.

6-17 Compute the angle of twist in an aluminum tube, having an outside diameter of 80 mm and an inside diameter of 60 mm, when subjected to a torque of 2250 N·m. The tube is 1200 mm long.

6-18 A steel rod, 8 ft long and 0.625 in. in diameter, is used as a long wrench to unscrew a plug at the bottom of a pool of water. If it requires 40 lb·ft of torque to loosen the plug, compute the angle of twist of the rod.

6-19 For the rod described in Problem 6-18, what must the diameter be if only 2.0 deg of twist is desired when 40 lb·ft of torque is applied?

6-20 Compute the angle of twist of the free end relative to the fixed end of the steel bar shown in Figure 6-24.

$\theta = .0756 \text{ rads}$

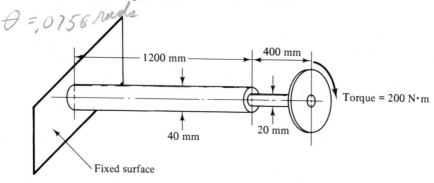

Figure 6-24

6-21   A meter for measuring torque uses the angle of twist of a shaft to indicate torque. The shaft is to be 150 mm long and made of 6061-T6 aluminum alloy. Determine the required diameter of the shaft if it is desired to have a twist of 10.0 deg when a torque of 5.0 N·m is applied to the meter. For the shaft thus designed, compute the torsional shear stress and then compute the resulting design factor for the shaft. Is it satisfactory? If not, what would you do?

6-22   A beryllium copper wire having a diameter of 1.50 mm and a length of 40 mm is used as a small torsion bar in an instrument. Determine what angle of twist would result in the wire when it is stressed to 250 MPa.

6-23   A fuel line in an aircraft is made of a titanium alloy. The tubular line has an outside diameter of 18 mm and an inside diameter of 16 mm. Compute the stress in the tube if a length of 1.65 m must be twisted through an angle of 40 deg during installation. Determine the design factor based on the yield strength in shear if the tube is titanium 6-Al-4V, aged at 900°F.

6-24   For the shaft shown in Figure 6-25, compute the angle of twist of pulleys B and C relative to A. The steel shaft has a diameter of 35 mm throughout its length. The torques are: $T_1 = 1500$ N·m, $T_2 = 1000$ N·m, $T_3 = 500$ N·m. The lengths are: $L_1 = 500$ mm, $L_2 = 800$ mm.

$\theta_{AC} = 0975$

rad

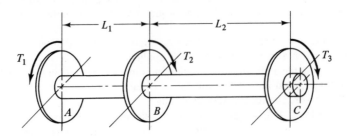

Figure 6-25

6-25   A torsion bar in a light truck suspension is to be 820 mm long and made of steel. It is subjected to a torque of 1360 N·m and must be limited to 2.2 deg of twist. Determine the required diameter of the solid round bar. Then compute the stress in the bar.

6-26   A steel drive shaft for an automobile is a hollow tube 1525 mm long. Its outside diameter is 75 mm, and its inside diameter is 55 mm. If the shaft transmits 120 kW of power at a speed of 225 rad/s, compute the torsional shear stress in the shaft and the angle of twist of one end relative to the other.

6-27   A rear axle of an automobile is a solid steel shaft having the configuration shown in Figure 6-26. Considering the stress concentration due to the shoulder, compute the torsional shear stress in the axle when it rotates at 70.0 rad/s, transmitting 60 kW of power.

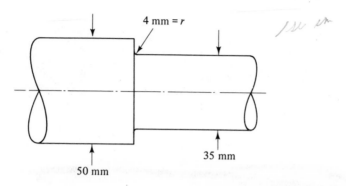

4 mm = r

35 mm

50 mm

Figure 6-26

6-28 An output shaft from an automotive transmission has the configuration shown in Figure 6-27. If the shaft is transmitting 105 kW at 220 rad/s, compute the maximum torsional shear stress in the shaft. Account for the stress concentration at the place where the speedometer gear is located.

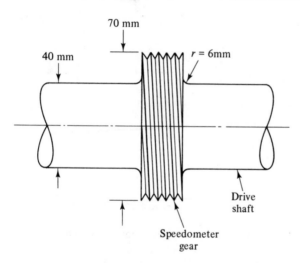

70 mm

40 mm

r = 6mm

Drive shaft

Speedometer gear

Figure 6-27

6-29 The shaft shown in Figure 6-28 is steel. Determine the maximum shear stress at points *A, B, C, D, E,* and *F*.

6-30 For the shaft shown in Figure 6-28, compute the angle of twist of each gear *A, D,* and *F* relative to gear *C*.

$\theta_{AC} = .0049$ rads

$\theta_{CD} = .018$ rads

$\theta_{CF} = .02255$ rads

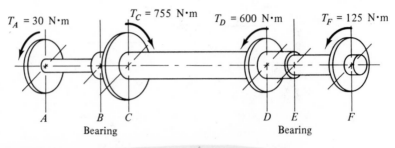

CW T and CCW T must be equal

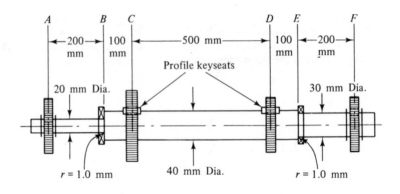

Retaining Ring Grooves at A and F
At A – Groove Dia. = 16 mm
At F – Groove Dia. = 25 mm
Assume K_t = 3.0 for Grooves

Figure 6-28

6-31 Figure 6-29 shows a sketch of a gear motor in which the motor shaft speed is reduced twice to produce the output shaft speed. The motor supplies 1.50 kW of power and rotates at 1725 rpm. The speed of shaft 1 is three times that of shaft 2. Shaft 2 rotates five times faster than shaft 3. Use a design factor of 4 based on the yield strength in shear of AISI 1141 OQT 1300 steel for all shafts, and determine the required diameter. Consider only torsional shear stresses.

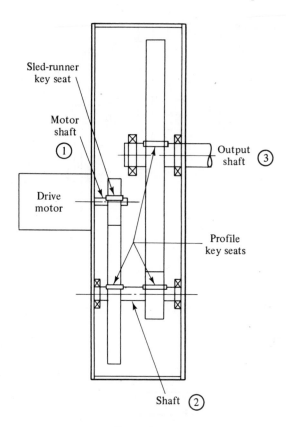

Sled-runner key seat

Motor shaft ①

Drive motor

Output shaft ③

Profile key seats

Shaft ②

Figure 6-29

6-32   Figure 6-30 shows part of a drive for a coal-processing machine. The motor delivers 30 kW of power to the center gear on shaft 2. At the left end, a gear delivers 18 kW of power to a crushing device. At the right end, a third gear delivers 12 kW of power to a conveyor which removes the crushed coal. The detailed design of shaft 2 is shown in part (b) of Figure 6-30. Compute the maximum stress due to torsion in shaft 2, which rotates at 24 rad/s.

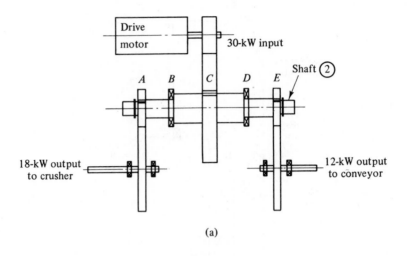

Drive motor

30-kW input

A B C D E Shaft ②

18-kW output
to crusher

12-kW output
to conveyor

(a)

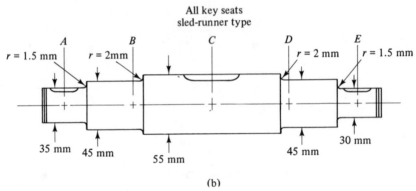

All key seats
sled-runner type

A   B   C   D   E

r = 1.5 mm   r = 2mm   r = 2 mm   r = 1.5 mm

35 mm   45 mm   45 mm

55 mm   30 mm

(b)

Figure 6-30

7

SHEARING FORCES
AND BENDING MOMENTS
IN BEAMS

7-1 BEAM LOADING

Much of the discussion in the next six chapters will deal with beams. A *beam* is a member which carries loads transversely, that is, perpendicular to its long axis. Figures 7-1 to 7-4 show several examples of beams. Part (a) of Figure 7-1 shows a beam, supported at its ends by rods, which carries four pipes in a utility tunnel supplying water, steam, compressed air, and natural gas to production processes in a factory. Such a beam is called a *simple beam* because of the manner of support. Note that the rods at the end provide only vertical supporting forces and supply these only at the ends of the beam. Also, the pipes on the beams are examples of *concentrated loads*. That is, the forces exerted on the beam act essentially at a single point or a very small part of the length of the beam. For purposes of analyzing this beam, the loads due to the pipes would be shown as arrows acting downward, as shown in Figure 7-1(b).

A second kind of loading is the *uniformly distributed load*, as shown in Figure 7-2. This shows a roof assembly made up of several beams running across the width of a structure. The roofing materials are then supported on the beams. Snow piled on top of the roof applies additional loading. Considering that the total roof load is shared evenly by the several beams, each

137

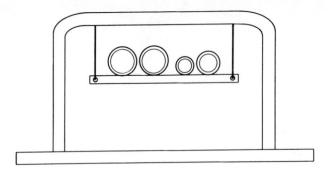

(a) Pictorial representation
of beam and loads

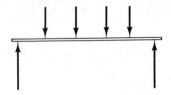

(b) Schematic representation
of beam and loads

Figure 7-1 Simple Beam with Concentrated Loads.

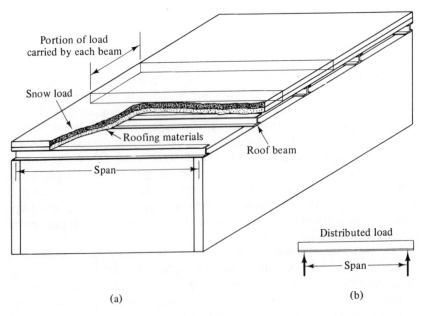

Portion of load
carried by each beam

Snow load

Roofing materials

Roof beam

Span

Distributed load

Span

(a)

(b)

Figure 7-2

beam can be considered to carry the uniformly distributed load as sketched in Figure 7-2(b). The load is usually described as being so much force per unit length. For example, the loading could be 20.3 kN/m or 1800 lb/ft.

An *overhanging beam* is one in which part of the loaded beam extends outside the supports. Figure 7-3 shows a part of a building structure in which the upper columns apply loads to the beam supported by the two lower columns. Figure 7-3(b) shows the schematic diagram of the beam.

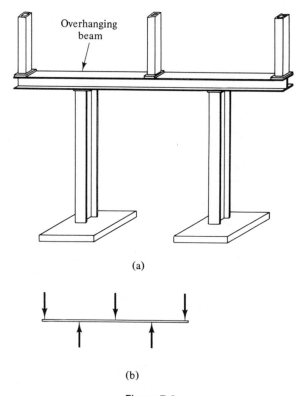

(a)

(b)

Figure 7-3

A *cantilever beam* is held fixed at the support and extends out to carry either concentrated or distributed loads. An example is the boom of a crane, such as the one shown in Figure 7-4. Part (b) of the figure shows the standard manner of sketching a cantilever beam.

For the four types of beams described so far, all the support reaction forces can be computed from the equations of equilibrium studied in statics or physics mechanics. Thus they are said to be *statically determinate*. Most of the work in this book will be for beams of this type. Statically indeterminate beams are discussed in Chapter 12.

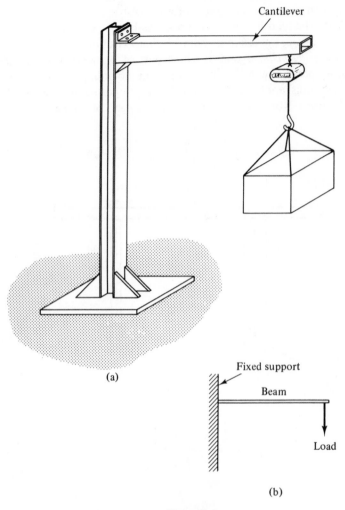

Cantilever

Fixed support

Beam

Load

(a)

(b)

Figure 7-4

7-2 BEAM SUPPORTS AND REACTIONS AT SUPPORTS

The first step in analyzing a beam in order to determine its safety under a given loading arrangement is to show completely the loads and support reactions on a free-body diagram. It is very important to be able to construct free-body diagrams from the physical picture or description of the loaded beam. This was done in each case for Figures 7-1 to 7-4, where part (b) of each figure is the free-body diagram.

After constructing the free-body diagram, it is necessary to compute the magnitude of all support reactions. It is assumed that the methods used to find reactions were studied previously. Therefore, only a few examples are shown here as a review and as an illustration of the techniques used throughout this book.

The following general procedure is recommended for solving for reactions on simple or overhanging beams.

1. Draw the free-body diagram.
2. Use the equilibrium equation $\sum M = 0$ by summing moments about the point of application of one support reaction. The resulting equation can then be solved for the other reaction.
3. Use $\sum M = 0$ by summing moments about the point of application of the second reaction to find the first reaction.
4. Use $\sum F = 0$ to check the accuracy of your calculations.

Example Problem 7-1

Figure 7-5 shows the free-body diagram for the beam carrying pipes, which was originally shown in Figure 7-1. Compute the reaction forces in the support rods.

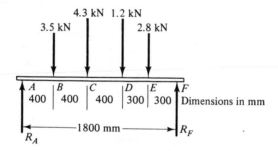

Figure 7-5

Solution

Since the free-body diagram is given, step 2 of the solution procedure will now be applied. To find the reaction R_F, sum moments about point A.

$$\sum M_A = 0 = 3.5(400) + 4.3(800) + 1.2(1200)$$
$$+ 2.8(1500) - R_F(1800)$$

Note that all forces are in kilonewtons and distances in millimetres. Now solve for R_F.

$$R_F = \frac{3.5(400) + 4.3(800) + 1.2(1200) + 2.8(1500)}{1800} = 5.82 \text{ kN}$$

Now, to find R_A, sum moments about point F.

$$\sum M_F = 0 = 2.8(300) + 1.2(600) + 4.3(1000) + 3.5(1400)$$
$$- R_A(1800)$$

$$R_A = \frac{2.8(300) + 1.2(600) + 4.3(1000) + 3.5(1400)}{1800} = 5.98 \text{ kN}$$

Now use $\sum F = 0$ for the vertical direction as a check.

Downward forces: $(3.5 + 4.3 + 1.2 + 2.8) \text{ kN} = 11.8 \text{ kN}$

Upward reactions: $(5.82 + 5.98) \text{ kN} = 11.8 \text{ kN}$ (check)

Remember to show the reaction forces R_A and R_F at their proper points on the beam.

Example Problem 7-2

Compute the reactions on the beam shown in Figure 7-6. Note that the distributed load covers only part of the beam.

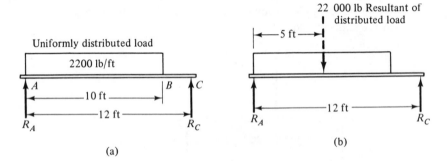

(a)

(b)

Figure 7-6

Solution

For purposes of finding reactions, it is convenient to work with the resultant of a distributed load by considering that the total load acts at the centroid of the load. This is shown in part (b) of Figure 7-6. Then the same procedure as before can be used to compute reactions.

$$\sum M_A = 0 = 22\ 000 \text{ lb } (5 \text{ ft}) - R_C(12 \text{ ft})$$

$$R_C = \frac{22\ 000 \text{ lb } (5 \text{ ft})}{12 \text{ ft}} = 9167 \text{ lb}$$

$$\sum M_C = 0 = 22\ 000 \text{ lb } (7 \text{ ft}) - R_A(12 \text{ ft})$$

$$R_A = \frac{22\ 000 \text{ lb } (7 \text{ ft})}{12 \text{ ft}} = 12\ 833 \text{ lb}$$

Finally, as a check, in the vertical direction,

$$\Sigma F = 0$$

Downward forces: 22 000 lb

Upward forces: $R_A + R_C = 12\ 833 + 9167$

$$= 22\ 000\ \text{lb} \qquad \text{(check)}$$

Example Problem 7-3

Compute the reactions for the overhanging beam shown in Figure 7-7.

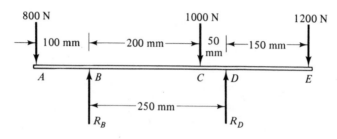

Figure 7-7

Solution

First, summing moments about point B,

$$\Sigma M_B = 0 = 1000(200) - R_D(250) + 1200(400) - 800(100)$$

Notice that forces that tend to produce clockwise moments about B are considered positive in this calculation. Now solving for R_D,

$$R_D = \frac{1000(200) + 1200(400) - 800(100)}{250} = 2400\ \text{N}$$

Summing moments about D will allow computation of R_B.

$$\Sigma M_D = 0 = 1000(50) - R_B(250) + 800(350) - 1200(150)$$

$$R_B = \frac{1000(50) + 800(350) - 1200(150)}{250} = 600\ \text{N}$$

Checking with $\Sigma F = 0$ in the vertical direction,

Downward forces: $(800 + 1000 + 1200)\ \text{N} = 3000\ \text{N}$

Upward forces: $R_B + R_D = (600 + 2400)\ \text{N}$

$$= 3000\ \text{N} \qquad \text{(check)}$$

Example Problem 7-4

Compute the reactions at the wall required to support the cantilever beam shown in Figure 7-8.

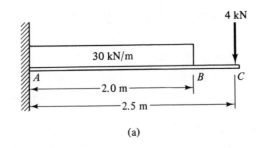

(a)

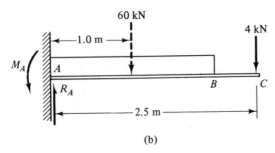

(b)

Figure 7-8

Solution

In the case of cantilever beams, the reactions at the wall are composed of an upward force R_A which must balance all downward forces on the beam and a reaction moment M_A which must balance the tendency for the applied loads to rotate the beam. These are shown in part (b) of the figure. Also shown is the resultant, 60 kN, of the distributed load. Then, by summing forces in the vertical direction,

$$R_A = 60 \text{ kN} + 4 \text{ kN} = 64 \text{ kN}$$

By summing moments about point A,

$$M_A = 60 \text{ kN } (1.0 \text{ m}) + 4 \text{ kN } (2.5 \text{ m}) = 70 \text{ kN} \cdot \text{m}$$

7-3 SHEARING FORCES

It will be shown later that the two kinds of stresses developed in a beam are shearing stresses and bending stresses. In order to compute these stresses, it will be necessary to know the magnitude of shearing forces and bending moments at all points in the beam. Therefore, although you may not yet understand the ultimate use of these factors, it is necessary to learn how to determine the variation of shearing forces and bending moments in beams for many types of loading and support combinations.

Shearing forces are internal forces developed in the material of a beam to balance externally applied forces in order to ensure equilibrium in all parts of the beam. The presence of these shearing forces can be visualized by considering a section of the beam and looking at all external forces. For example, Figure 7-9 shows a simple beam carrying a concentrated load at its center. The beam as a whole is in equilibrium under the action of the 1000-N load and the two 500-N reaction forces at the supports. If the whole beam is in equilibrium, then so is any part of it. Consider a part of the beam 0.5 m long, as shown in part (b) of Figure 7-9. In order for this part of the beam to be in equilibrium, there must be a force of 500 N inside the beam acting downward.* This internal force is called the *shearing force* and will be denoted by the symbol V. Notice that the shearing force would be of the same magnitude if the beam were cut anywhere between A and B.

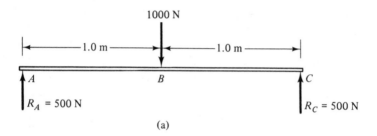

(a)

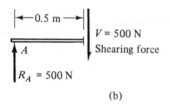

(b)

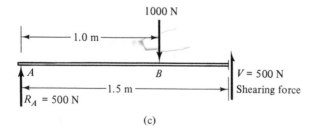

(c)

Figure 7-9

*Also, a moment would have to exist inside the beam to keep it from rotating. This is called the bending moment and will be discussed in the next section.

Now consider a part of the beam 1.5 m long, as shown in Figure 7-9(c). In order for that part to be in equilibrium, an internal shearing force of 500 N upward must exist in the beam. This would be the same if the beam were cut anywhere between B and C.

Shear Diagrams. The free-body diagram could be drawn for any part of any beam with any loading to determine the magnitude of the shearing force in the beam. However, it is a time-consuming process, and a simpler method is available to obtain a plot of the variation of shearing force versus position on the beam for the entire beam. The name given to such a plot is the *shear diagram*. The fundamental principle involved in creating a shear diagram is that *the magnitude of the shearing force in any part of the beam is equal to the algebraic sum of all external forces acting to the left of the section of interest.* In drawing the diagram, upward forces and reactions will be considered positive, and downward forces will be negative.

Figure 7-10 illustrates the shear diagram for the beam considered earlier. Construction should start at the left end, point A, where the reaction force of 500 N is encountered immediately. Then the shear diagram starts at 500 N. Between points A and B, there are no other forces applied to the beam. Therefore, the shear remains at 500 N all the way from A to B. At point B, the 1000-N concentrated load is encountered, causing an abrupt change in shear from positive 500 N to negative 500 N. Typically, the application of a concentrated load will cause an abrupt change in the magnitude of the shear in the beam. Then between B and C, no other loads are applied, resulting in no change in shear. At point C, the 500-N reaction R_C brings the plot abruptly back to zero.

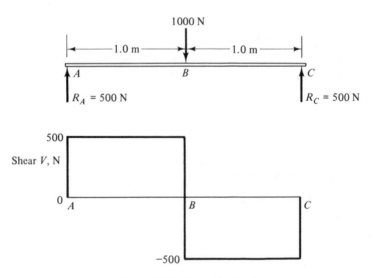

Figure 7-10   Shear Diagram

Looking at the complete shear diagram of Figure 7-10, it can be seen that the greatest value of shear is 500 N. Notice that even though there is an applied load of 1000 N, the maximum shearing force in the beam is only 500 N.

Another beam carrying concentrated loads will now be considered in an example problem. The general approach used above will be applicable to any beam carrying concentrated loads.

Example Problem 7-5

Draw the complete shear diagram for the beam shown in Figure 7-11.

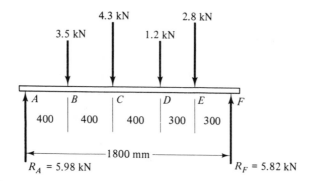

Figure 7-11

Solution

This is the same beam which was analyzed to determine the reactions R_A and R_F in an earlier problem. In general, the calculation of the reaction forces would be the first step in the process of drawing the shear diagram.

The completed diagram is shown in Figure 7-12. The process used to determine each part of the diagram is described below.

Point A: Starting at the left end of the beam, the reaction R_A is encountered immediately. Thus, the shear starts abruptly at 5.98 kN.

Point B: Since no loads are applied between A and B, the shear is constant at 5.98 kN. At B, the 3.5-kN downward load causes an abrupt decrease in shear to 2.48 kN.

Point C: No change between B and C. At C, shear decreases by 4.3 kN to -1.82 kN.

Point D: No change between C and D. At D, shear decreases by 1.2 kN to -3.02 kN.

Point E: No change from D to E. At E, shear decreases by 2.8 kN to -5.82 kN.

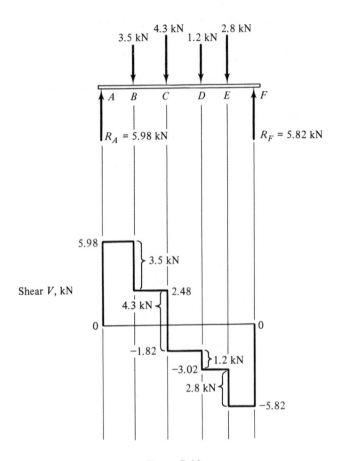

Figure 7-12

Point F: No change between E and F. At F, the upward reaction $R_F = 5.82$ kN brings the shear back to zero.

Notice some general characteristics of shear diagrams for beams having concentrated loads, as illustrated in this example.

1. The shear diagram starts and ends at zero at the ends of the beam.
2. At each concentrated load or reaction, the value of the shear changes abruptly by an amount equal to the load or reaction force.
3. Between concentrated loads, there is no change in shear, and the shear curve plots as a straight horizontal line.
4. The symbol V is used to identify the shearing force axis, and the units for force are indicated.

Example Problem 7-6

Draw the complete shear diagram for the beam shown in Figure 7-13. Work the problem completely yourself before looking at the solution in Figure 7-14.

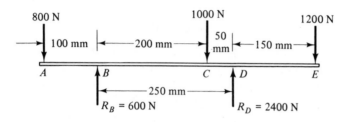

Figure 7-13

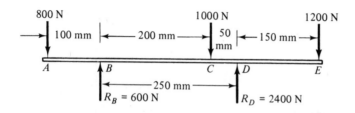

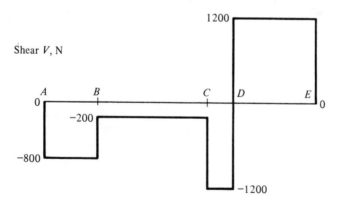

Figure 7-14

Shear Diagrams for Distributed Loads. The variation of shearing force with position on the beam for distributed loads is different from that for concentrated loads. The free-body diagram method is useful to help visualize these variations.

Consider the beam in Figure 7-15, carrying a uniformly distributed load of 1500 N/m over part of its length. It is desired to determine the magnitude

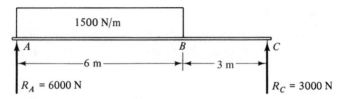

Figure 7-15

of the shear at several points in the beam in order to draw a shear diagram. Let's choose points at intervals of 2 m across the beam, including the two ends of the beam. At the left end of the beam, just to the right of point A, the shearing force in the beam must be 6000 N in order to balance the reaction force R_A. Now, if a segment of the beam 2 m long was considered to be cut from the rest of the beam, the free-body diagram shown in Figure 7-16(a) would be obtained. There must be a shearing force in the beam to balance

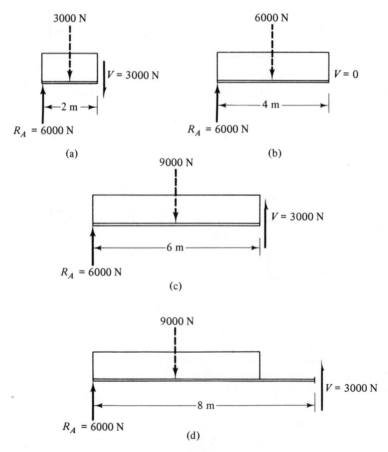

Figure 7-16

the external forces in order for the segment to be in equilibrium. The reaction R_A of 6000 N acts upward, while the total distributed load of 3000 N acts downward. The shearing force V must be 3000 N downward.

Analyzing a segment 4 m long in a similar manner would produce the free-body diagram shown in Figure 7-16(b). In this case $V = 0$, since the external loads themselves are balanced.

For a segment 6 m long, Figure 7-16(c) would be the free-body diagram. Now a shearing force V of 3000 N upward must exist. At 8 m, Figure 7-16(d) shows that $V = 3000$ N upward again. This would be true throughout the last 3 m of the beam's length, since there are no external loads applied in this part.

In summary, the shearing forces calculated were:

At point A:	$V = 6000$ N downward
At 2 m:	$V = 3000$ N downward
At 4 m:	$V = 0$
At 6 m:	$V = 3000$ N upward
Between B and C:	$V = 3000$ N upward

By convention, downward shearing forces are considered positive, and upward shearing forces are negative. If these values are plotted on a graph of shearing force versus position on the beam, the shear diagram shown in Figure 7-17 would be produced. Notice that for the portion of the beam carrying the uniformly distributed load, the shear curve is a straight line. This is typical

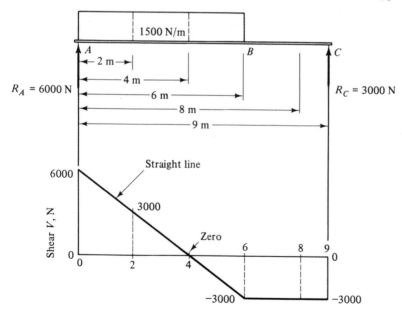

Figure 7-17

for such loads. Also, the following general rules can be derived from this example. For the part of a beam carrying a uniformly distributed load:

1. The *change in shear* between any two points is equal to the area under the load diagram between those points.
2. The slope of the straight-line shear curve is equal to the rate of loading on the beam, that is, the load per unit length.

In the present example, rule 1 is illustrated by noting that between points *A* and *B* the shear decreases by 9000 N, from positive 6000 N to negative 3000 N. This is the area under the load curve as computed from

$$(1500 \text{ N/m})(6 \text{ m}) = 9000 \text{ N.}$$

Rule 2 states that the shear decreases 1500 N over each metre of length of the beam.

The following example problem shows a beam having a uniformly distributed load plus two concentrated loads. This may occur in the case of a roof beam which carries the roofing material and snow over its entire length, plus a piece of equipment whose legs apply the concentrated loads. The general principles developed for both concentrated loads and distributed loads must be applied in the solution of the problem. It would be of the greatest benefit to you if you would work the problem in steps yourself before looking at the given solution. The following steps should be applied:

1. Solve for the reaction forces at the supports.
2. Make a sketch of the beam. It helps to make the sketch approximately to scale.
3. Draw lines vertically down from key points on the loaded beam to the area below, where the shear diagram will be drawn.
4. Draw the horizontal axis of the shear diagram equal to the length of the beam. Label the vertical shear axis, showing the units for the shearing forces to be plotted.
5. Starting at the left end of the beam, plot the variation in shear across the entire beam. Remember that:
   a. The shear changes abruptly at each point where a concentrated load is applied. The change in shear is equal to the load.
   b. The shear curve is a straight, horizontal line between points where no loads are applied.
   c. The shear curve is a straight, inclined line between points where uniformly distributed loads are applied. The slope of the line is equal to the rate of loading.
   d. The change in shear between points equals the area under the load curve between the points.
6. Show the value of the shear at each point where major changes occur, such as concentrated loads and at the beginning and end of distributed loads.

Example Problem 7-7

Draw the complete shear diagram for the beam shown in Figure 7-18 using the six steps just outlined. The solution is shown in three parts, with Part 1 giving the results of step 1; Part 2 covering steps 2, 3, and 4; and Part 3 completing steps 5 and 6. Complete each part yourself before looking at the solution.

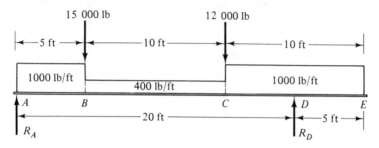

Figure 7-18

Solution

Part 1: Reactions at supports:

$$R_A = 20\ 625\ \text{lb}$$
$$R_D = 25\ 375\ \text{lb}$$

Figure 7-19 was drawn to aid in computing reactions. The resultant of each distributed load is shown acting at the middle of the loaded section. These resultants are used in the following computations.

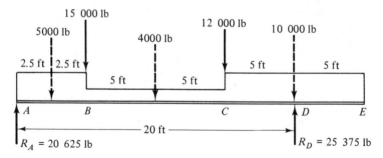

Figure 7-19

Summing moments about point D,

$$\sum M_D = 0 = 12\ 000(5) + 15\ 000(15) + 4000(10) + 5000(17.5)$$
$$- R_A(20)$$

$$R_A = \frac{60\ 000 + 225\ 000 + 40\ 000 + 87\ 500}{20} = 20\ 625\ \text{lb}$$

Notice that the 10 000-lb resultant of the distributed load at the right of the beam acts right at point D and, therefore, has no moment about that point.

Now summing moments about point A,

$$\Sigma M_A = 0 = 15\ 000(5) + 12\ 000(15) + 5000(2.5) + 4000(10)$$
$$+ 10\ 000(20) - R_D(20)$$

$$R_D = \frac{75\ 000 + 180\ 000 + 12\ 500 + 40\ 000 + 200\ 000}{20}$$

$$R_D = 25\ 375\ \text{lb}$$

As a check, we can sum forces in the vertical direction.

Downward forces: $5000 + 15\ 000 + 4000 + 12\ 000 + 10\ 000$
$$= 46\ 000\ \text{lb}$$

Upward forces: $20\ 625 + 25\ 375 = 46\ 000\ \text{lb}$ (check)

Now complete Steps 2, 3, and 4.

Part 2: Figure 7-20 shows the base on which the shear diagram will be drawn. Don't forget the units for the shear on the vertical axis. Now do steps 5 and 6 of the shear diagram procedure.

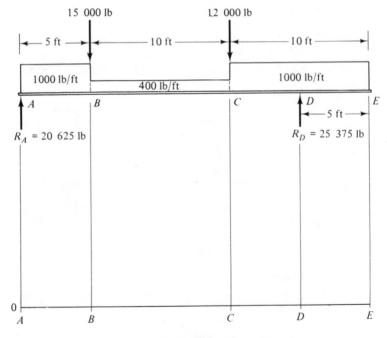

Figure 7-20

Part 3: Figure 7-21 shows the completed shear diagram. The key points were computed as shown below:

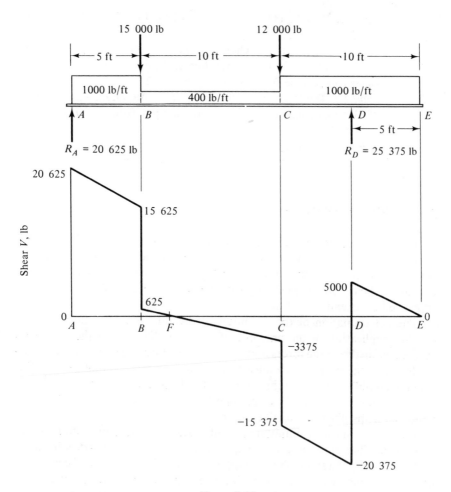

Figure 7-21

Point A: The reaction R_A causes the shear to increase abruptly to 20 625 lb.

Between points A and B: The distributed load results in the downward-sloped straight line. The total change in shear from A to B is 5000 lb, the area under the load curve between A and B. The slope of the line is 1000 lb/ft. The shear at B is 20 625 lb — 5000 lb = 15 625 lb.

Point B: The concentrated load of 15 000 lb causes the shear to decrease from 15 625 lb to 625 lb.

Between points B and C: The change in shear is 4000 lb, decreasing at the rate of 400 lb/ft between B and C. Then at C, the shear is 625 lb − 4000 lb = −3375 lb.

Point C: The 12 000-lb concentrated load decreases the shear to −15 375 lb.

Between C and D: The change in shear is 5000 lb, the area of the load curve between C and D. The slope of the line is 1000 lb/ft, causing the shear to decrease from −15 375 lb to −20 375 lb.

Point D: The large upward reaction R_D causes an increase in shear by 25 375 lb. Then at D, the shear changes from −20 375 lb to +5000 lb.

Between D and E: The change in shear is 5000 lb, causing the curve to decrease to zero at E, where the beam ends.

7-4 BENDING MOMENTS

Bending moments, in addition to shearing forces, are developed in beams as a result of the application of loads acting perpendicular to the beam. It is these bending moments which cause the beam to assume its characteristic curved, or "bent," shape. Pushing on the middle of a thin stick, such as a ruler supported at its ends, illustrates this.

The determination of the magnitude of bending moments in a beam is another application of the principle of static equilibrium. In the preceding section, we analyzed the forces in the vertical direction to determine the shearing forces in the beam which must be developed to maintain all parts of the beam in equilibrium. To do this, it was helpful to consider parts of the beam as free bodies in order to visualize what happens inside the beam. A similar approach helps illustrate bending moments.

Figure 7-22 shows a simply supported beam carrying a concentrated load at its center. The entire beam is in equilibrium, and so is any part of it. Look at the free-body diagrams shown in parts (a), (b), (c), and (d) of Figure 7-22. By summing moments about the point where the beam is cut, the magnitude of the bending moment inside required to keep the segment in equilibrium can be found. The first 0.5-m segment is shown in Figure 7-22(a). Summing moments about point B gives.

$$M_B = 500 \text{ N } (0.5 \text{ m}) = 250 \text{ N} \cdot \text{m}$$

The shearing force, determined earlier in Section 7-3, is also shown.

In part (b) of Figure 7-22, a segment 1.0 m long, which includes the left half of the beam but *not* the 1000-N load, is drawn as a free body. Summing moments about C gives

$$M_C = 500 \text{ N } (1.0 \text{ m}) = 500 \text{ N} \cdot \text{m}$$

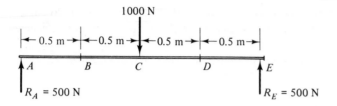

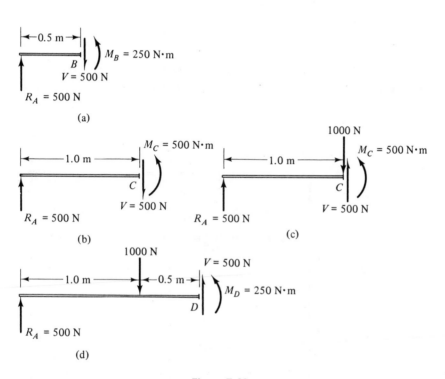

Figure 7-22

If the 1000-N load had been considered as shown in Figure 7-22(c), the result would be the same, since the load acts right at point C and, therefore, has no moment about that point.

Part (d) of Figure 7-22 shows 1.5 m of the beam isolated as a free body. Summing moments about point D gives

$$M_D = 500 \text{ N} (1.5 \text{ m}) - 1000 \text{ N} (0.5 \text{ m}) = 250 \text{ N} \cdot \text{m}$$

If we consider the entire beam as a free body and sum moments about point E at the right end of the beam, we will get

$$M_E = 500 \text{ N} (2.0 \text{ m}) - 1000 \text{ N} (1.0 \text{ m}) = 0$$

A similar result would be obtained for point *A* at the left end. In fact, a general rule is that *the bending moments at the ends of a simply supported beam are zero.*

In summary, for the beam in Figure 7-22, the bending moments are:

Point A: 0
Point B: 250 N·m
Point C: 500 N·m
Point D: 250 N·m
Point E: 0

Figure 7-23 shows these values plotted on a bending moment diagram below the shear diagram developed earlier for the same beam. Notice that between

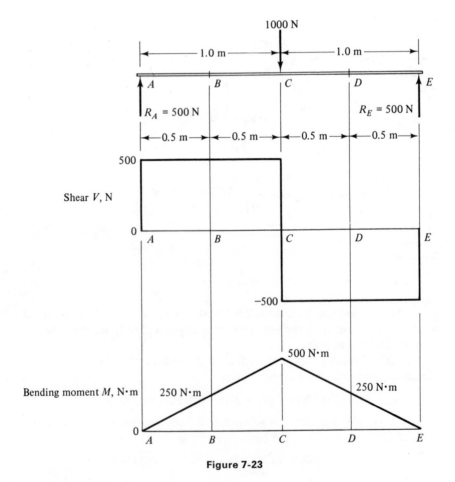

Figure 7-23

A and *C* the bending moment values fall on a straight line. Likewise, between *C* and *E*, the points fall on a straight line. This is typical for segments of beams carrying only concentrated loads.

Figure 7-23 also illustrates another general rule. The *change in moment* between two points on a beam is equal to the *area under the shear curve* between the same two points. Between *A* and *B*, under the shear curve, is the area

$$(500 \text{ N})(0.5 \text{ m}) = 250 \text{ N} \cdot \text{m}$$

Since the moment at *A* is zero, this would make $M_B = 250 \text{ N} \cdot \text{m}$, as shown before. Between *C* and *D*, the area under the shear curve is

$$(-500 \text{ N})(0.5 \text{ m}) = -250 \text{ N} \cdot \text{m}$$

This is the change in moment from *C* to *D*. Then

$$M_D = M_C - 250 \text{ N} \cdot \text{m} = 500 \text{ N} \cdot \text{m} - 250 \text{ N} \cdot \text{m}$$
$$M_D = 250 \text{ N} \cdot \text{m}$$

These same concepts are illustrated in the following example problem. The shear diagram for this beam was developed in Example Problem 7-5, with the result shown in Figure 7-12.

Example Problem 7-8

Draw the complete bending moment diagram for the beam shown in Figure 7-24. Show the values for the moment at points *A*, *B*, *C*, *D*, *E*, and *F*. The shear diagram shown was developed in Example Problem 7-5.

Solution

It is most convenient to start from the left end of the beam and work toward the right, considering each segment of the beam separately. The segments chosen should be those between points where the shear changes value. We can first note that at the ends of the beam, points *A* and *F*, the moment must be zero. Now, considering segment *AB*, the change in bending moment from *A* to *B* is equal to the area under the shear curve between *A* and *B*. That is,

$$M_B = M_A + 5.98 \text{ kN} (0.4 \text{ m}) = 0 + 2.39 \text{ kN} \cdot \text{m}$$
$$M_B = 2.39 \text{ kN} \cdot \text{m}$$

This value is plotted on the bending moment diagram, as shown in Figure 7-24, directly below the shear diagram. The straight line from M_A to M_B shows the variation of moment with position on the beam.

For the segment *BC*,

$$M_C = M_B + 2.48 \text{ kN} (0.4 \text{ m}) = 2.39 \text{ kN} \cdot \text{m} + 0.99 \text{ kN} \cdot \text{m}$$
$$M_B = 3.38 \text{ kN} \cdot \text{m}$$

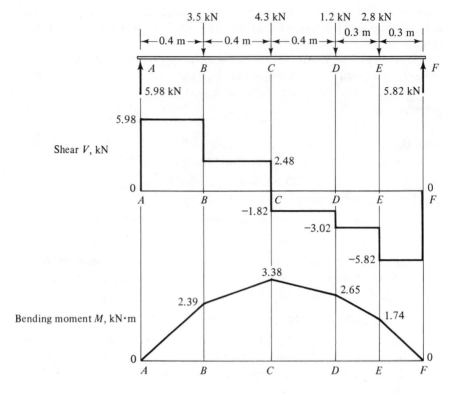

Figure 7-24

For segment CD,

$$M_D = M_C - 1.82 \text{ kN } (0.4 \text{ m}) = 3.38 \text{ kN} \cdot \text{m} - 0.73 \text{ kN} \cdot \text{m}$$
$$M_D = 2.65 \text{ kN} \cdot \text{m}$$

Note that the area between C and D on the shear curve is *negative*.
For segment DE,

$$M_E = M_D - 3.02 \text{ kN } (0.3 \text{ m}) = 2.65 \text{ kN} \cdot \text{m} - 0.91 \text{ kN} \cdot \text{m}$$
$$M_E = 1.74 \text{ kN} \cdot \text{m}$$

For segment EF,

$$M_F = M_D - 5.82 \text{ kN } (0.3 \text{ m}) = 1.74 \text{ kN} \cdot \text{m} - 1.74 \text{ kN} \cdot \text{m}$$
$$M_F = 0$$

The values of the bending moment are plotted in Figure 7-24. The fact that M_F was computed to be zero is a check on the computations, since the moment at the end of a simply supported beam *must* be zero.

If this were not the result, some error would have been made, and the computations would have to be checked.

Bending Moment Diagrams for Distributed Loads. The preceding examples showed the computation of bending moments and the plotting of bending moment diagrams for beams carrying only concentrated loads. Now we will consider distributed loads. The free-body diagram approach will be used again as an aid to visualizing the variation in bending moment as a function of position on the beam.

The beam shown in Figure 7-25 will be used to illustrate the typical results for distributed loads. This is the same beam for which the shearing force was

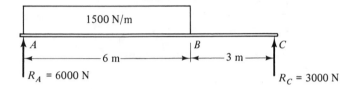

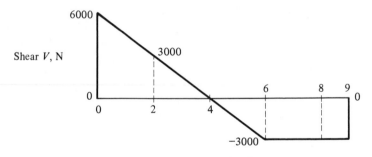

Figure 7-25

determined, as shown in Figures 7-16 and 7-17. The free-body diagrams for segments of the beam taken in increments of 2 m will be used to compute the bending moments. Refer to Figure 7-26.

For a segment at the left end of the beam, 2 m long, we can determine the bending moment in the beam by summing moments about that point due to all external loads to the left of the section, as shown in Figure 7-26(a). Notice that the resultant of the distributed load is shown to be acting at the middle of the 2-m segment. Then because the segment is in equilibrium,

$$M_2 = 6000 \text{ N } (2 \text{ m}) - 3000 \text{ N } (1 \text{ m}) = 9000 \text{ N·m}$$

The symbol M_2 is being used to indicate the bending moment at the point 2 m out on the beam.

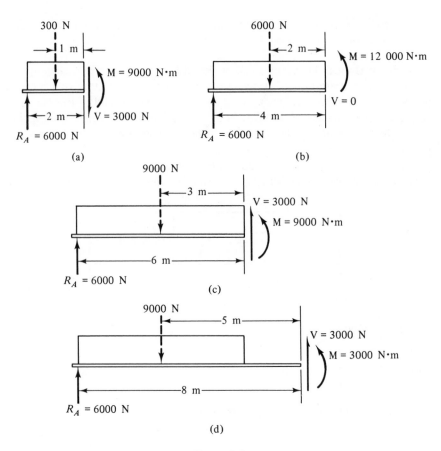

Figure 7-26

Using a similar method at points 4 m, 6 m, and 8 m out on the beam, as shown in Figure 7-26(b), (c), and (d), we would get

$$M_4 = 6000 \text{ N (4 m)} - 6000 \text{ N (2 m)} = 12 \ 000 \text{ N} \cdot \text{m}$$

$$M_6 = 6000 \text{ N (6 m)} - 9000 \text{ N (3 m)} = 9000 \text{ N} \cdot \text{m}$$

$$M_8 = 6000 \text{ N (8 m)} - 9000 \text{ N (5 m)} = 3000 \text{ N} \cdot \text{m}$$

Remember that at each end of the beam the bending moment is zero. Now we have several points which can be plotted on a bending moment diagram below the shear diagram, as shown in Figure 7-27. First look at that section of the beam where the distributed load is applied, the first 6 m. By joining the points plotted for bending moment with a smooth curve, the typical shape of a bending moment curve for a distributed load is obtained. For the last 3 m, where no loads are applied, the curve is a straight line, as was the case in earlier examples.

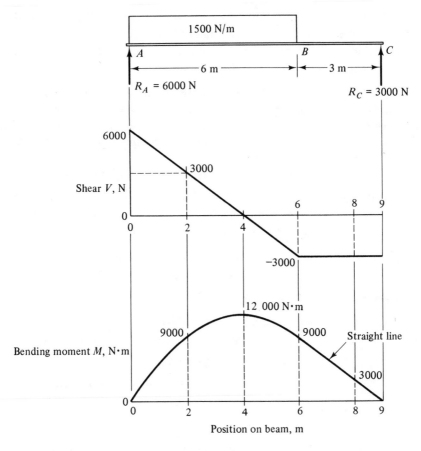

Figure 7-27

Some important observations can be made from Figure 7-27, which can be generalized into rules for drawing bending moment diagrams.

1. At the ends of a simply supported beam, the bending moment is zero.
2. The *change in bending moment* between two points on a beam is equal to the area under the shear curve between those two points. Thus, when the area under the shear curve is positive (above the axis), the bending moment is increasing, and vice versa.
3. The maximum bending moment occurs at a point where the shear curve crosses its zero axis.
4. On a section of the beam where distributed loads act, the bending moment diagram will be curved.

5. On a section of the beam where no loads are applied, the bending moment diagram will be a straight line.

6. The *slope* of the bending moment curve at any point is equal to the magnitude of the shear at that point.

Consider these rules as they apply to the beam in Figure 7-27. *Rule 1* is obviously satisfied since the moment at each end is zero. *Rule 2* can be used to check the points plotted in the moment diagram at the 2-m intervals. For the first 2 m, the area under the shear curve is composed of a rectangle and a triangle. Then the area is

$$A_{0-2} = 3000 \text{ N } (2 \text{ m}) + \tfrac{1}{2}(3000 \text{ N})(2 \text{ m}) = 9000 \text{ N} \cdot \text{m}$$

This is the change in moment from point 0 to point 2 on the beam. For the segment from 2 to 4, the area under the shear curve is a triangle. Then

$$A_{2-4} = \tfrac{1}{2}(3000 \text{ N})(2 \text{ m}) = 3000 \text{ N} \cdot \text{m}$$

Since this is the change in moment from point 2 to point 4,

$$M_4 = M_2 + A_{2-4} = 9000 \text{ N} \cdot \text{m} + 3000 \text{ N} \cdot \text{m} = 12\ 000 \text{ N} \cdot \text{m}$$

Similarly for the remaining segments,

$$A_{4-6} = \tfrac{1}{2}(-3000 \text{ N})(2 \text{ m}) = -3000 \text{ N} \cdot \text{m}$$

$$M_6 = M_4 + A_{4-6} = 12\ 000 \text{ N} \cdot \text{m} - 3000 \text{ N} \cdot \text{m} = 9000 \text{ N} \cdot \text{m}$$

$$A_{6-8} = (-3000 \text{ N})(2 \text{ m}) = -6000 \text{ N} \cdot \text{m}$$

$$M_8 = M_6 + A_{6-8} = 9000 \text{ N} \cdot \text{m} - 6000 \text{ N} \cdot \text{m} = 3000 \text{ N} \cdot \text{m}$$

$$A_{8-9} = (-3000 \text{ N})(1 \text{ m}) = -3000 \text{ N} \cdot \text{m}$$

$$M_9 = M_8 + A_{8-9} = 3000 \text{ N} \cdot \text{m} - 3000 \text{ N} \cdot \text{m} = 0$$

Here the fact that $M_9 = 0$ is a check on the process because *rule 1* must be satisfied.

Rule 3 is illustrated at point 4. Where the maximum bending moment occurs, the shear curve crosses the zero axis.

Rule 6 will probably take some practice to get familiar with, but it is extremely helpful in the process of sketching moment diagrams. Usually sketching is sufficient. The use of the six rules stated above will allow you to quickly sketch the shape of the diagram and compute key values.

In applying *rule 6*, remember the basic concepts about the slope of a curve or line, as illustrated in Figure 7-28. Seven different segments are shown, with curves for both the shear diagram and the moment diagram, including those most often encountered in constructing these diagrams. Thus, if you are drawing a portion of a diagram in which a particular shape is observed in the shear curve, the corresponding shape for the moment curve should be as illustrated in Figure 7-28. Applying this approach to the moment diagram in Figure 7-27, we can see that the curve from point 0 to point 4 is like that of

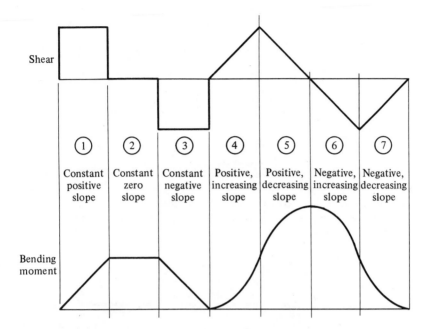

Figure 7-28 General Shapes for Moment Curves.

type 5 in Figure 7-28. Between points 4 and 6, the curve is like type 6. Finally, between points 6 and 9, the straight line with a negative slope, type 3, is used.

7-5 SHEAR AND BENDING MOMENTS FOR CANTILEVER BEAMS

The manner of support of a cantilever beam causes the analysis of its shear and bending moments to be somewhat different from that for simply supported beams. The most notable difference is that, at the place where the beam is supported, it is fixed and can therefore resist moments. Thus, at the fixed end of the beam, the bending moment is not zero, as it was for simply supported beams. In fact, the bending moment at the fixed end of the beam is the *maximum*.

Consider the cantilever beam shown in Figure 7-29. Earlier, in Example Problem 7-4, it was shown that the support reactions at point A are a vertical force $R_A = 64$ kN and a moment $M_A = 70$ kN·m. These are equal to the values of the shearing force and bending moment at the left end of the beam. According to convention, the upward reaction force R_A is positive, and the counterclockwise moment M_A is negative, giving the starting values for

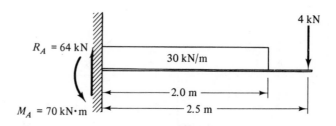

Figure 7-29

the shear diagram and bending moment diagram shown in Figure 7-30. The general rules developed in earlier sections about shear and bending moment diagrams can then be used to complete the diagrams.

The shear decreases in a straight-line manner from 64 kN to 4 kN in the

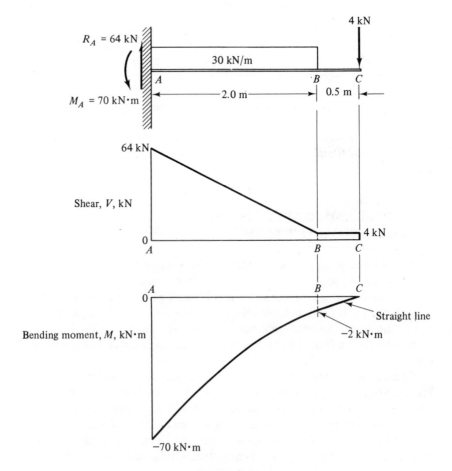

Figure 7-30

interval A to B. Note that the change in shear is equal to the amount of the distributed load. The shear remains constant from B to C, where no loads are applied. The 4-kN load at C returns the curve to zero.

The bending moment diagram starts at -70 kN·m because of the reaction moment M_A. Between points A and B, the curve has a positive but decreasing slope (type 5 in Figure 7-28). The change in moment between A and B is equal to the area under the shear curve between A and B. The area is

$$A_{A\text{-}B} = 4 \text{ kN } (2 \text{ m}) + \tfrac{1}{2}(60 \text{ kN})(2 \text{ m}) = 68 \text{ kN·m}$$

Then the moment at B is

$$M_B = M_A + A_{A\text{-}B} = -70 \text{ kN·m} + 68 \text{ kN·m} = -2 \text{ kN·m}$$

Finally, from B to C,

$$M_C = M_B + A_{B\text{-}C} = -2 \text{ kN·m} + 4 \text{ kN } (0.5 \text{ m}) = 0$$

Since point C is a *free* end of the beam, the moment there must be zero.

7-6 SUMMARY EXAMPLE

An example problem will now be shown for drawing bending moment diagrams; it includes most of the concepts developed thus far. Many steps are required to complete the problem. The solution is presented in several parts so you can check your progress toward the final result along the way. It is recommended that you work each part yourself before checking the solution. The beam to be considered is the one used in Example Problem 7-7 to illustrate shear diagrams. Review that problem before going on.

Example Problem 7-9

For the beam shown in Figure 7-31, draw the complete bending moment diagram. Show the magnitude of the bending moment at points A, B, C, D, and E. Also show the maximum positive and maximum negative bending moments.

Solution

Before making any computations, you should be able to tell a few things about the general character of the moment curve. For example:

1 What is the moment at points A and E?
2 Where will the maximum moment likely occur?
3. What is the general shape of the moment curve in each segment, AB, BC, CD, and DE?

Refer to the rules for moment diagrams discussed in Section 7-4 and see Figure 7-28 before looking at the first part of the solution.

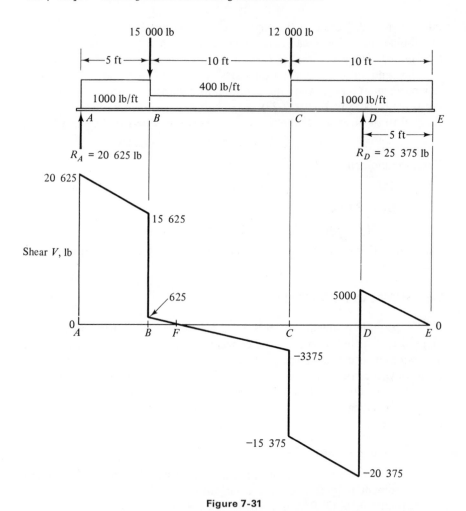

Figure 7-31

Part 1. The answers to the questions posed above are given here.

1. Since this is a simply supported beam, $M_A = 0$, and $M_E = 0$ (rule 1).

2. Remember that the maximum bending moment will occur at a point where the shear curve crosses its axis (rule 3). Thus either the point F in segment BC, or point D, is where the maximum moment will occur. Because of this observation, it will be necessary to locate point F where the shear curve crosses the axis. The shear curve between B and F has been

redrawn in Figure 7-32 as an aid to finding point *F*. We want to find the horizontal distance *BF* required for the shear curve to decrease by 625 lb.

The slope of this portion of the shear curve is −400 lb/ft because of the distributed load on this segment of the beam. Therefore, the following proportion can be set up:

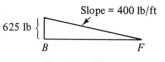

Figure 7-32

$$\frac{625 \text{ lb}}{BF} = \frac{400 \text{ lb}}{1 \text{ ft}}$$

Then

$$BF = \frac{625 \text{ lb}}{400 \text{ lb}} (1\text{ft}) = 1.563 \text{ ft}$$

The distance *BF* locates point *F*.

3. The general shape of the bending moment curve is shown in Figure 7-33.

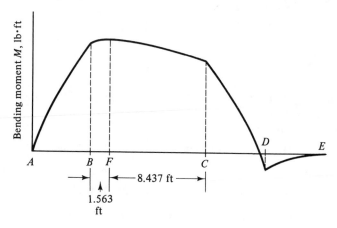

Figure 7-33

Now, applying rule 2 to find the values of the moment at the points of interest requires the computation of the areas under the shear curve. Do this now and show your results on the moment diagram. Remember to consider the area between *B* and *F* to be *positive* and the area between *F* and *C* to be *negative*.

Part 2. The results for the areas under the shear curve are summarized as follows:

$$A_{AB} = 90\ 625\ \text{lb·ft}, \qquad A_{CD} = -89\ 375\ \text{lb·ft}$$

$$A_{BF} = 488\ \text{lb·ft} \qquad A_{DE} = 12\ 500\ \text{lb·ft}$$

$$A_{FC} = -14\ 238\ \text{lb·ft}$$

These calculations are carried out to several places of accuracy in order to minimize round-off errors since the areas will be accumulated to find bending moments.

Figure 7-34 shows the completed bending moment diagram, with magnitudes shown at key points. From the figure, it can be seen that

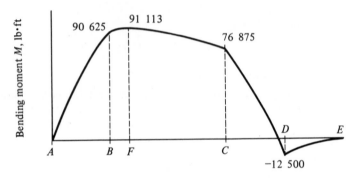

Figure 7-34

the maximum positive moment is $M_F = 91\ 113\ \text{lb·ft}$ and the maximum negative moment is $M_D = -12\ 500\ \text{lb·ft}$. The values were computed as follows:

$$M_B = M_A + A_{AB} = 0 + 90\ 625\ \text{lb·ft} = 90\ 625\ \text{lb·ft}$$

$$M_F = M_B + A_{BF} = 90\ 625\ \text{lb·ft} + 488\ \text{lb·ft} = 91\ 113\ \text{lb·ft}$$

$$M_C = M_F + A_{FC} = 91\ 113\ \text{lb·ft} - 14\ 238\ \text{lb·ft} = 76\ 875\ \text{lb·ft}$$

$$M_D = M_C + A_{CD} = 76\ 875\ \text{lb·ft} - 89\ 375\ \text{lb·ft} = -12\ 500\ \text{lb·ft}$$

$$M_E = M_D + A_{DE} = -12\ 500\ \text{lb·ft} + 12\ 500\ \text{lb·ft} = 0$$

PROBLEMS

For each of the beams which are shown in Figures 7-35 to 7-48, perform the following:

1. Compute the reactions at the supports.
2. Draw the complete shearing force and bending moment diagrams.
3. Determine the magnitude and location of the maximum absolute value of the shearing force and bending moment.

Note that the symbol K is used to indicate *kip*, which equals 1000 lb. For example, 25 K means 25 000 lb. In the figures, the arrows without labels indicate the locations of the supports.

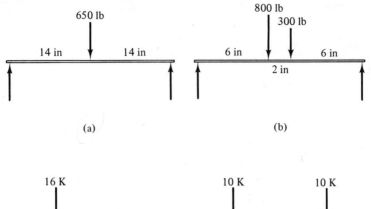

(a)

(b)

(c)

(d)

Figure 7-35

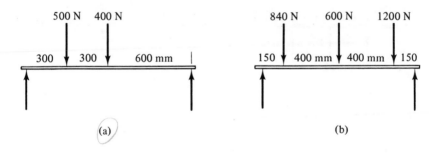

(a)

(b)

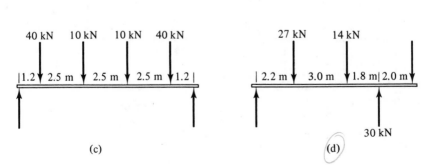

(c)

(d)

Figure 7-36

171

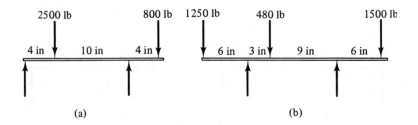

(a)

(b)

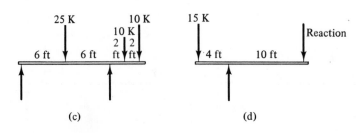

(c)

(d)

Figure 7-37

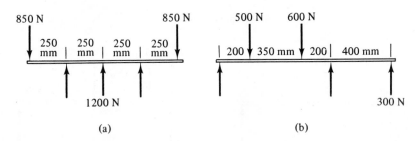

(a)

(b)

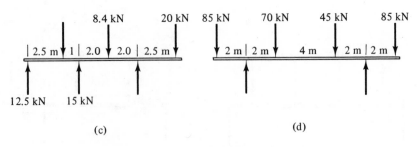

(c)

(d)

Figure 7-38

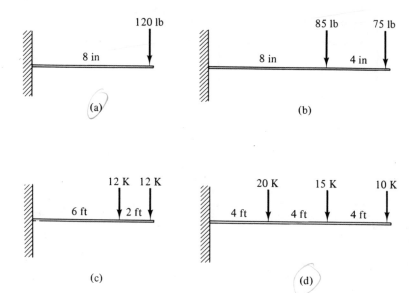

Figure 7-39

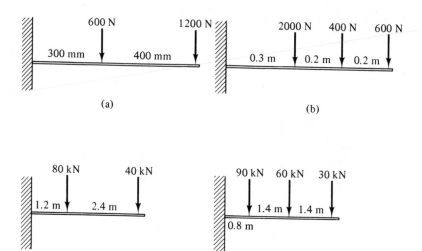

Figure 7-40

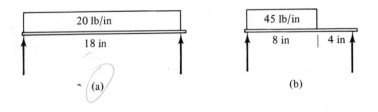

(a)

(b)

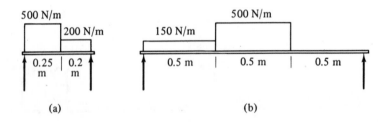

(c)

(d)

Figure 7-41

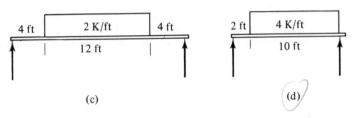

(a)

(b)

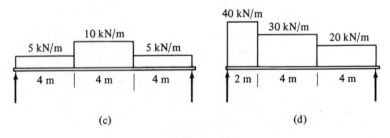

(c)

(d)

Figure 7-42

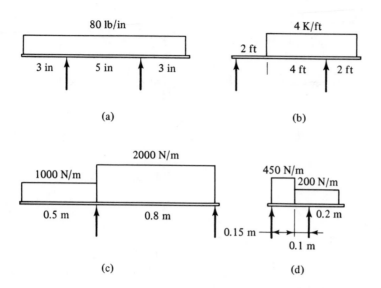

(a)

(b)

(c)

(d)

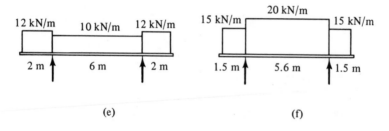

(e)

(f)

Figure 7-43

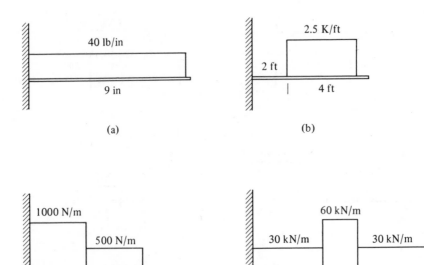

(a)

40 lb/in

9 in

(b)

2.5 K/ft

2 ft

4 ft

(c)

1000 N/m

500 N/m

0.4 m   0.4 m

(d)

60 kN/m

30 kN/m   30 kN/m

2 m   1 m   2 m

Figure 7-44

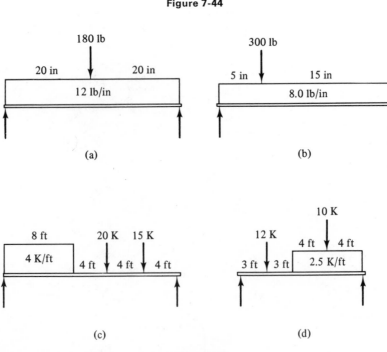

(a)

180 lb

20 in   20 in

12 lb/in

(b)

300 lb

5 in   15 in

8.0 lb/in

(c)

8 ft

4 K/ft

20 K   15 K

4 ft   4 ft   4 ft

(d)

10 K

12 K

4 ft   4 ft

3 ft   3 ft   2.5 K/ft

Figure 7-45

176

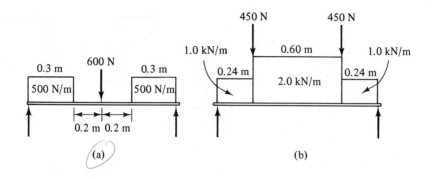

(a)

(b)

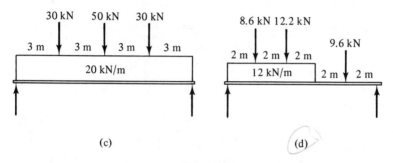

(c)

(d)

Figure 7-46

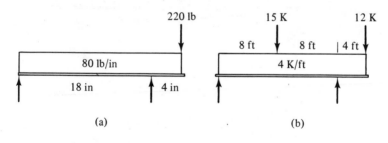

(a)

(b)

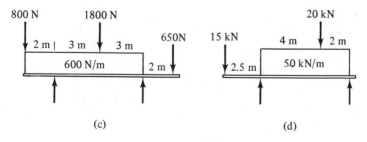

(c)

(d)

Figure 7-47

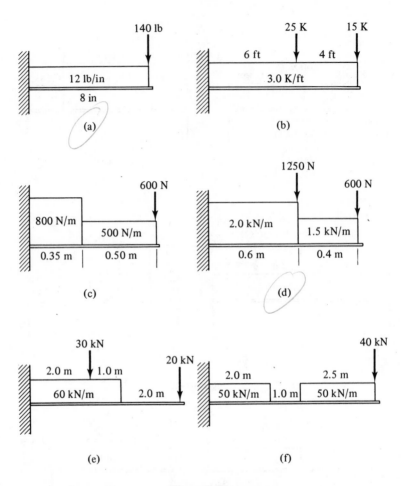

Figure 7-48

8

CENTROIDS
AND MOMENTS
OF INERTIA OF AREAS

8-1 THE CONCEPT OF CENTROID—
SIMPLE SHAPES

The *centroid* of an area is the point about which the area could be balanced if it was supported from that point. The word is derived from the word *center*, and it can be thought of as the geometrical center of an area. For three-dimensional bodies, the term *center of gravity*, or *center of mass*, is used to define a similar point.

For simple areas, such as the circle, the square, the rectangle, and the triangle, the location of the centroid is easy to visualize. Figure 8-1 shows the locations. If these shapes were carefully made and the location for the centroid carefully found, the shapes could be balanced on a pencil point at the centroid. Of course a steady hand is required. How's yours?

8-2 CENTROID OF COMPLEX SHAPES

Most complex shapes can be considered to be made up by combining several simple shapes together. This can be used to facilitate the location of the centroid, as will be demonstrated later.

179

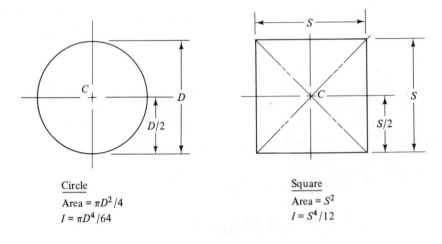

Circle
Area = $\pi D^2/4$
$I = \pi D^4/64$

Square
Area = S^2
$I = S^4/12$

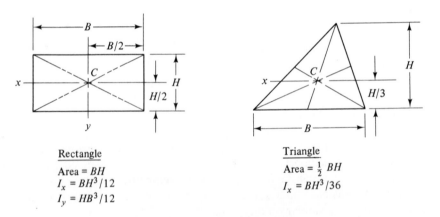

Rectangle
Area = BH
$I_x = BH^3/12$
$I_y = HB^3/12$

Triangle
Area = $\frac{1}{2}BH$
$I_x = BH^3/36$

Figure 8-1   Properties of Simple Areas.

Another concept which aids in the location of centroids is that if the area has an axis of symmetry, the centroid will be on that axis. Some complex shapes have two axes of symmetry, and therefore the centroid is at the intersection of these two axes. Figure 8-2 shows examples where this occurs.

Where two axes of symmetry do not occur, the method of composite areas can be used. If a complex area can be considered to be a composite of two or more simple areas, the centroid can be found by applying the principle that the product of the total area times the distance to the centroid of the total area is equal to the sum of the products of the area of each component part

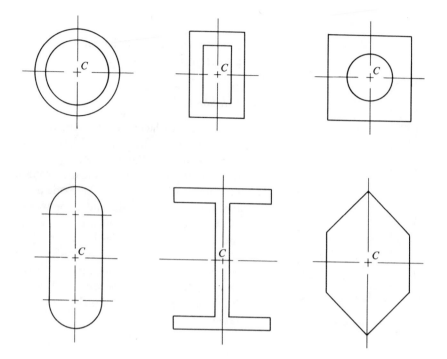

Figure 8-2   Composite Shapes Having Two Axes of Symmetry.

times the distance to its centroid. Stating this mathematically,

$$A_T \bar{Y} = \sum (A_i y_i) \tag{8-1}$$

where A_T = total area of the composite shape
  $\bar{Y}$ = distance to the centroid of the composite shape measured from some reference axis
  A_i = area of one component part of the shape
  y_i = distance to the centroid of the component part from the reference axis

The subscript i indicates that there may be several component parts, and the product $A_i y_i$ for each must be formed and then summed together, as called for in Equation (8-1). Since our objective is to compute $\bar{Y}$, Equation (8-1) can be solved:

$$\bar{Y} = \frac{\sum (A_i y_i)}{A_T} \tag{8-2}$$

A tabular form of writing the data helps keep track of the parts of the calculations called for in Equation (8-2). An example problem will illustrate the method.

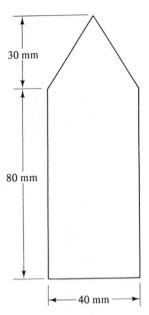

30 mm

80 mm

|←————40 mm————→|

Figure 8-3

Example Problem 8-1

Find the location of the centroid of the area shown in Figure 8-3.

Solution

Since the area has a vertical axis of symmetry, we know the centroid lies on that line. Then only the distance up to the horizontal centroidal axis must be found. Consider the total area to be made up of the sum of a triangle and a rectangle. Each of these is a simple area for which the centroid can be easily found. (See Figure 8-1.) The centroids are shown in Figure 8-4, with their locations measured from the bottom of the section. Now to implement Equation (8-2), the following table is formed.

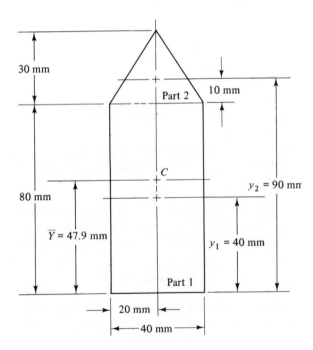

Figure 8-4

Part	A_i	y_i	$A_i y_i$
1	3200 mm²	40 mm	128 000 mm³
2	600 mm²	90 mm	54 000 mm³
$A_T = 3800$ mm²			182 000 mm³ $= \sum (A_i y_i)$

Now $\bar{Y}$ can be computed.

$$\bar{Y} = \frac{\sum (A_i y_i)}{A_T} = \frac{182\ 000 \text{ mm}^3}{3800 \text{ mm}^2} = 47.9 \text{ mm}$$

This locates the centroid as shown in Figure 8-4.

The composite area method works also for sections where parts are removed as well as added. In this case the removed area is considered negative, illustrated as follows.

Example Problem 8-2

Find the location of the centroid of the area shown in Figure 8-5.

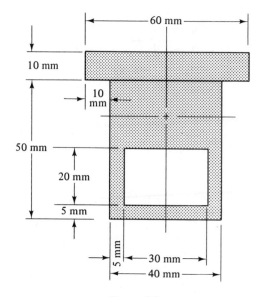

Figure 8-5

Solution

Again this area has a vertical axis of symmetry. The centroid then falls at the intersection of the horizontal centroidal axis, to be found, and the vertical axis of symmetry. Consider the composite area to be composed of three parts, as labeled in Figure 8-6. Notice that part 2

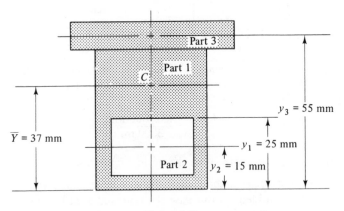

Figure 8-6

is *removed* from the composite area. The areas and individual centroidal distances are shown in the following table.

Part	A_i	y_i	A_iy_i
1	2000 mm²	25 mm	50 000 mm³
2	−600 mm²	15 mm	−9 000 mm³
3	600 mm²	55 mm	33 000 mm³
	$A_T = 2000$ mm²		74 000 mm³ $= \sum (A_iy_i)$

Then

$$\bar{Y} = \frac{\sum (A_iy_i)}{A_T} = \frac{74\ 000 \text{ mm}^3}{2000 \text{ mm}^2} = 37.0 \text{ mm}$$

8-3 THE CONCEPT OF MOMENT OF INERTIA—SIMPLE SHAPES

In the study of strength of materials, the property of moment of inertia is an indication of the stiffness of a particular shape. That is, a shape having a higher moment of inertia would deflect less when subjected to bending moments than one having a lower moment of inertia. The stress developed in a beam is also affected by the moment of inertia of its cross section, as will be shown in Chapter 9. For these reasons, the ability to compute the magnitude of the moment of inertia is very important.

The physical meaning of the moment of inertia is difficult to grasp since it is a derived property. That is, it is defined to be a property because it is important to the calculation of stress and deflection of beams and other processes. But it is not possible to describe it in physical terms, as we can an area or a volume. With practice, you should be able to identify sections which have inherently high moments of inertia because of the effective use of material. The discussion in Chapters 9 and 12 will emphasize this.

The moment of inertia of an area with respect to a particular axis is defined as the sum of the products obtained by multiplying each element of the area by the square of its distance from the axis. This definition was applied to mathematically determine the formulas for moment of inertia listed in Figure 8-1 for simple areas. The symbol *I* is used to denote moment of inertia. If a subscript is shown, it indicates the axis about which the moment of inertia is taken. If no subscript is used, it should be assumed that the axis is the horizontal centroidal axis of the section. That is the one of interest in most strength and deflection calculations. For the simple areas, then, the formulas can be applied directly to compute moment of inertia. The units for *I* will be *length to the fourth power*, usually mm⁴ in the SI system and in.⁴ in the English system.

8-4 MOMENT OF INERTIA OF COMPLEX SHAPES

The method of composite areas can be used to determine the moment of inertia for a complex shape which is made up of two or more simple areas joined together.

If the component parts of a composite area all have the same centroidal axis, the total moment of inertia can be found by adding or subtracting the moments of inertia of the component parts with respect to the centroidal axis. The following example problems show the process.

Example Problem 8-3

The cross-shaped section shown in Figure 8-7 has its centroid at the intersection of its axes of symmetry. Compute the moment of inertia of the section with respect to the horizontal axis *x-x*.

Solution

The cross can be considered to be made up of the vertical central stem plus the two horizontal arms. All three sections have as their horizontal controidal axes the axis *x-x*. Therefore the total moment of inertia I_x is the sum of the moments of inertia of each part with respect to the axis *x-x*. That is,

$$I_x = I_1 + I_2 + I_3$$

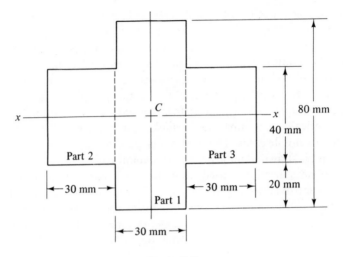

Figure 8-7

Referring to Figure 8-1,

$$I_1 = \frac{BH^3}{12} = \frac{30(80)^3}{12} = 1.28 \times 10^6 \text{ mm}^4$$

$$I_2 = I_3 = \frac{30(40)^3}{12} = 0.16 \times 10^6 \text{ mm}^4$$

Then

$$I_x = 1.28 \times 10^6 \text{ mm}^4 + 2(0.16 \times 10^6 \text{ mm}^4) = 1.60 \times 10^6 \text{ mm}^4$$

Example Problem 8-4

Compute the moment of inertia for the section shown in Figure 8-8 with respect to the axis x-x.

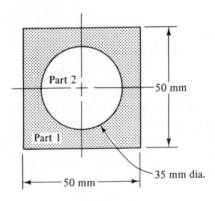

Figure 8-8

Solution

Since the square and the circle both have their centroidal axes coincident with the axis *x-x*,

$$I_x = I_1 - I_2$$

Notice that since the circular area is removed from the section, its moment of inertia is subtracted. Now

$$I_1 = \frac{S^4}{12} = \frac{(50)^4}{12} = 520.8 \times 10^3 \text{ mm}^4$$

$$I_2 = \frac{\pi D^4}{64} = \frac{\pi (35)^4}{64} = 73.7 \times 10^3 \text{ mm}^4$$

For the composite section,

$$I_x = I_1 - I_2 = 447.1 \times 10^3 \text{ mm}^4$$

Transfer-of-Axis Theorem. When a composite section is composed of parts whose centroidal axes do not lie on the centroidal axis of the entire section, the process of summing the values of *I* for the parts *cannot* be used. It can be said that the effect of material added to or removed from the section at a point far away from the centroid of the composite section is greater than if it were closer to the centroid. This is illustrated in Figure 8-9. Let's start with the simple rectangular section in (a) and add material to it in an attempt to increase its moment of inertia. Two ideas are tried. In (b), an area of 4 in.² is added at the centroidal axis. In (c), the same area is added at the top of the section. The tee section formed in (c) is stiffer than that in (b). That is, it has

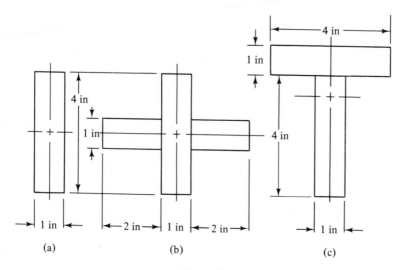

(a)   (b)   (c)

Figure 8-9

a higher moment of inertia. The reason is that the added area was placed far away from the original centroidal axis of the section. To show how much stiffer (c) is, we will compute the moment of inertia of all three sections. Remember, the higher the value of *I*, the stiffer the section is. The value of *I* for sections (a) and (b) can be computed using the basic methods already developed since the centroidal axes of all parts are the same. Section (c) will be used as an example to illustrate the transfer-of-axis theorem.

For section (a):

$$I_a = \frac{BH^3}{12} = \frac{1(4)^3}{12} = 5.33 \text{ in.}^4$$

For section (b):

$$I_b = I_1 + 2I_2 = \frac{1(4)^3}{12} + 2\left[\frac{2(1)^3}{12}\right] = 5.67 \text{ in.}^4$$

Very little increase in *I* was obtained from (a) to (b). Now I_c will be found.

Example Problem 8-5

Compute the moment of inertia of the tee section shown in Figure 8-9(c), and compare it with the values found for sections (a) and (b).

Solution

The process for finding the moment of inertia of a section whose component parts have centroidal axes different from that for the entire section is outlined as follows:

1. Divide the composite section into component parts which are simple areas.
2. Locate the centroid for the composite section.
3. Determine the distances d_1, d_2, etc., from the centroid of the composite section to the centroid of each part.
4. Compute the moment of inertia of each part with respect to its own centroidal axis. Call these I_1, I_2, etc.
5. Compute the *transfer term* for each part by multiplying the area of the part by the square of the distances d_1, d_2, etc., found in step 3. The transfer terms will then be $A_1d_1^2$, $A_2d_2^2$, etc.
6. Compute the total moment of inertia of the composite section from

$$I_T = I_1 + A_1d_1^2 + I_2 + A_2d_2^2 + \cdots \qquad (8\text{-}3)$$

Equation (8-3) is called the transfer-of-axis theorem because it defines how to transfer the moment of inertia of an area from one axis to any parallel axis. For each part of a composite section, the sum $I + Ad^2$ is a measure of its contribution to the total moment of inertia.

Now let's apply the above six-step procedure to compute the value of I for the section in Figure 8-9(c).

Step 1:

Let the vertical stem be part 1 and the flange at the top be part 2. Both are simple rectangles.

Step 2:

The procedure outlined in Section 8-2 can be used to locate the centroid of the composite section.

Part	A_i	y_i	$A_i y_i$
1	4.0 in.^{2}	2.0 in.	8.0 in.^{3}
2	4.0 in.^{2}	4.5 in.	18.0 in.^{3}
$A_T = 8.0$ in.^{2}			26.0 in.$^3 = \sum (A_i y_i)$

$$\bar{Y} = \frac{\sum (A_i y_i)}{A_T} = \frac{26.0 \text{ in.}^3}{8.0 \text{ in.}^2} = 3.25 \text{ in.}$$

Figure 8-10 shows these dimensions.

Step 3:

Figure 8-10 and the data from step 2 are helpful in finding the required distances.

$$d_1 = \bar{Y} - y_1 = 3.25 \text{ in.} - 2.0 \text{ in.} = 1.25 \text{ in.}$$
$$d_2 = y_2 - \bar{Y} = 4.5 \text{ in.} - 3.25 \text{ in.} = 1.25 \text{ in.}$$

Step 4:

For part 1,

$$I_1 = \frac{BH^3}{12} = \frac{1(4)^3}{12} = 5.33 \text{ in.}^4$$

For part 2,

$$I_2 = \frac{4(1)^3}{12} = 0.33 \text{ in.}^4$$

Step 5:

For part 1, the transfer term is

$$A_1 d_1^2 = (4.0 \text{ in.}^2)(1.25 \text{ in.})^2 = 6.25 \text{ in.}^4$$

For part 2,

$$A_2 d_2^2 = (4.0 \text{ in.}^2)(1.25 \text{ in.})^2 = 6.25 \text{ in.}^4$$

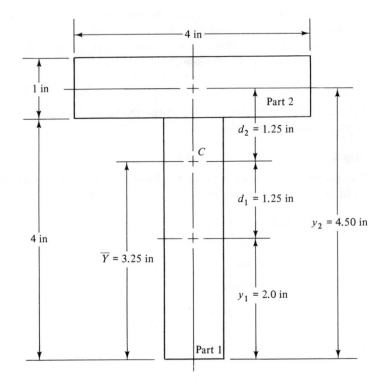

Figure 8-10

It is just coincidence that the transfer terms for each part are the same in this problem.

Step 6:

The total moment of inertia is

$$I_c = I_T = I_1 + A_1 d_1^2 + I_2 + A_2 d_2^2$$
$$= 5.33 \text{ in.}^4 + 6.25 \text{ in.}^4 + 0.33 \text{ in.}^4 + 6.25 \text{ in.}^4$$
$$I_c = I_T = 18.16 \text{ in.}^4$$

Let's compare the values of I for sections (a), (b), and (c) in Figure 8-9:

$$I_a = 5.33 \text{ in.}^4$$
$$I_b = 5.67 \text{ in.}^4 \qquad (1.06 \text{ times greater than } I_a)$$
$$I_c = 18.16 \text{ in.}^4 \qquad (3.4 \text{ times greater than } I_a \text{ and}$$
$$3.2 \text{ times greater than } I_b)$$

8-5 COMMERCIALLY AVAILABLE SHAPES FOR BEAMS

Steel and aluminum manufacturers provide a large variety of standard shapes for use as beams. Some are also used as columns, as will be shown in a later chapter. It is important to know the standard terminology for these shapes and how to use the tables of data which report their geometrical properties.

For steel shapes, the standard symbols used are listed in Figure 8-11, along with a sketch of the shape and an example of how it is designated. In the designation, the first letter or letters give the basic shape symbol. The first numbers give the nominal size, and the last set of numbers gives the weight per foot of length of a beam having this shape and sige. For example, the W12 × 16.5 shape would be a wide flange shape with a nominal depth (vertical height) of 12 in., and a 1-ft length would weigh 16.5 lb. This format is typical for all sections listed in Figure 8-11 except the angles and the construction pipe. For angles, the designation L4 × 3 × $\frac{1}{2}$ would be for an unequal leg angle whose legs are 4 in. and 3 in. and whose thickness is $\frac{1}{2}$ in. For the construction pipe, the designation TS4 × 0.188 would be for a pipe with a nominal size of 4 in. and a wall thickness of 0.188 in. It must be noted that the actual dimensions of standard shapes are not always the same as the nominal sizes. Reference must be made to standard tables of dimensions for actual sizes.

The properties of several standard shapes are listed in the Appendix. The data include such items as area of the section, depth, flange width and thickness, web thickness, and moment of inertia with respect to major axes. Other properties such as section modulus and radius of gyration, used in later chapters, are also listed. For channels and angles, the dimensions to the centroidal axes are also given.

The properties for Aluminum Association standard structural I-beams and channels are also listed in the Appendix. These are extruded shapes having flanges with uniform thickness. In contrast, rolled shapes have tapered flanges, making joining somewhat more difficult. The proportions for flange width and thickness are also somewhat different for the extruded sections. The designations for aluminum shapes are similar to those for steel shapes. For example, I8 × 6.181 would be a standard I-beam with a depth of 8.00 in., and a 1-ft length would weigh 6.181 lb. A C4 × 1.738 would be a standard channel with a depth of 4.00 in., and a 1-ft length would weigh 1.738 lb. For these aluminum shapes, the actual depth is the same as the nominal depth.

Combining Structural Shapes. When used separately, the properties for designing with standard shapes can be read directly from the tables. However, frequently two or more shapes are joined together to obtain a special

191

Name of Shape	Shape	Symbol	Example Designation
Wide flange beam		W	W12x16.5
American Standard beam		S	S10x35
Channel		C	C15x50
Angle		L	L4x3x$\frac{1}{2}$
Tees, cut from W shape		WT	WT5x8.5
Tees, cut from S shape		ST	ST6x15.9
Zees		Z	Z5x16.4
Construction pipe	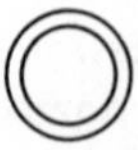	TS	TS4x0.188

Figure 8-11   Designations for Steel Shapes.

shape or to make a stronger section. The properties of the combined section must then be computed.

The principles developed earlier in this chapter are used to analyze combined sections. The special nature of the data given for standard shapes will be illustrated in the following examples.

Example Problem 8-6

In order to increase the stiffness of a standard aluminum I-beam, a 0.50-in. thick by 6.00-in. wide plate is welded to both the top and bottom flanges, as shown in Figure 8-12. Compute the moment of inertia of the combined section.

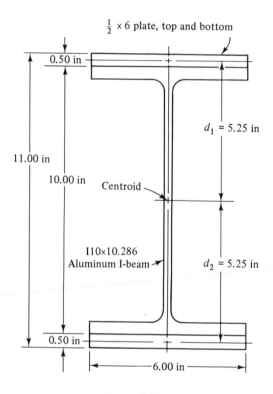

Figure 8-12

Solution

The centroid of the combined section is the same as the centroid for the beam itself since the added plates are placed symmetrically on the beam. Still, the transfer-of-axis theorem will have to be used to compute the moment of inertia since the centroid of each plate is away from that for the entire section. Calling the beam part 1, the top plate part 2, and the bottom plate part 3, the total moment of inertia will be

$$I_T = I_1 + I_2 + A_2 d_2^2 + I_3 + A_3 d_3^2$$

From the Appendix, $I_1 = 155.79$ in.^{4}. Now

$$I_2 = I_3 = \frac{BH^3}{12} = \frac{(6.0)(0.5)^3}{12} = 0.063 \text{ in.}^4$$

$$A_1 = A_2 = BH = (6.0)(0.5) = 3.0 \text{ in.}^2$$

$$d_1 = d_2 = 5.25 \text{ in.}$$

Then

$$I_T = 155.79 + 0.063 + (3.0)(5.25)^2 + 0.063 + (3.0)(5.25)^2$$

$$I_T = 321.29 \text{ in.}^4$$

The moment of inertia was more than doubled by the addition of the plates.

Example Problem 8-7

A built-up beam is made by fastening four L4×4×$\frac{1}{2}$ steel angles to a $\frac{1}{2}$ in. × 16 in. plate, as shown in Figure 8-13. Compute the moment of inertia of the beam.

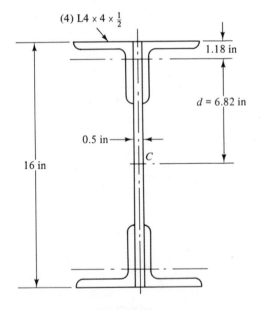

Figure 8-13

Solution

The beam has two axes of symmetry, so its centroid is at the middle of the plate. The centroid of the angles is away from the centroid of the entire section, so the transfer-of-axis theorem must be used. Notice that all four angles are identical and that all are the same distance from the beam's centroid. Therefore, the total moment of inertia will be

$$I_T = I_1 + 4(I_2 + A_2 d_2^2)$$

where the subscript 1 refers to the plate and the subscript 2 refers to one of the angles. Now

$$I_1 = \frac{BH^3}{12} = \frac{(0.50)(16)^3}{12} = 170.67 \text{ in}^4.$$

$$I_2 = 5.56 \text{ in.}^4 \qquad \text{(from Appendix)}$$

$$A_2 = 3.75 \text{ in.}^2$$

$$d_2 = 8.0 \text{ in.} - 1.18 \text{ in.} = 6.82 \text{ in.}$$

The computation of d_2 required the location of the centroidal axis of the angle, which is 1.18 in. down from its top leg, as shown in Table A-15. The calculation can now be completed.

$$I_T = 170.67 + 4[5.56 + 3.75\,(6.82)^2]$$

$$I_T = 170.67 + 719.93 = 890.60 \text{ in.}^4$$

PROBLEMS

For each of the shapes in Figures 8-14 to 8-20, determine the location of the horizontal centroidal axis and the magnitude of the moment of inertia of the shape with respect to that axis. Figures 8-17 through 8-18 are composed of standard steel or aluminum sections for which the properties are listed in the Appendix. The sections in Figure 8-19 are made of standard wood forms. The sections in Figure 8-20 are typical aluminum or plastic extrusions or molded shapes found in business machines, toys, coat hangers, and similar products.

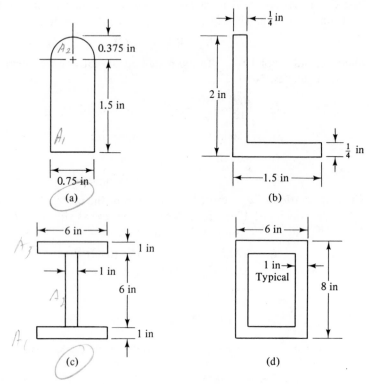

Figure 8-14

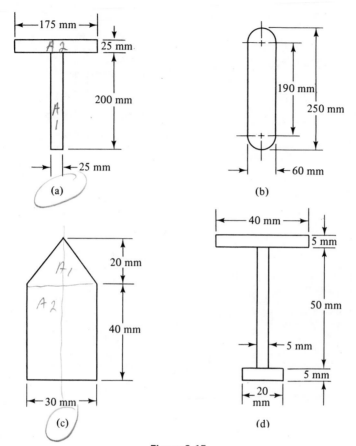

Figure 8-15

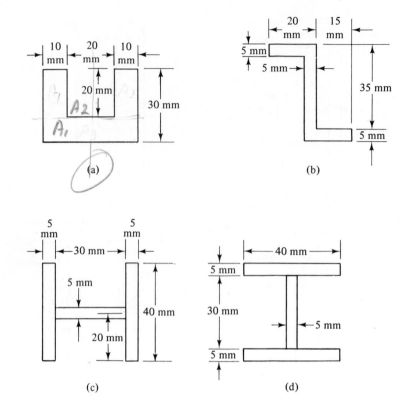

Figure 8-16

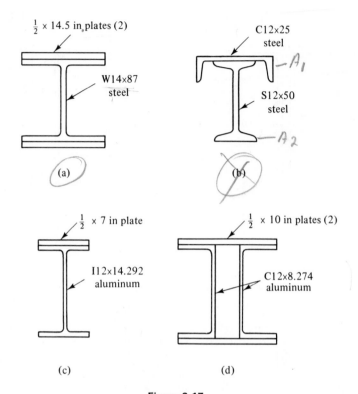

$\frac{1}{2}$ × 14.5 in plates (2)

W14×87 steel

(a)

C12×25 steel

A_1

S12×50 steel

A_2

(b)

$\frac{1}{2}$ × 7 in plate

I12×14.292 aluminum

(c)

$\frac{1}{2}$ × 10 in plates (2)

C12×8.274 aluminum

(d)

Figure 8-17

199

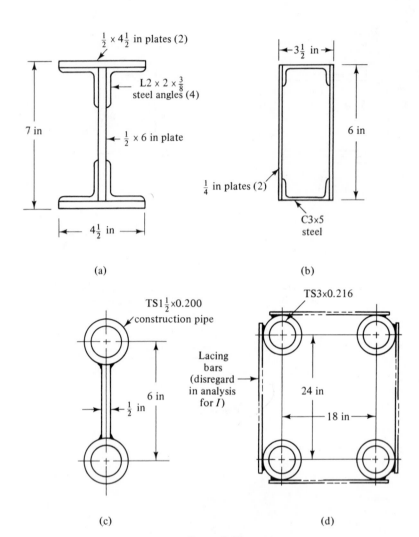

$\frac{1}{2} \times 4\frac{1}{2}$ in plates (2)

L2 × 2 × $\frac{3}{8}$
steel angles (4)

7 in

$\frac{1}{2}$ × 6 in plate

$4\frac{1}{2}$ in

(a)

$3\frac{1}{2}$ in

$\frac{1}{4}$ in plates (2)

6 in

C3×5
steel

(b)

TS1$\frac{1}{2}$×0.200
construction pipe

6 in

$\frac{1}{2}$ in

(c)

TS3×0.216

Lacing
bars
(disregard
in analysis
for I)

24 in

18 in

(d)

Figure 8-18

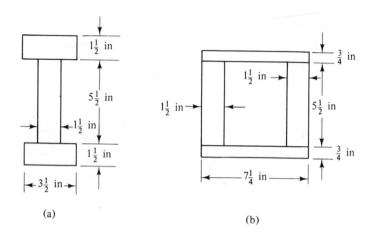

(a)

(b)

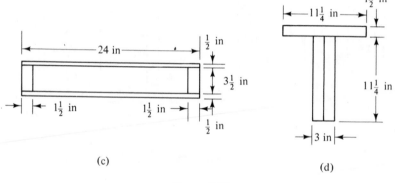

(c)

(d)

Figure 8-19

201

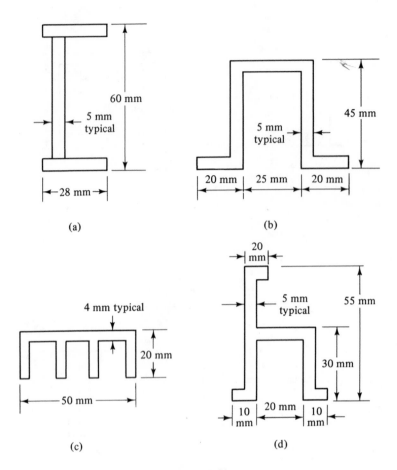

(a)

(b)

(c)

(d)

Figure 8-20

9

STRESS DUE
TO BENDING

9-1   THE FLEXURE FORMULA

Beams must be designed to be safe. When loads are applied perpendicular to the long axis of a beam, bending moments are developed inside the beam, causing it to bend. By observing a thin beam, the characteristically curved shape shown in Figure 9-1 is evident. The fibers of the beam near its top surface are shortened and placed in compression. Conversely, the fibers near the bottom surface are stretched and placed in tension.

In designing or analyzing beams, it is usually the objective to determine what the maximum tensile or compressive stress is. The maximums will occur at the outer surfaces of the beam. The flexure formula is used to com-

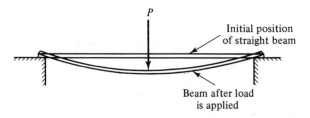

Figure 9-1

pute the maximum stress due to bending. The typical form for the formula is

$$s = \frac{Mc}{I} \tag{9-1}$$

where s = maximum stress at the outermost fiber of the beam
M = bending moment at the section of interest
c = distance from the centroidal axis of the beam to the outermost fiber
I = moment of inertia of the cross section with respect to its centroidal axis

More about the flexure formula will be discussed later. A simple problem will now be shown to illustrate the application of the formula.

Example Problem 9-1

For the beam shown in Figure 9-1, compute the maximum stress due to bending. The cross section of the beam is a rectangle 100 mm high and 25 mm wide. The load at the middle of the beam is 1500 N, and the length of the beam is 3.40 m.

Solution

The method of solving any beam stress problem involves the following steps:

1. Determine the maximum bending moment on the beam by drawing the shear and bending moment diagrams.
2. Locate the centroid of the cross section of the beam.
3. Compute the moment of inertia of the cross section with respect to its centroidal axis.
4. Compute the distance c from the centroidal axis to the top or bottom of the beam, whichever is greater.
5. Compute the stress from the flexure formula,

$$s = \frac{Mc}{I} \quad or \quad S = \frac{m}{2}$$

For this problem, refer to Figure 9-2 when completing steps 1–4.

Step 1:

The maximum bending moment is 1275 N·m at the middle of the beam.

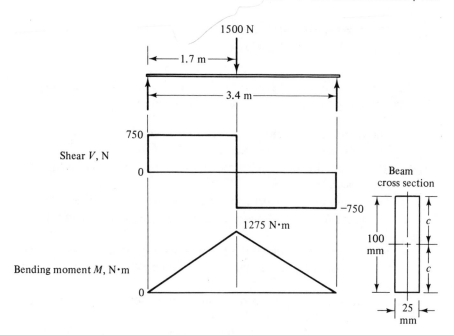

Figure 9-2

Step 2:

The centroid of the rectangular section is at the intersection of its two axes of symmetry.

Step 3:

For the rectangular section,

$$I = \frac{BH^3}{12} = \frac{25(100)^3}{12} = 2.08 \times 10^6 \text{ mm}^4$$

Step 4:

$$c = 50 \text{ mm}$$

Step 5:

The maximum stress due to bending is

$$s = \frac{Mc}{I} = \frac{(1275 \text{ N} \cdot \text{m})(50 \text{ mm})}{2.08 \times 10^6 \text{mm}^4} \times \frac{10^3 \text{ mm}}{\text{m}}$$

$$s = 30.6 \text{ N/mm}^2 = 30.6 \text{ MPa}$$

$m = \times 1000^2$

9-2 CONDITIONS ON THE USE
OF THE FLEXURE FORMULA

The proper application of the flexure formula requires the understanding of the conditions under which it is valid, listed as follows:

1. The beam must be straight or very nearly so.
2. The cross section of the beam must be uniform.
3. All loads and support reactions must act perpendicular to the axis of the beam.
4. The beam must not twist while the loads are being applied.
5. The beam must be relatively long and narrow in proportion to its depth.
6. The material from which the beam is made must be homogeneous, and it must have an equal modulus of elasticity in tension and compression.
7. The stress resulting from the loading must not exceed the proportional limit of the material.
8. No part of the beam may fail from instability, that is, from the buckling or crippling of thin sections.

Although the list of conditions appears to be long, the flexure formula still applies to a wide variety of real cases. Beams violating some of the conditions can be analyzed by using a modified formula or by using a combined stress approach. For example, for condition 2, a change in cross section will cause stress concentrations which can be handled as described in Section 9-5. The combined bending and axial stress or bending and torsional stress produced by violating condition 3 will be discussed in Chapter 11. If the other conditions are violated, special analyses are required which are not covered in this book.

Condition 4 is important, and attention must be paid to the shape of the cross section to ensure that twisting does not occur. In general, if the beam has a vertical axis of symmetry and if the loads are applied through that axis, no twisting will result. Figure 9-3 shows some typical shapes used for beams which satisfy condition 4. Conversely, Figure 9-4 shows several which do not satisfy condition 4; in each of these cases, the beam would tend to twist as well as bend as the load is applied in the manner shown. Of course, these sections can support some load, but the actual stress condition in them is different from that which would be predicted from the flexure formula. More about these kinds of beams is presented in Section 9-6.

Condition 8 is important because long, thin members and, sometimes, thin sections of members tend to buckle at stress levels well below the yield

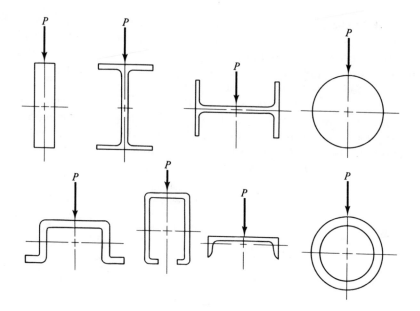

Figure 9-3

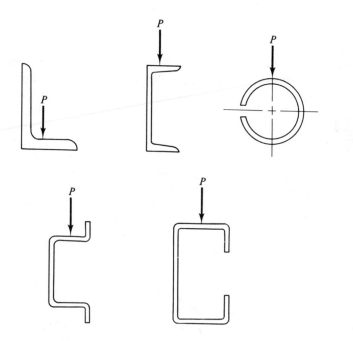

Figure 9-4

strength of the material. Such failures are called *instability* and are to be avoided. Frequently, cross braces or local stiffeners are added to beams to relieve the problem of instability. An example can be seen in the wood joist floor construction of many homes and commercial buildings. The relatively slender wood joists are braced near their midpoints to avoid buckling.

One of the basic assumptions made in the derivation of the flexure formula is that a plane section through the beam remains straight as bending occurs. During the action of bending, the plane section will rotate as indicated in Figure 9-5. The result is that fibers near the bottom will be placed in tension

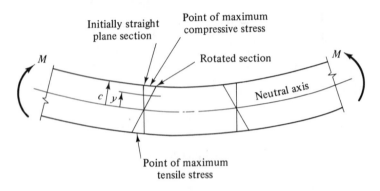

Figure 9-5

and fibers near the top will be in compression, as we have already observed. Furthermore, the plane section will rotate about an axis where neither tension nor compression is developed. This axis is called the *neutral axis*, and it is coincident with the *centroidal axis* of the cross section. We discussed how to locate the centroidal axis in Chapter 8. If no tension or compression exists at the neutral axis, then no stress exists. Another observation can be made from Figure 9-5. There is a linear variation in the strain or deformation of the material as a function of distance away from the neutral axis. Since stress is proportional to strain in a material obeying Hooke's law, stress, too, varies linearly with distance from the neutral axis. The maximum stress occurs at the outer surface at a distance c from the neutral axis. Then at any other distance y,

$$s = s_{max} \frac{y}{c} \tag{9-2}$$

This analysis shows that the stress distribution on a section subjected to bending would be as shown in Figure 9-6. The same distribution would occur for any section having a centroidal axis equidistant from the top and bottom surfaces. For this case, the maximum compressive stress would equal the maximum tensile stress.

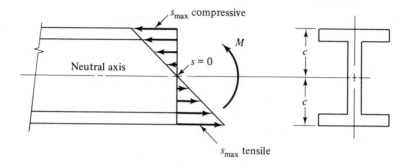

Figure 9-6

If the centroidal axis of the section is not the same distance from both the top and bottom surfaces, the stress distribution shown in Figure 9-7 would occur. Still the stress at the neutral axis would be zero. Still the stress

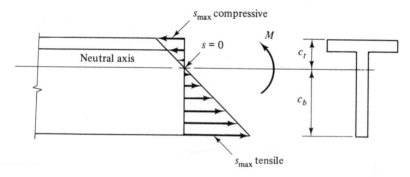

Figure 9-7

would vary linearly with distance from the neutral axis. But now the maximum stress at the bottom of the section is greater than that at the top because it is farther from the neutral axis. Using the distances c_b and c_t as indicated in Figure 9-7, the stresses would be

$$s_{max} = \frac{Mc_b}{I} \quad \text{(tension at the bottom)}$$

$$s_{max} = \frac{Mc_t}{I} \quad \text{(compression at the top)}$$

9-3   APPLICATIONS—BEAM ANALYSIS

Beam analysis refers to those cases where the description of the beam, its loading, and its manner of support are completely defined and the objective of the computation is to determine the stresses which exist. In performing the analyses, keep in mind the five-step procedure outlined in Example Problem 9-1.

Example Problem 9-2

The shape shown in Figure 9-8 is the cross section of a beam carrying a bending moment of 3850 N·m. Compute the stress due to bending at the bottom surface and also at the axis *a-a*. The moment of inertia for the section is 1.60×10^6 mm⁴.

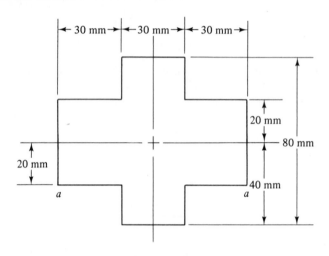

Figure 9-8

Solution

The maximum stress occurs at the bottom surface, and Equation (9-1) applies directly. Using $c = 40$ mm,

$$s_{max} = \frac{Mc}{I} = \frac{(3850 \text{ N·m})(40 \text{ mm})}{1.60 \times 10^6 \text{ mm}^4} \times \frac{10^3 \text{ mm}}{\text{m}}$$

$$s_{max} = 96.25 \text{ N/mm}^2 = 96.25 \text{ MPa}$$

At axis *a-a*, Equation (9-2) applies,

$$s = s_{max} \frac{y}{c}$$

Using $y = 20$ mm,

$$s = 96.25 \text{ MPa} \times \frac{20 \text{ mm}}{40 \text{ mm}} = 48.13 \text{ MPa}$$

Example Problem 9-3

The tee section shown in Figure 9-9 is from a simply supported beam which carries a bending moment of 100 000 lb·in. due to a load on

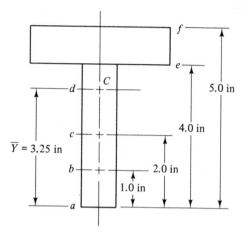

Figure 9-9

the top surface. Earlier, this section was analyzed to determine that $I = 18.16$ in.^{4}. The centroid of the section is 3.25 in. up from the bottom of the beam. Compute the stress due to bending in the beam at the six axes a to f indicated in the figure. Then plot a graph of stress versus position in the cross section.

Solution

Let's start at *axis a*. The maximum stress would occur there because it is the farthest point from the centroidal axis of the beam. Using $c = 3.25$ in.,

$$s_a = \frac{Mc}{I} = \frac{(100\ 000\ \text{lb}\cdot\text{in.})(3.25\ \text{in.})}{18.16\ \text{in.}^4}$$

$$s_a = 17\ 900\ \text{psi} \qquad (\text{tension})$$

The remaining stresses will be computed from Equation (9-2).

At axis b: $\qquad s_b = s_{max}\dfrac{y}{c} = s_a\dfrac{y}{c}$

$$y = 3.25\ \text{in.} - 1.0\ \text{in.} = 2.25\ \text{in.}$$

Then

$$s_b = 17\ 900\ \text{psi} \times \frac{2.25}{3.25} = 12\ 400\ \text{psi} \qquad (\text{tension})$$

At axis c: $\quad y = 3.25\ \text{in.} - 2.0\ \text{in.} = 1.25\ \text{in.}$

$$s_c = 17\ 900\ \text{psi} \times \frac{1.25}{3.25} = 6900\ \text{psi} \qquad (\text{tension})$$

At axis d: At the centroid $y = 0$ and $s_d = 0$.

At axis e: $y = 4.0 \text{ in.} - 3.25 \text{ in.} = 0.75 \text{ in.}$

$$s_e = 17\ 900 \text{ psi} \times \frac{0.75}{3.25} = 4100 \text{ psi} \qquad \text{(compression)}$$

At axis f: $y = 5.0 \text{ in.} - 3.25 \text{ in.} = 1.75 \text{ in.}$

$$s_f = 17\ 900 \text{ psi} \times \frac{1.75}{3.25} = 9600 \text{ psi} \qquad \text{(compression)}$$

The graph of these data is shown in Figure 9-10. Notice the linear variation of stress with distance from the neutral axis.

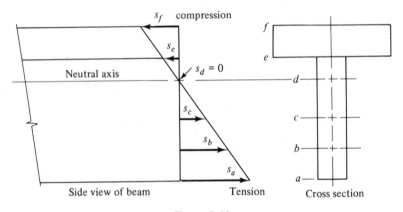

Figure 9-10

Example Problem 9-4

A beam 25 ft long supports loads which cause the bending moments indicated in Figure 9-11. The beam is a W14×38. Compute the maximum stress due to bending in the beam.

Solution

Figure 9-11 is the bending moment diagram for the beam, showing the variation in M as a function of position on the beam. The maximum stress would occur at point F, where the bending moment is 91 113 lb·ft. Equation (9-1),

$$s = \frac{Mc}{I}$$

will be used to compute the stress. The Table A-12, listing properties of wide flange shapes, gives $I = 386 \text{ in.}^4$ and the depth of the beam to

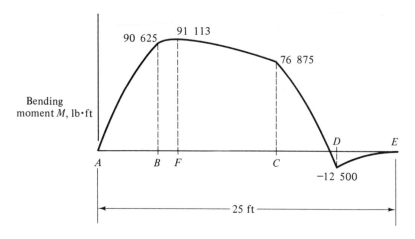

Figure 9-11

be 14.12 in. Therefore,

$$c = 7.06 \text{ in.}$$

Then

$$s = 91 \ 113 \text{ lb·ft} \times \frac{12 \text{ in.}}{\text{ft}} \times \frac{7.06 \text{ in.}}{386 \text{ in.}^4} = 20 \ 000 \text{ psi}$$

9-4   APPLICATIONS—BEAM DESIGN

To design a beam, its material, length, placement of loads, placement of supports, and the size and shape of its cross section must be specified. Normally, the length, placement of loads, and placement of supports are given by the requirements of the application. Then the material specification and the size and shape of the cross section are determined by the designer. The primary duty of the designer is to ensure the safety of the design. This requires a stress analysis of the beam and a decision concerning the allowable or design stress which the chosen material may be subjected to. The examples presented here will concentrate on these items. Also of interest to the designer are cost, appearance, physical size, weight, compatibility of the design with other components of the machine or structure, and the availability of the material or beam shape.

Two basic approaches will be shown for beam design. One involves the specification of the *material* from which the beam will be made and the general *shape* of the beam (circular, rectangular, W-beam, etc.), with the subsequent determination of the required dimensions of the cross section of the beam. The second requires specifying the *dimensions* and *shape* of the beam and then computing the required strength of a material from which to make the beam. Then the actual material is specified.

In both approaches described above, some representation of *design stresses* must be made. We will use the same approach described in Section 3-5, with design factors selected from Table 3-1 according to the type of load and the type of material. In general, then, the design stress will be computed from

$$s_d = \frac{s_y}{N} \qquad \text{based on yield strength}$$

or

$$s_d = \frac{s_u}{N} \qquad \text{based on ultimate strength}$$

These equations for design stress apply because the kind of stresses developed in a beam are tensile and compressive and the modes of failure are similar to those described in Section 3-5.

The stress analysis will require the use of the flexure formula,

$$s = \frac{Mc}{I}$$

But a modified form is desirable for cases where the determination of the dimensions of a section is to be done. Notice that both the moment of inertia I and the distance c are geometrical properties of the cross-sectional area of the beam. Therefore, the quotient I/c is also a geometrical property. For convenience, we can define a new term, *section modulus*, denoted by the letter Z.

$$Z = \frac{I}{c} \tag{9-3}$$

The flexure formula then becomes

if looking for dimen-
sions

$$s = \frac{M}{Z} \tag{9-4}$$

This is the most convenient form for use in design. Example problems will demonstrate the use of the section modulus. It should be noted that some designers and manufacturers use the symbol S in place of Z for section modulus. In this book, Z is used to eliminate possible confusion with stress.

Example Problem 9-5 *To find height*

A beam is to be designed to carry the load shown in Figure 9-12. A rectangular cross section will be used, with the beam being cut from a $1\frac{1}{4}$-in. thick plate of A36 structural steel. (See Table A-3.) The required height of the cross section is to be determined. Assume the forces will be dead loads.

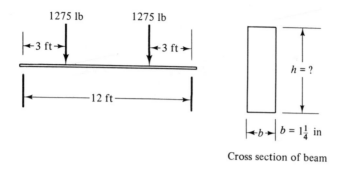

Cross section of beam

Figure 9-12

Solution

The following steps will be used to determine the required height of the cross section.

1. The shear and bending moment diagrams will be drawn to determine the maximum bending moment M.
2. The design stress s_d for the A36 steel will be computed.
3. The flexure formula $s = M/Z$ will be solved for Z,

$$Z = \frac{M}{s}$$

4. Letting $s = s_d$, the required section modulus Z will be computed.
5. The required height h of the rectangular section will be computed.

Step 1:

Figure 9-13 shows the completed diagrams. The maximum bending moment is 45 900 lb·in.

Step 2:

From Table A-3, $s_y = 36\ 000$ psi for A36 steel. For a dead load, a design factor of $N = 2$ based on yield strength is reasonable. Then

$$s_d = \frac{s_y}{N} = \frac{36\ 000\ \text{psi}}{2} = 18\ 000\ \text{psi}$$

Steps 3 and 4:

The required Z is

$$Z = \frac{M}{s_d} = \frac{45\ 900\ \text{lb·in.}}{18\ 000\ \text{lb/in.}^2} = 2.55\ \text{in.}^3$$

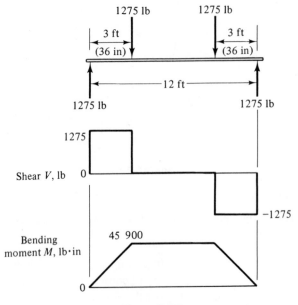

Figure 9-13

Step 5:

The formula for the section modulus for a rectangular section with a height h and a thickness b is

$$Z = \frac{I}{c} = \frac{bh^3}{12(h/2)} = \frac{bh^2}{6}$$

For the beam in this design problem, b will be $1\frac{1}{4}$ in. Then solving for h,

$$Z = \frac{bh^2}{6}$$

$$h = \sqrt{\frac{6Z}{b}} = \sqrt{\frac{6(2.55 \text{ in.}^3)}{1.25 \text{ in.}}}$$

$$h = 3.50 \text{ in.}$$

The beam will then be rectangular in shape, with dimensions of $1\frac{1}{4}$ in. by $3\frac{1}{2}$ in.

Example Problem 9-6

A circular shaft is to be designed to carry the three loads shown in Figure 9-14. The shaft rotates at 65 rad/s. The bearings at B and D provide simple supports for the shaft. Considering bending stress only, and assuming that the entire shaft will have the same diameter with no stress concentrations, determine the required diameter of the shaft. Use AISI 1137 OQT 1300 steel for the shaft.

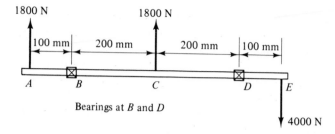

Figure 9-14

Solution

The basic steps to be used are similar to those used in the preceding problem. Figure 9-15 shows the shear and bending moment diagrams, giving the maximum bending moment of 400 N·m. The fact that the moment is negative is of no concern here, since the maximum absolute value is desired. The shaft will see the maximum tensile stress on one

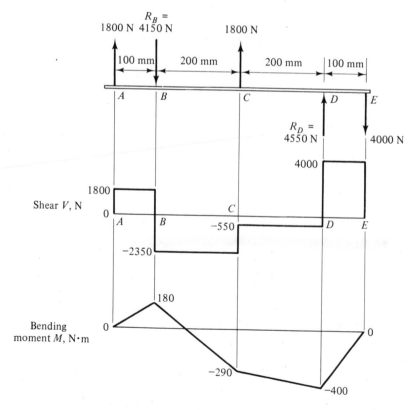

Figure 9-15

side and the maximum compressive stress on the other side. Also, since the shaft rotates, different parts of the shaft will experience the stresses throughout one revolution.

To determine design stress, it must be observed that a *repeated* load condition exists. Consider a single point on the surface of the shaft where the 400 N·m bending moment occurs. As the shaft rotates, the point will experience alternately the maximum tensile and maximum compressive stresses. For this reason, the design stress should be computed from

$$s_d = \frac{s_u}{N}$$

with $N = 8$, as described in Section 3-5. For the AISI 1137 OQT 1300, (Table A-1), $s_u = 600$ MPa. Then

$$s_d = \frac{600 \text{ MPa}}{8} = 75 \text{ MPa}$$

Now the required section modulus can be found.

$$Z = \frac{M}{s_d} = \frac{400 \text{ N·m}}{75 \times 10^6 \text{ N/m}^2} = 5.33 \times 10^{-6} \text{ m}^3$$

The section modulus is related to the shaft diameter D by

$$Z = \frac{I}{c} = \frac{\pi D^4}{64(D/2)} = \frac{\pi D^3}{32}$$

Solving for D,

$$D = \left(\frac{32Z}{\pi}\right)^{1/3}$$

To obtain a value of $Z = 5.33 \times 10^{-6} \text{ m}^3$, the minimum required diameter would be

$$D = \left[\frac{(32)(5.33 \times 10^{-6} \text{ m}^3)}{\pi}\right]^{1/3} = 0.0379 \text{ m} = 37.9 \text{ mm}$$

The next preferred size, 40 mm, should be specified.

Example Problem 9-7 *suitable mat'l*

In order to satisfy space limitations, a beam in a machine structure has been specified to be a 60-mm square bar. The forces on the bar are as shown in Figure 9-16. All are dead loads. An aluminum alloy will be used because of its corrosion resistance. Specify a suitable aluminum alloy from Table A-4.

Solution

Figure 9-17 shows the shear and bending moment diagrams. The maximum bending moment occurs at the supports, 6592 N·m. Since the cross-sectional dimensions are already specified, the stress in the

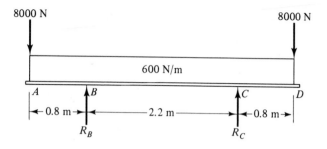

Figure 9-16

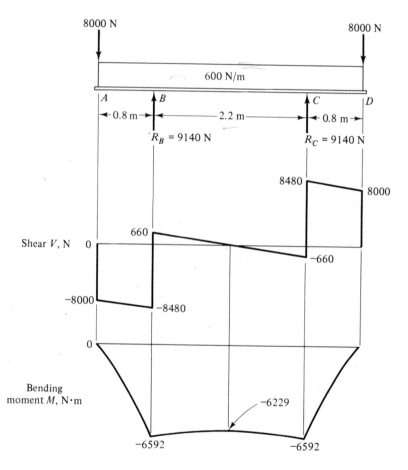

Figure 9-17

bar can be computed directly from the flexure formula,

$$s = \frac{Mc}{I}$$

For the 60-mm square bar,

$$I = \frac{(60 \text{ mm})^4}{12} = 1.08 \times 10^6 \text{ mm}^4$$

$$c = \frac{60 \text{ mm}}{2} = 30 \text{ mm}$$

Then

$$s = \frac{(6592 \text{ N} \cdot \text{m})(30 \text{ mm})}{1.08 \times 10^6 \text{ mm}^4} \times \frac{10^3 \text{ mm}}{\text{m}} = 183 \text{ N/mm}^2 = 183 \text{ MPa}$$

In selecting a material, the stress just computed can be considered to be the allowable stress or design stress. The required yield strength of the aluminum then would be, using $N = 2$ because of the dead load,

$$s_d = \frac{s_y}{N}$$

$$s_y = N s_d = 2(183 \text{ MPa}) = 366 \text{ MPa}$$

Referring to Table A-4, we find that alloys 2014-T6 or 7075-T6 could be used. No others listed there have sufficient strength.

Example Problem 9-8

The floor of an industrial building is to be supported by wide flange beams spaced 4 ft on centers across a 20-ft span, as sketched in Figure 9-18. The floor will be a poured concrete slab, 4 in. thick. The design

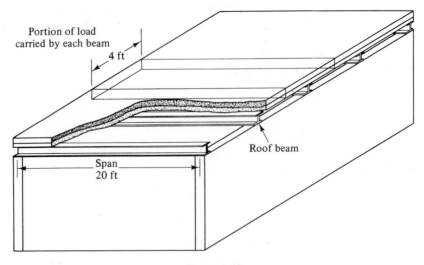

Portion of load carried by each beam

4 ft

Roof beam

Span 20 ft

Figure 9-18

live load on the floor is 200 lb/ft². Specify a suitable wide flange beam which will limit the stress in the beam to 22 000 psi.

Solution

We must first determine the load on each beam of the floor structure. Dividing the load evenly among adjacent beams would result in each beam carrying a 4-ft wide portion of the floor load. In addition to the 200 lb/ft² live load, the weight of the concrete slab offers a sizeable load. Assuming that the concrete weighs 150 lb/ft³, each square foot of the floor, 4.0 in. thick, would weigh 50 lb. This is called the dead load. Then the total loading due to the floor is 250 lb/ft². Now, notice that each foot of length of the beam carries 4 ft² of the floor. Therefore, the load on the beam is a uniformly distributed load of 1000 lb/ft. Figure 9-19 shows the loaded beam and the shear and bending moment diagrams. The maximum bending moment is 50 000 lb·ft.

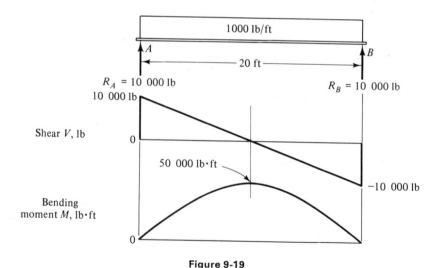

Figure 9-19

In order to select a wide flange beam, the required section modulus must be calculated.

$$s = \frac{M}{Z}$$

$$Z = \frac{M}{s_d} = \frac{50\ 000\ \text{lb·ft}}{22\ 000\ \text{lb/in.}^2} \times \frac{12\ \text{in.}}{\text{ft}} = 27.3\ \text{in.}^3$$

A beam must be found from Table A-12 which has a value of Z greater than 27.3 in.³ In considering alternatives, of which there are many, you should search for the lightest beam which will be safe, since the cost of the beam is based on its wieght. Some possible beams are

$$W14 \times 22: \quad Z = 28.9 \text{ in.}^3, \text{ 22 lb/ft}$$

$$W12 \times 27: \quad Z = 34.2 \text{ in.}^3, \text{ 27 lb/ft}$$

$$W10 \times 29: \quad Z = 30.8 \text{ in.}^3, \text{ 29 lb/ft}$$

$$W8 \times 31: \quad Z = 27.4 \text{ in.}^3, \text{ 31 lb/ft}$$

Of these, the $W14 \times 22$ would be preferred because it is the lightest.

9-5 STRESS CONCENTRATIONS

The conditions specified for valid use of the flexure formula in Section 9-2 included the statement that the beam must have a uniform cross section. Changes in cross section result in higher local stresses than would be predicted from the direct application of the flexure formula. Similar observations were made in earlier chapters concerning direct axial stresses and torsional shear stresses. The use of *stress concentration factors* will allow the analysis of beams which do include changes in cross section.

In the design of round shafts for carrying power-transmitting elements, the use of steps in the diameter is encountered frequently. Examples were shown in Chapter 6, where torsional shear stresses were discussed. Figure 9-20 shows such a shaft. Considering the shaft as a beam subjected to bending moments, there would be stress concentrations at the shoulder (2), the key

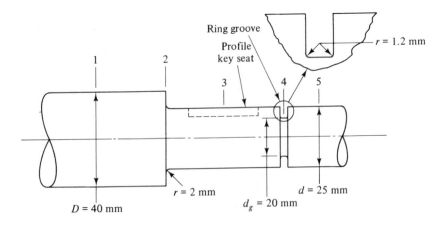

Figure 9-20

seat (3), and the groove (4). At sections where stress concentrations occur, the stress due to bending would be calculated from a modified form of the flexure formula,

$$s = \frac{McK_t}{I} = \frac{MK_t}{Z} \qquad (9\text{-}5)$$

The stress concentration factor K_t is found experimentally, with the values reported in graphs such as those in Figures 9-21 to 9-23. Table 9-1 shows recommended values for K_t for key seats.

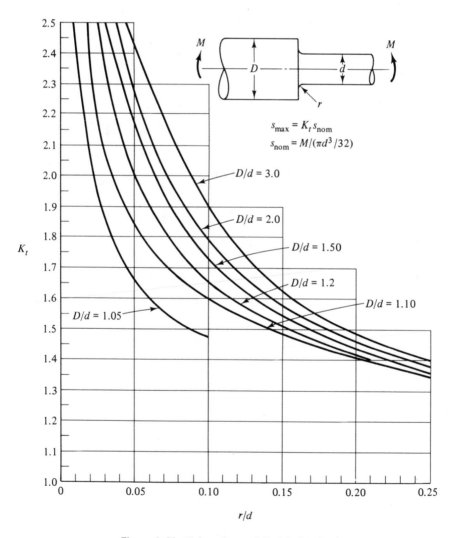

Figure 9-21 K_t for a Stepped Shaft in Bending

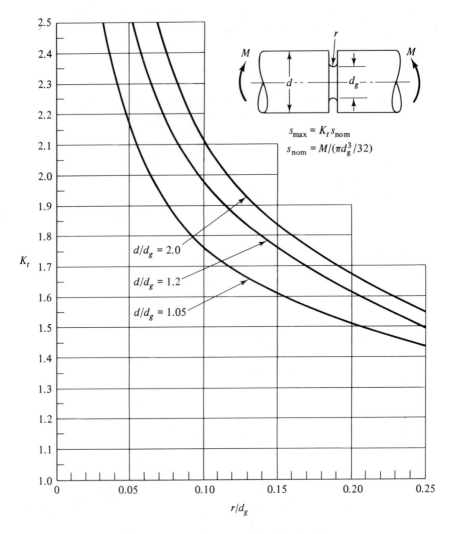

$$s_{max} = K_t s_{nom}$$

$$s_{nom} = M/(\pi d_g^3/32)$$

Figure 9-22 K_t for a Grooved Shaft in Bending

TABLE 9-1 Stress Concentration Factors for Key Seats

Type of Key Seat	K_t for Bending*
Sled-runner	1.6
Profile	2.0

*K_t is to be applied to the stress computed for the full nomi-nal diameter of the shaft where the key seat is located.

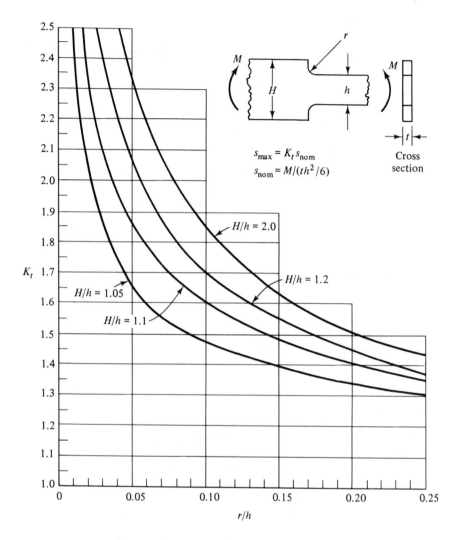

Figure 9-23 K_t for a Stepped Flat Bar in Bending

Similarly, a flat bar with a rectangular cross section would experience a stress concentration at a place where the section size changed. Figure 9-23 shows the stress concentration factors for this case.

Example Problem 9-9

Figure 9-20 shows a portion of a round shaft where a gear is mounted. A bending moment of 30 N·m is applied at this location. Compute the stress due to bending at sections 2, 3, and 4.

Solution

At *section 2*, the step in the shaft causes a stress concentration to occur. Then the stress is

$$s_2 = \frac{MK_t}{Z}$$

The smaller of the diameters at section 2 is used to compute Z.

$$Z = \frac{\pi d^3}{32} = \frac{\pi (25 \text{ mm})^3}{32} = 1534 \text{ mm}^3$$

The value of K_t depends on the ratios r/d and D/d.

$$\frac{r}{d} = \frac{2 \text{ mm}}{25 \text{ mm}} = 0.08$$

$$\frac{D}{d} = \frac{40 \text{ mm}}{25 \text{ mm}} = 1.60$$

From Figure 9-21, $K_t = 1.87$. Then

$$s_2 = \frac{MK_t}{Z} = \frac{(30 \text{ N} \cdot \text{m})(1.87)}{1534 \text{ mm}^3} \times \frac{10^3 \text{ mm}}{\text{m}} = 36.6 \text{ N/mm}^2$$

$$s_2 = 36.6 \text{ MPa}$$

At *section 3*, the key seat causes a stress concentration factor of 2.0. Then

$$s_3 = \frac{MK_t}{Z} = \frac{(30 \text{ N} \cdot \text{m})(2.0)}{1534 \text{ mm}^3} \times \frac{10 \text{ mm}^3}{\text{m}} = 39.1 \text{ N/mm}^2$$

$$s_3 = 39.1 \text{ MPa}$$

The groove at *section 4* requires the use of Figure 9-22 to find K_t. Note that the nominal stress is based on the root diameter of the groove, d_g. For the groove,

$$\frac{r}{d_g} = \frac{1.2 \text{ mm}}{20 \text{ mm}} = 0.06$$

$$\frac{d}{d_g} = \frac{25 \text{ mm}}{20 \text{ mm}} = 1.25$$

Then $K_t = 2.45$. The section modulus at the root of the groove is

$$Z = \frac{\pi d_g^3}{32} = \frac{\pi (20 \text{ mm})^3}{32} = 785 \text{ mm}^3$$

Now the stress at section 4 is

$$s_4 = \frac{MK_t}{Z} = \frac{(30 \text{ N} \cdot \text{m})(2.45)}{785 \text{ mm}^3} \times \frac{10^3 \text{ mm}}{\text{m}} = 93.6 \text{ N/mm}^2$$

$$s_4 = 93.6 \text{ MPa}$$

Example Problem 9-10

Figure 9-24 shows a lever used as an emergency brake actuator for a piece of construction machinery. The operator pulls at the top of the lever with 15 lb of force. Compute the stress which is produced at the place where the cross section of the lever changes.

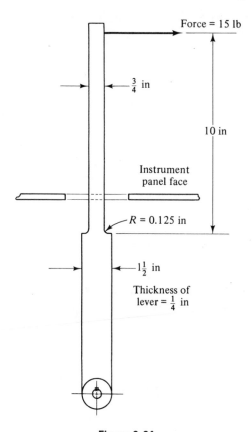

Figure 9-24

Solution

The lever acts like a cantilever beam. Then the moment at the section of interest is

$$M = (15 \text{ lb})(10 \text{ in.}) = 150 \text{ lb} \cdot \text{in.}$$

The lever has a cross section of $\frac{1}{4}$ in. by $\frac{3}{4}$ in. Then

$$Z = \frac{th^2}{6} = \frac{(0.25 \text{ in.})(0.75 \text{ in.})^2}{6} = 0.0234 \text{ in.}^3$$

at the fillet between the $\frac{3}{4}$-in. section and the $1\frac{1}{2}$-in. section,

$$\frac{r}{h} = \frac{0.125}{0.75} = 0.167$$

$$\frac{H}{h} = \frac{1.50}{0.75} = 2.0$$

Then from Figure 9-23, $K_t = 1.58$. The stress can now be computed.

$$s = \frac{MK_t}{Z} = \frac{150 \text{ lb}\cdot\text{in. } (1.58)}{0.0234 \text{ in.}^3} = 10\ 100 \text{ psi}$$

9-6   FLEXURAL CENTER (SHEAR CENTER)

The flexure formula is valid for computing the stress in a beam provided the applied loads pass through a point called the *flexural center*, or sometimes, the *shear center*. If a section has an axis of symmetry and if the loads pass through that axis, then they also pass through the flexural center. The beam sections shown in Figure 9-3 are of this type. For sections loaded away from an axis of symmetry, the position of the flexural center, indicated by Q, must be found. Such sections were identified in Figure 9-4. In order to result in pure bending, the loads must pass through Q, as shown in Figure 9-25. If they don't, then a condition of *unsymmetrical bending* occurs and other analyses would have to be performed which are not discussed in this book. The sections of the type shown in Figure 9-25 are used frequently in structures. Some lend themselves nicely to production by extrusion and are therefore very economical. But because of the possibility of producing unsymmetrical bending, care must be taken in their application.

Example Problem 9-11

Determine the location of the shear center for the two sections shown in Figure 9-26.

Solution

Channel section (a)

From Figure 9-25, the distance e to the shear center is

$$e = \frac{b^2 h^2 t}{4I_x}$$

Note that the dimensions b and h are measured to the middle of the flange or web. Then $b = 40$ mm and $h = 50$ mm. Because of symmetry about the X axis, I_x can be found by the difference between

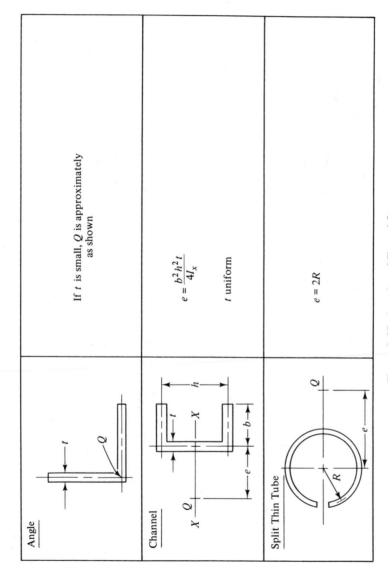

Angle

If t is small, Q is approximately as shown

Channel

$$e = \frac{b^2 h^2 t}{4 I_x}$$

t uniform

Split Thin Tube

$$e = 2R$$

Figure 9-25 Location of Flexural Center

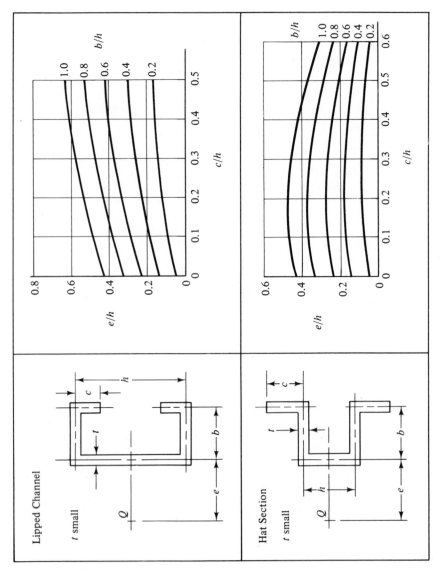

Figure 9-25 (cont'd)

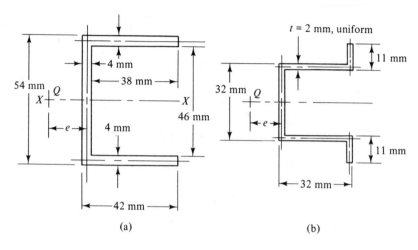

Figure 9-26

the value of I for the large outside rectangle (54 mm by 42 mm) and the smaller rectangle removed (46 mm by 38 mm).

$$I_X = \frac{(42)(54)^3}{12} - \frac{(38)(46)^3}{12} = 0.243 \times 10^6 \text{ mm}^4$$

Then

$$e = \frac{(40)^2(50)^2(4)}{4(0.243 \times 10^6)} \text{ mm} = 16.5 \text{ mm}$$

This dimension is drawn to scale in Figure 9-26(a).

Hat section (b)

Here the distance e is a function of the ratios c/h and b/h.

$$\frac{c}{h} = \frac{10}{30} = 0.3$$

$$\frac{b}{h} = \frac{30}{30} = 1.0$$

Then from Figure 9-25, $e/h = 0.45$. Solving for e,

$$e = 0.45h = 0.45(30 \text{ mm}) = 13.5 \text{ mm}$$

This dimension is drawn to scale in Figure 9-26(b).

Now, can you devise a design for using either section as a beam and provide for the application of the load through the flexural center Q in order to produce pure bending?

PROBLEMS

9-1 A square bar, 30 mm on a side, is used as a simply supported beam subjected to a bending moment of 425 N·m. Compute the maximum stress due to bending in the bar.

9-2 Compute the maximum stress due to bending in a round rod, 20 mm in diameter, if it is subjected to a bending moment of 120 N·m.

9-3 A bending moment of 5800 lb·in. is applied to a rectangular beam made as a rectangle, 0.75 in. by 1.50 in. in cross section. Compute the maximum bending stress in the beam (a) if the 1.50-in. side is set vertical, and (b) if the 0.75-in. side is vertical.

9-4 A wood beam carries a bending moment of 15 500 lb·in. It has a rectangular cross section, 1.50 in. wide by 7.25 in. high. Compute the maximum stress due to bending in the beam.

9-5 Compute the required diameter of a round bar used as a beam to carry a bending moment of 240 N·m with a stress no greater than 125 MPa.

9-6 A rectangular bar is to be used as a beam subjected to a bending moment of 145 N·m. If its height is to be three times its width, compute the required dimensions of the bar to limit the stress to 55 MPa.

9-7 The tee section shown in Figure 8-15(a) is to carry a bending moment of 28.0 kN·m. It is to be made of steel plates welded together. If the load on the beam is a dead load, would AISI 1020 hot-rolled steel be satisfactory for the plates?

9-8 The modified I-section shown in Figure 8-15(d) is to be extruded aluminum. Specify a suitable aluminum alloy if the beam is to carry a repeated load resulting in a bending moment of 275 N·m.

9-9 A standard steel pipe is to be used as a chinning bar for personal exercise. The bar is to be 42 in. long and simply supported at its ends. Specify a suitable size pipe if the bending stress is to be limited to 10 000 psi and a 280-lb man hangs by one hand in the middle.

9-10 A pipeline is to be supported above ground on horizontal beams, 14 ft long. Consider each beam to be simply supported at its ends. Each beam carries the combined weight of 50 ft of 48-in. diameter pipe and the oil flowing through it, about 42 000 lb. Assuming the load acts at the center of the beam, specify the required section modulus of the beam to limit the bending stress to 20 000 psi. Then specify a suitable wide flange or American Standard beam.

9-11 A wood platform is to be made of standard plywood and finished lumber using the cross section shown in Figure 8-19(c). Would the platform be safe if four men, weighing 250 lb each, were to stand 2 ft apart, as shown in Figure 9-27? Consider only bending stresses. (See Chapter 10 for shear stresses.)

9-12 A diving board has a hollow rectangular cross section 30 in. wide and 3.0 in. thick and is supported as shown in Figure 9-28. Compute the maximum stress due to bending in the board if a 300-lb person stands at the end. Would the board be safe if it was made of extruded 6061-T4 aluminum and the person landed at the end of the board with an impact?

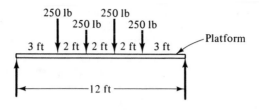

Figure 9-27

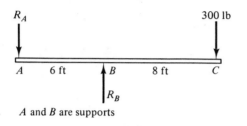

(a)

(b)

Figure 9-28

9-13   The loading shown in Figure 7-35(d) is to be carried by a W10×21 steel beam. Compute the stress due to bending.

9-14   An American Standard beam, S12×40.8, carries the load shown in Figure 7-37(c). Compute the stress due to bending.

9-15   The 24-in. long beam shown in Figure 7-37(b) is an aluminum channel, C4×2.331, positioned with the legs down so that the flat 4-in. surface can carry the applied loads. Compute the maximum tensile and maximum compressive stresses in the channel.

9-16   The 650-lb load at the center of the 28-in. long bar shown in Figure 7-35(a) is carried by a standard steel construction pipe, TS1½×0.145. Compute the stress in the pipe due to bending.

9-17   The loading shown in Figure 7-36(b) is to be carried by an extruded aluminum hat-section beam having the cross section shown in Figure 8-20(b). Compute the maximum stress due to bending in the beam. If it is made of extruded 6061-T4 aluminum and the loads are dead loads, would the beam be safe?

9-18   The extruded shape shown in Figure 8-20(c) is to be used to carry the loads shown in Figure 7-36(a), which is a part of a business machine frame. The loads are due to a motor mounted on the frame and can be considered dead loads. Specify a suitable aluminum alloy for the beam.

9-19 The loading shown in Figure 7-36(c) is to be carried by the fabricated beam shown in Figure 8-17(d). Compute the stress due to bending in the beam.

9-20 An aluminum I-beam, I9 × 8.361, carries the load shown in Figure 7-36(d). Compute the stress due to bending in the beam.

$S = 116$
MPA
$(16,800 PSI)$

9-21 A beam is being designed to support the loads shown in Figure 9-29. The four shapes proposed are: (a) round bar, (b) square bar, (c) rectangular bar

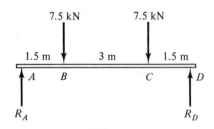

Figure 9-29

with the height made four times the thickness, and (d) lightest American Standard beam. Determine the required dimensions of each proposed shape to limit the maximum stress due to bending to 80 MPa. Then compare the magnitude of the cross-sectional areas of the four shapes. Since the weight of the beam is proportional to its area, the one with the smallest area will be the lightest.

9-22 A rack is being designed to support large sections of pipe, as shown in Figure 9-30. Each pipe exerts a force of 2500 lb on the support arm. The height of

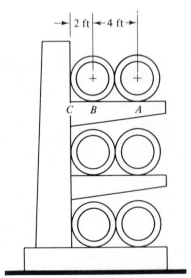

Figure 9-30

the arm is to be tapered as suggested in the figure, but the thickness will be a constant 1.50 in. Determine the required height of the arm at sections *B* and *C*, considering only bending stress. Use AISI 1040 hot-rolled steel for the arms and a design factor of 4 based on yield strength.

9-23   A children's play gym includes a cross beam carrying four swings, as shown in Figure 9-31. Assume that each swing carries 300 lb. It is desired to use a standard steel pipe for the beam, keeping the stress due to bending below 10 000 psi. Specify the suitable size pipe for the beam.

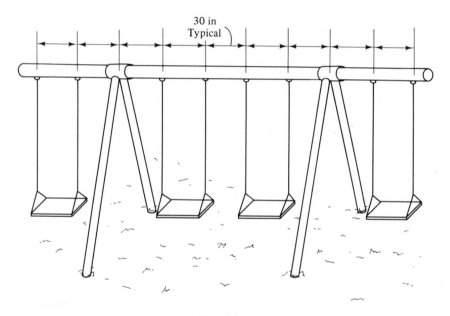

Figure 9-31

9-24   A part of a truck frame is composed of two channel-shaped members, as shown in Figure 9-32. If the moment at the section is 60 000 lb·ft, compute the bending stress in the frame. Assume that the two channels act as a single beam.

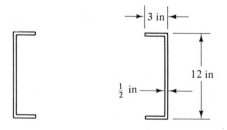

Figure 9-32

9-25 The loading shown in Figure 7-43(e) represents the load on a floor beam of a commercial building. Determine the maximum bending moment on the beam, and then specify a wide flange shape which will limit the stress to 150 MPa.

9-26 Figure 7-43(c) represents the loading on a motor shaft; the two supports are bearings in the motor housing. The larger load between the supports is due to the rotor plus dynamic forces. The smaller, overhung load is due to externally applied loads. Using AISI 1141 OQT 1300 for the shaft, specify a suitable diameter based on bending stress only. Use a design factor of 8 based on ultimate strength.

9-27 The cantilever beam loading shown in Figure 7-39(c) is applied to the structure shown in Figure 9-33. The 12-in. pipe mates smoothly with its support, so that no stress concentration exists at B. At C, the 10-in. pipe is placed inside the 12-in. pipe with a spacer ring to provide a good fit. Then a $\frac{3}{8}$-in. fillet weld is used to secure the section together. Accounting for the stress concentration at the joint, determine how far out point C must be to limit the stress to 35 500 psi.

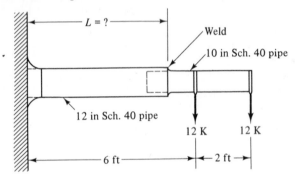

Figure 9-33

9-28 Figure 9-34 shows a round shaft from a gear transmission. Gears are mounted at points A, C, and E. Supporting bearings are at B and D. The forces trans-

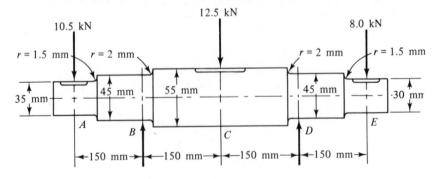

Figure 9-34

mitted from the gears to the shaft are shown, all acting downward. Compute the maximum stress due to bending in the shaft, accounting for stress concentrations.

9-29 The forces shown on the shaft in Figure 9-35 are due to gears mounted at *B* and *C*. Compute the maximum stress due to bending in the shaft.

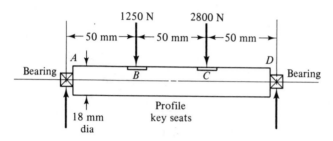

Figure 9-35

9-30 Compute the location of the flexural center of a split, thin tube if it has an outside diameter of 50 mm and a wall thickness of 4 mm.

9-31 For an aluminum channel C2×0.577 with its web oriented vertically, compute the location of its flexural center. Neglect the effect of the fillets between the flanges and the web.

9-32 If the hat section shown in Figure 8-20(b) were turned 90 deg from the position shown, compute the location of its flexural center.

9-33 A simple, solid axle for a trailer suspension is made as shown in Figure 9-36. A load of 3600 N is applied through each spring. Assume that the supports at each wheel are simple supports. Design the axle, specifying the configuration, the material, the dimensions, and the design factor used.

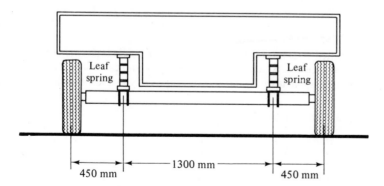

Figure 9-36

9-34 A cantilevered roof over a shipping dock is 15 ft wide by 30 ft long, as shown in Figure 9-37. Three main beams extend out from the building. Cross beams connect the main beams and carry 25 lb per square foot of distributed load

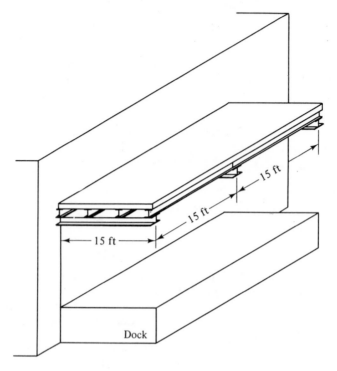

Figure 9-37

from the roof. Consider each cross beam to be simply supported at its ends and carrying a uniformly distributed load equal to its share of the roof load. Consider the cantilever beams to be loaded with a series of concentrated loads from each cross beam. Specify standard W- or S-beams for each component of the roof structure which will limit the stress to 0.6 times the yield strength of ASTM A36 steel.

9-35 A simplified version of the loading on a truck frame is shown in Figure 9-38. The two forces at the left represent the engine and cab loads. The two forces at the right are loads transferred from the van-type body at points of attach-

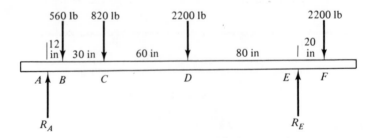

Figure 9-38

ment. Draw the shear and bending moment diagrams for the frame. Then design the frame members, assuming that the entire frame will have the same cross section. Specify the material, the basic configuration, the dimensions, and the design factor used.

9-36 A cover for a service station is to be designed in the form shown in Figure 9-39. Its area is 40 ft by 50 ft. With maximum snow load on the cover, the load is 25 lb per square foot. A suggested framing plan is shown, but others are possible.

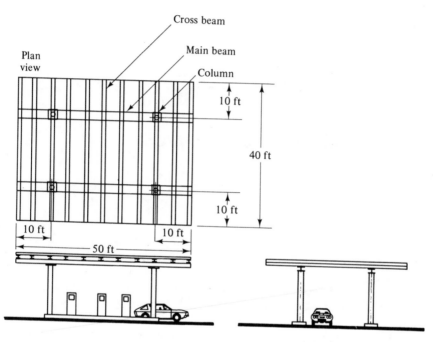

Figure 9-39

10

STRESSES
DUE TO VERTICAL
SHEARING FORCE

10-1 NEED TO CONSIDER SHEAR IN BEAMS

The vertical shearing force applied to a beam causes shear stresses to be developed inside the beam. The shear stresses must provide an internal resisting force equal to the external vertical shearing force in order to maintain equilibrium. In beams, however, there are also *horizontal* shear stresses of the same magnitude as the *vertical* shear stresses. The existence of these horizontal shear stresses can be visualized by considering several flat strips of wood to be stacked together to form a beam. As shown in Figure 10-1, when a bending load is applied to the beam, the strips tend to slide over each other. Since the strips do not act as a unit, the beam is relatively weak. A stronger beam can be made by gluing the strips together. Now, the tendency for the strips to slide is resisted by the glue, and a *shear stress* is developed in the glue.

A similar condition exists in a solid beam. Here the tendency for horizontal sliding of one part of the beam cross section relative to the part above it is resisted by the *material of the beam*. Therefore, a shear stress is developed on any horizontal plane.

Horizontal shear is one of the primary modes of failure for wood beams. The grain usually runs parallel to the long axis of the beam. The shear stress causes separation of the adjacent "layers" of the wood.

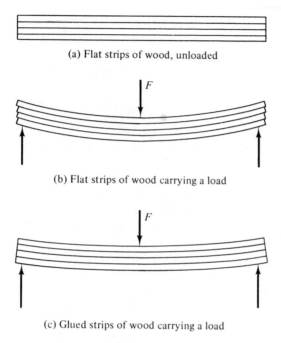

(a) Flat strips of wood, unloaded

(b) Flat strips of wood carrying a load

(c) Glued strips of wood carrying a load

Figure 10-1

In metal beams the shear stress is usually of negligible importance as compared with the stress due to bending. Failure will occur by yielding in tension or compression before the shear stress builds to a level at which shear failure would occur. Exceptions would be beams which are very short as compared with their height. For a metal beam whose length is shorter than about two times its height, shear stresses may be significant. Otherwise, the beam should be analyzed for bending rather than for shear.

Built-up beams made by gluing, nailing, riveting, welding, or bolting sections together should also be analyzed with respect to shear to ensure that the fastening method is safe. The number, size, and placement of nails, rivets, or bolts must be such that they collectively resist the shearing forces. The shear strength of glue and the area over which it acts must be chosen to ensure that the built-up beam acts as a unit. The weld pattern and the thickness of welds must be specified to provide sufficient resistance to shearing forces.

10-2   THE GENERAL SHEAR FORMULA

The magnitudes of shear stresses in beams are very much dependent on the geometry of the cross section of the beam. There is an inverse proportion between the shear stress and the *moment of inertia I* of the cross section.

Likewise, the stress at a particular point is inversely proportional to the thickness t of the section at that point. A third factor, called the *statical moment Q*, is directly proportional to the shear stress.

Of course, the shear stress is also dependent on the magnitude of the vertical shearing force V. The *general shear formula* is usually written as

$$s_s = \frac{VQ}{It} \tag{10-1}$$

The terms V, I, and t all have the same meaning as in previous discussions. Remember:

1. V is the vertical shearing force at the *section of interest*. The shear diagram developed in Chapter 7 is a plot of the variation of shearing force with position on the beam. If the maximum value of V is required, it can be read from the shear diagram. Generally, the maximum absolute value, positive or negative, is used.
2. I is the moment of inertia of the *entire* cross section of the beam with respect to its centroidal axis. This is the same value of I used in the flexure formula ($s = Mc/I$) to compute bending stresses.
3. The thickness t is taken at the point where the shear stress is to be calculated. There is a considerable variation of shear stress with position in the cross section.

The statical moment Q must be evaluated carefully. By definition,

$$Q = A_p \bar{y} \tag{10-2}$$

where A_p = area of that *part* of the total cross section which lies *to the outside* of the place where the shear stress is to be calculated

$\bar{y}$ = distance to the centroid of A_p from the centroid of the entire area

Figure 10-2 shows the proper identification of A_p and $\bar{y}$ for three cases in which the objective would be the computation of the shear stress at the axis *a-a*. The term *statical moment* means the moment of an area, that is, an area times a distance. This is consistent with $Q = A_p \bar{y}$ as the definition of Q.

Example Problem 10-1

Compute the shear stress at the axis *a-a* in the rectangular beam whose cross section is shown in Figure 10-2(a). Use $t = 1.50$ in. and $h = 7.50$ in. Axis *a-a* is 0.75 in. above the centroid of the cross section. The beam is 14 ft long, simply supported at its ends, and carries a uniformly distributed load of 100 lb per foot. Compute the stress at the place where the maximum shearing force occurs. The beam is of the type used in the floor or roof of a wood-frame structure.

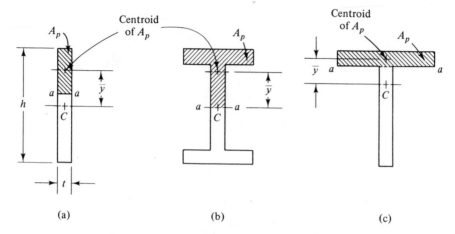

(a)   (b)   (c)

Figure 10-2

Solution

We will use the general shear formula,

$$s_s = \frac{VQ}{It}$$

Let's first find the maximum value of V. Figure 10-3 shows the shear diagram. The largest value of V, 700 lb, occurs at the supports. This is where the maximum shear stress would occur.

The moment of inertia I for the rectangular cross section is

$$I = \frac{th^3}{12} = \frac{(1.50)(7.50)^3}{12} = 52.7 \text{ in.}^4$$

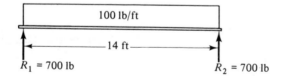

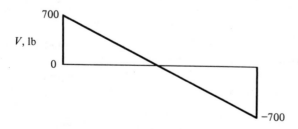

Figure 10-3

The thickness at all sections is 1.50 in. Now Q can be found for the area above a-a. Since a-a is 0.75 in. above the centroid,

$$A_p = \left(\frac{h}{2} - 0.75\ \text{in.}\right)(t) = \left(\frac{7.50}{2} - 0.75\right)(1.50)$$

$$A_p = (3.0)(1.50) = 4.50\ \text{in.}^2$$

Also,

$$\bar{y} = \frac{h}{2} - \frac{3.0}{2} = 3.75 - 1.50 = 2.25\ \text{in.}$$

Finally,

$$Q = A_p\bar{y} = (4.50\ \text{in.}^2)(2.25\ \text{in.}) = 10.13\ \text{in.}^3$$

Now the shear stress can be computed.

$$s_s = \frac{VQ}{It} = \frac{(700\ \text{lb})(10.13\ \text{in.}^3)}{(52.7\ \text{in.}^4)(1.50\ \text{in.})} = 89.7\ \text{psi}$$

Since the allowable stress for most woods is 100 psi or greater, this is an acceptable level of stress.

Example Problem 10-2

Compute the shear stress for the same beam as in Example Problem 10-1, except use an axis through the centroid of the entire section.

Solution

The values of V, I, and t would be the same as before. Only Q changes, being evaluated for the upper half of the cross-sectional area above the centroidal axis as shown in Figure 10-4.

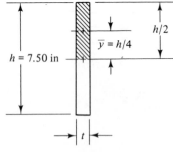

Figure 10-4

$$A_p = \left(\frac{h}{2}\right)(t) = \left(\frac{7.50}{2}\right)(1.50)$$
$$= 5.625\ \text{in.}^2$$

$$\bar{y} = \frac{h}{4} = \frac{7.50}{4} = 1.875\ \text{in.}$$

$$Q = A_p\bar{y}$$
$$= (5.625\ \text{in.}^2)(1.875\ \text{in.})$$
$$= 10.55\ \text{in.}^3$$

Now the shear stress is

$$s_s = \frac{VQ}{It} = \frac{(700\ \text{lb})(10.55\ \text{in.}^3)}{(52.7\ \text{in.}^4)(1.50\ \text{in.})} = 93.4\ \text{psi}$$

This is larger than that found at axis *a-a* but still less than 100 psi, so the beam should be safe from shear failure.

For most sections used for beams, the maximum shear stress occurs at the centroidal axis of the section. This is true provided the thickness at the centroidal axis is not greater than that at some other axis. Figure 10-2 shows three examples of sections for which the maximum shear stress does occur at the centroidal axis. Figure 10-5 shows three where it does *not*.

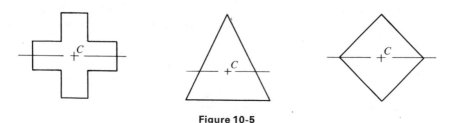

Figure 10-5

Example Problem 10-3

For the rectangular beam cross section shown in Figure 10-6, compute the shear stresses which occur at the axes *a* through *e* and plot the variation of stress with position on the section. The shearing force is 1000 lb.

Solution

The general shear formula will be used.

$$s_s = \frac{VQ}{It} = \frac{VA_p\bar{y}}{It}$$

In this case *V*, *I*, and *t* are the same for all axes.

$$V = 1000 \text{ lb}$$

$$I = \frac{th^3}{12} = \frac{(2)(8)^3}{12} = 85.33 \text{ in.}^4$$

$$t = 2.0 \text{ in.}$$

The value of *Q* varies with the position of the axis. The following table shows the calculations for all five axes.

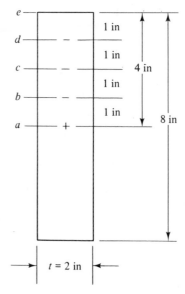

Figure 10-6

Axis	A_p, in.^{2}	$\bar{y}$, in.	$Q = A_p\bar{y}$, in.^{3}	$s_s = VQ/It$, psi
a-a	8	2.0	16	93.8
b-b	6	2.5	15	87.9
c-c	4	3.0	12	70.3
d-d	2	3.5	7	41.0
e-e	0	4	0	0

Notice that because of the symmetry of the cross section, the shear stresses on corresponding axes below the centroid would be the same. The shear stresses computed above are plotted in Figure 10-7. The maximum shear stress occurs at the centroidal axis.

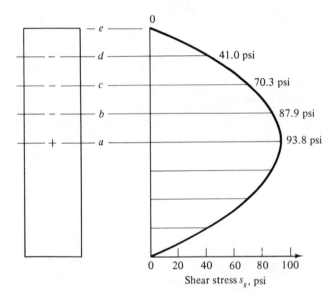

Figure 10-7

Example Problem 10-4

For the triangular beam cross section shown in Figure 10-8, compute the shear stress which occurs at the axes a through g, each 50 mm apart. Plot the variation of stress with position on the section. The shearing force is 50 kN.

Solution

In the general shear formula, the values of V and I will be the same for all computations.

$$I = \frac{bh^3}{36} = \frac{(300)(300)^3}{36} = 225 \times 10^6 \text{ mm}^4$$

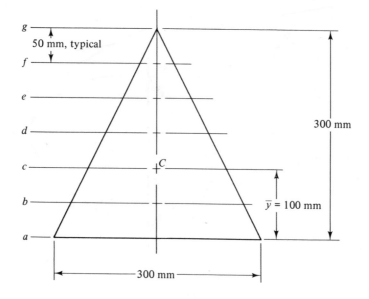

Figure 10-8

The following table shows the remaining computations.

Axis	A_p, mm^2	$\bar{y}$, mm	$Q = A_p\bar{y}$, mm^3	t, mm	s_s, MPa
a-a	0	100	0	300	0
b-b	13 750	75.8	1.042 × 10⁶	250	0.92
c-c	20 000	66.7	1.333 × 10⁶	200	1.48
d-d	11 250	100.0	1.125 × 10⁶	150	1.67
e-e	5 000	133.3	0.667 × 10⁶	100	1.48
f-f	1 250	166.7	0.208 × 10⁶	50	0.92
g-g	0	200	0	0	0

Figure 10-9 shows a plot of these stresses. The maximum shear stress occurs at half the height of the section, and the stress at the centroid (at $h/3$) is lower. This illustrates the general statement made earlier that for sections whose minimum thickness does not occur at the centroidal axis, the maximum shear stress may occur at some axis other than the centroidal axis.

One further note can be made about the computations shown for the triangular section. For the axis b-b, the partial area A_p was taken as that area *below* b-b. The resulting section is the trapezoid between b-b and the bottom of the beam. For all other axes, the partial area A_p was taken as the triangular area *above* the axis. The area below the axis could have been used, but the computations would have been more difficult.

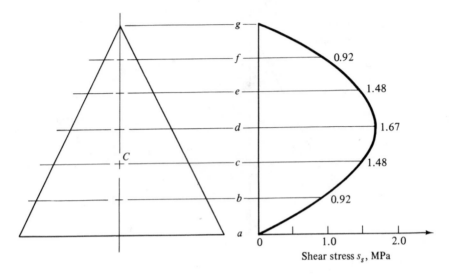

Figure 10-9

10-3 SPECIAL SHEAR FORMULAS

The general shear formula can be used to compute the shear stress at any part of any cross section. However, frequently it is desired to know only the

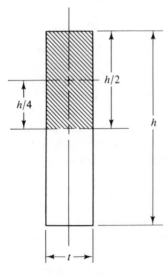

Figure 10-10

maximum shear stress. For several common shapes, it is possible to derive a special, simplified formula which will give the maximum stress quickly. The rectangle, the circle, and thin-webbed shapes can be analyzed this way.

Rectangular Shape. Figure 10-10 shows a typical rectangular cross section having a thickness t and a height h. The three geometrical terms in the general shear formula can be expressed in terms of t and h.

$$I = \frac{th^3}{12}$$

$$t = t$$

$$Q = A_p \bar{y} \quad \text{(for area above centroidal axis)}$$

$$Q = \frac{th}{2} \cdot \frac{h}{4} = \frac{th^2}{8}$$

Putting these terms in the general shear formula gives

$$S_s = \frac{VQ}{It} = V \cdot \frac{th^2}{8} \cdot \frac{12}{th^3} \cdot \frac{1}{t} = \frac{3}{2}\frac{V}{th}$$

But since th is the total area of the section,

$$S_s = \frac{3V}{2A} \tag{10-3}$$

Equation (10-3) can be used to compute exactly the maximum shear stress in a rectangular beam at its centroidal axis.

Circular Shape. For a circular bar (shaft) used as a beam, the maximum shear stress also occurs at the centroidal axis, even though other axes have thinner sections. In terms of the diameter D (see Figure 10-11),

$$t = D$$

$$I = \frac{\pi D^4}{64}$$

$$Q = A_p\bar{y} \qquad \text{(for the semicircle above the centroid)}$$

$$Q = \frac{\pi D^2}{8} \cdot \frac{2D}{3\pi} = \frac{D^3}{12}$$

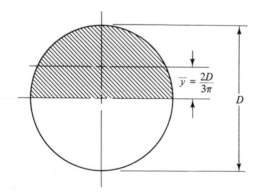

Figure 10-11

Then the maximum shear stress is

$$S_s = \frac{VQ}{It} = V \cdot \frac{D^3}{12} \cdot \frac{64}{\pi D^4} \cdot \frac{1}{D} = \frac{64V}{12\pi D^2}$$

To refine the equation, factor out a 4 from the numerator and then note that the total area of the circular section is $A = \pi D^2/4$.

$$S_s = \frac{16(4)V}{12\pi D^2} = \frac{16V}{12A}$$

$$S_s = \frac{4V}{3A} \tag{10-4}$$

Thin-Webbed Shapes. Structural shapes such as W- and S-beams have relatively thin webs. The distribution of shear stress in such beams is typically like that shown in Figure 10-12. The maximum shear stress is at the centroidal axis. It decreases slightly in the rest of the web and then drastically

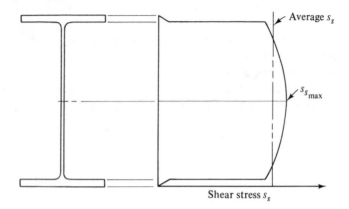

Figure 10-12

in the flanges. Thus most of the resistance to the vertical shearing force is provided by the web. Also, the average shear stress in the web would be just slightly smaller than the maximum shear stress. For these reasons, the *web shear formula* is often used to get a quick estimate of the shear stress in thin-webbed shapes.

$$s_s = \frac{V}{A_{web}} = \frac{V}{th} \tag{10-5}$$

The thickness of the web is t. The simplest approach would be to use the full height of the beam for h. This would result in a shear stress approximately 15 percent lower than the actual maximum shear stress at the centroidal axis for typical beam shapes. Using just the web height between the flanges would produce a closer approximation of the maximum shear stress, probably less than 10 percent lower than the actual value. In this book, problems using the web shear formula will use the full height of the cross section unless otherwise stated.

Example Problem 10-5

Compute the maximum shear stress which would occur in the rectangular cross section of a beam like that shown in Figure 10-6. The shearing force is 1000 lb.

Solution

Using Equation (10-3),

$$s_s = \frac{3V}{2A} = \frac{3(1000 \text{ lb})}{2(2 \text{ in.})(8 \text{ in.})} = 93.8 \text{ psi}$$

This is the same as that computed in Example Problem 10-3.

Example Problem 10-6

Compute the maximum shear stress which would occur in a circular shaft, 50 mm in diameter, if it is subjected to a vertical shearing force of 110 kN.

Solution

Equation (10-4) will give the maximum shear stress at the horizontal diameter of the shaft.

$$s_s = \frac{4V}{3A}$$

But

$$A = \frac{\pi D^2}{4} = \pi(50 \text{ mm})^2/4 = 1963 \text{ mm}^2$$

Then

$$s_s = \frac{4(110 \times 10^3 \text{ N})}{3(1963 \text{ mm}^2)} = 74.7 \text{ MPa}$$

Example Problem 10-7

Using the web shear formula, compute the shear stress in a W10×33 beam if it is subjected to a shearing force of 25 000 lb.

Solution

In Table A-12 for W-beams, it is found that the web thickness is 0.292 in. and the overall depth (height) of the beam is 9.75 in. Then using Equation (10-5),

$$s_s = \frac{V}{th} = \frac{25\ 000 \text{ lb}}{(0.292 \text{ in.})(9.75 \text{ in.})} = 8780 \text{ psi}$$

If it were desirable to get a closer approximation of the maximum shear stress, the web area between the flanges could be used. In this case, the 9.75-in. depth of the beam would have to be reduced by 2(0.433 in.) for the two flanges to get *h*. That is,

$$h = 9.75 \text{ in.} - 2(0.433 \text{ in.}) = 8.884 \text{ in.}$$

Then in the web shear formula,

$$s_s = \frac{V}{th} = \frac{25\ 000\ \text{lb}}{(0.292\ \text{in.})\ (8.884\ \text{in.})} = 9640\ \text{psi}$$

The actual maximum shear stress would probably be about 10 percent greater than this value, about 10 600 psi.

10-4   SHEAR FLOW

Built-up sections used for beams, such as those shown in Figure 10-13, must be analyzed to determine the proper size and spacing of fasteners. The discussion in preceding sections showed that horizontal shearing forces

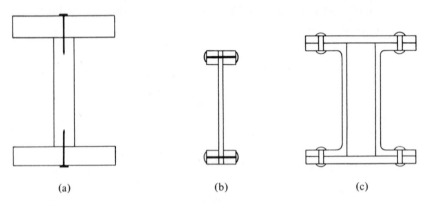

(a)   (b)   (c)

Figure 10-13

exist at the planes joined by the nails, bolts, and rivets. Thus the fasteners are subjected to shear. Usually, the size and material of the fastener will permit the specification of an allowable shearing force on each. Then the beam must be analyzed to determine a suitable spacing for the fasteners which will ensure that all parts of the beam act together.

The term *shear flow* is useful for analyzing built-up sections. Called q, the shear flow is found by multiplying the shear stress at a section by the thickness at that section. That is,

$$q = s_s t \tag{10-6}$$

But from the general shear formula,

$$s_s = \frac{VQ}{It}$$

Then

$$q = s_s t = \frac{VQ}{I} \qquad (10\text{-}7)$$

The units for q are *force per unit length*, such as N/m, N/mm, or lb/in. The shear flow is a measure of how much shearing force must be resisted at a particular section per unit length. Knowing the shearing force capacity of a fastener then allows the determination of a safe spacing for the fasteners.

Example Problem 10-8

Determine the proper spacing of nails used to secure the flange boards to the web of the built-up I-beam shown in Figure 10-13(a). All boards measure 1.50 in. thick and 7.50 in. wide. The vertical shearing force on the beam is 700 lb. The nails to be used can resist 150 lb of shearing force each. The moment of inertia of the entire section is 513 in.^{4}.

Solution

The shear flow must be calculated.

$$q = \frac{VQ}{I}$$

At the place where the nails join the boards, Q is evaluated for the area of the top (or bottom) flange board.

$$Q = A_p \bar{y} = (1.5 \text{ in.})(7.5 \text{ in.})(4.5 \text{ in.}) = 50.6 \text{ in.}^3$$

$$q = \frac{VQ}{I} = \frac{(700 \text{ lb})(50.6 \text{ in.}^3)}{513 \text{ in.}^4} = 69 \text{ lb/in.}$$

This means that 69 lb of force must be resisted along each inch of length of the beam at the point between the flange and the web boards. Since each nail can withstand 150 lb, the required spacing is

$$\frac{150 \text{ lb}}{69 \text{ lb/in.}} = 2.17 \text{ in.}$$

A spacing of 2.0 in. would be reasonable.

Example Problem 10-9

A fabricated beam is made by riveting square aluminum bars to a vertical plate, as shown in Figure 10-13(b). The bars are 20 mm square. The plate is 6 mm thick and 200 mm high. The rivets can withstand 800 N of shear force across one cross section. Determine the required spacing of the rivets if a shearing force of 5 kN is applied.

Solution

In the shear flow equation,

$$q = \frac{VQ}{I}$$

I is the moment of inertia of the entire cross section, and Q is the product of $A_p\bar{y}$ for the area *outside* of the section where the shear is to be calculated. In this case, the partial area A_p is the 20-mm square area *to the side* of the web. The shear flow equation is valid for shear across vertical cutting planes in flanges of beams as well as for shear across horizontal cutting planes, as done in previous problems. For the beam in Figure 10-13(b),

$$I = \frac{6(200)^3}{12} + 4\left[\frac{20^4}{12} + (20)(20)(90)^2\right]$$

$$I = 17.0 \times 10^6 \text{ mm}^4$$

$$Q = A_p\bar{y} \quad \text{(for one square bar)}$$

$$Q = (20)(20)(90) \text{ mm}^3 = 36\ 000 \text{ mm}^3$$

Then for $V = 5$ kN,

$$q = \frac{VQ}{I} = \frac{(5 \times 10^3 \text{ N})(36 \times 10^3 \text{ mm}^3)}{17.0 \times 10^6 \text{ mm}^4} = 10.6 \text{ N/mm}$$

Thus a shear force of 10.6 N is to be resisted for each millimetre of length of the beam. Since each rivet can withstand 800 N of shear force, the required spacing is

$$800 \text{ N} \times \frac{1 \text{ mm}}{10.6 \text{ N}} = 75.5 \text{ mm}$$

PROBLEMS

10-1   Use the general shear formula to compute the maximum shear stress in a rectangular beam, 50 mm wide by 200 mm high, when subjected to a vertical shearing force of 85 kN.

For Problems 10-2 through 10-8, refer to the figures at the end of Chapter 8 for the cross-sectional shape of the beam. Compute the shear stress at the neutral axis of each shape for the given shearing force.

10-2   Use Figure 8-14(a); $V = 1500$ lb.
10-3   Use Figure 8-14(c); $V = 850$ lb.
10-4   Use Figure 8-14(d); $V = 850$ lb.
10-5   Use Figure 8-15(a); $V = 112$ kN.
10-6   Use Figure 8-15(b); $V = 71.2$ kN.

10-7 Use Figure 8-15(c); $V = 1780$ N.

10-8 Use Figure 8-15(d); $V = 675$ N.

In Problems 10-9 through 10-12, for each beam cross section indicated, assume that it is made of wood having an allowable shearing stress of 100 psi. Determine how much vertical shearing force each could carry.

10-9 Use Figure 8-19(a).

10-10 Use Figure 8-19(b).

10-11 Use Figure 8-19(c).

10-12 Use Figure 8-19(d).

10-13 The loading shown in Figure 7-35(d) is to be carried by a W10×21 steel beam. Compute the shear stress in the beam using the web shear formula.

10-14 An American Standard beam, S12×40.8, carries the load shown in Figure 7-37(c). Compute the shear stress in the beam using the web shear formula.

10-15 An aluminum I-beam, I9×8.361, carries the load shown in Figure 7-36(d). Compute the shear stress in the beam using the web shear formula.

10-16 A wood floor joist in a home is 12 ft long and carries a uniformly distributed load of 80 lb/ft. The cross-sectional dimensions are $1\frac{1}{2}$ in. by $7\frac{1}{4}$ in. Compute the shear stress in the joist.

10-17 A steel beam is made as a rectangle, 0.50 in. wide by 4.00 in. high. Compute the shear stress in the beam if it carries the load shown in Figure 7-37(a).

10-18 A wood beam, rectangular in cross section, is 8.0 in. wide and 20.0 in. high and is made of construction grade Douglas fir. Would it be safe when carrying the load shown in Figure 7-37(c)?

10-19 A aluminum rectangular bar 8 mm wide by 80 mm high carries the load shown in Figure 7-36(b). Compute the shear stress in the bar. Compare the shear stress with the stress due to bending.

10-20 It is planned to use a rectangular bar to carry the load shown in Figure 7-46(a). Its thickness is to be 6 mm. Determine what its height should be to limit the stress due to bending to 60 MPa. Then compute the shear stress in the bar.

10-21 A round shaft, 40 mm in diameter, carries the load shown in Figure 7-46(b). Compute the shear stress in the shaft.

10-22 Compute the required diameter of a round bar, which carries the load shown in Figure 7-46(a), to limit the bending stress to 120 MPa. Then compute the resulting shear stress in the shaft.

10-23 Determine how much vertical shearing force can be carried by a wood dowel, 1.50 in. in diameter, if the maximum allowable shear stress is 80 psi.

10-24 The I-section shown in Figure 8-19(a) is fabricated from three wood boards by nailing through the top and bottom flanges into the web. Each nail can withstand 110 lb of shearing force. If the beam having this section carries a vertical shearing force of 500 lb, what spacing would be required between nails?

10-25 The built-up section shown in Figure 8-19(b) is nailed together by driving one nail through each side of the top and bottom boards into the $1\frac{1}{2}$-in. thick sides. If each nail can withstand 180 lb of shearing force, determine the required nail spacing when the beam carries a vertical shearing force of 1200 lb.

10-26   The platform whose cross section is shown in Figure 8-19(c) is glued together. How much force per unit length of the platform must the glue withstand if it carries a vertical shearing force of 500 lb?

10-27   The built-up section shown in Figure 8-17(a) is fastened together by passing two $\frac{3}{8}$-in. rivets through the top and bottom plates into the flanges of the beam. Each rivet will withstand 1650 lb in shear. Determine the required spacing of the rivets along the length of the beam if it carries the load shown in Figure 7-46(c).

10-28   A fabricated beam having the cross section shown in Figure 8-17(b) carries the loading shown in Figure 7-36(c). The channel is riveted to the S-beam with two $\frac{1}{4}$-in. diameter rivets which can withstand 750 lb each in shear. Determine the required rivet spacing.

11

COMBINED STRESSES

11-1 COMBINED BENDING AND DIRECT TENSION OR COMPRESSION

In earlier chapters, stress analyses were concerned with only one kind of stress at a time. You learned about direct tension, direct compression, direct shear, bending stress, vertical shear stress, and torsional shear stress. In many practical problems, two or more of these kinds of stresses may exist at the same time at the same place in a load-carrying member. Then the combined effect of all stresses must be determined.

The first combination to be considered is bending with direct tension or compression. In any combined stress problem, it is helpful to visualize the stress distribution caused by the various components of the total stress pattern. Notice that bending results in tensile and compressive stresses, as do both direct tension and direct compression. Since the same kind of stresses are produced, a simple algebraic sum of the stresses produced at any point is all that is required to compute the resultant stress at that point. This process is called *superposition*.

An example of a member in which both bending and direct tensile stresses are developed is shown in Figure 11-1. The two horizontal beams support a 10 000-lb load by means of the cable assembly. The beams are rigidly attached

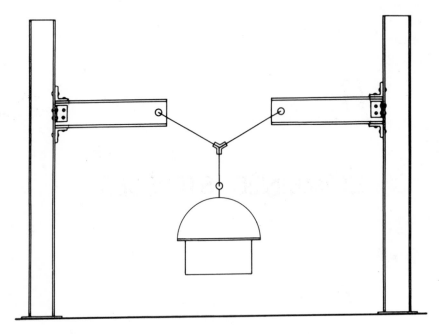

Figure 11-1

to columns, so that they act as cantilever beams. The load at the end of each beam is equal to the tension in the cable. Figure 11-2 shows that the vertical component of the tension in each cable must be 5000 lb. That is,

$$F \cos 60° = 5000 \text{ lb}$$

and the total tension in the cable is

$$F = \frac{5000 \text{ lb}}{\cos 60°} = 10\ 000 \text{ lb}$$

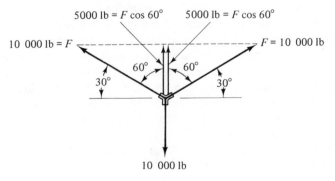

Figure 11-2

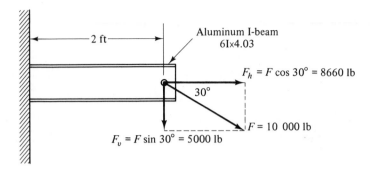

Figure 11-3

This is the force applied to each beam, as shown in Figure 11-3. For purposes of analyzing the stresses in the beam, it is convenient to resolve the applied force into horizontal and vertical components. The principle of superposition can now be applied by considering the effect of each component separately. The horizontal component F_h tends to produce a direct tensile stress over the entire cross section of the beam. The vertical component F_v produces bending in such a way that the upper part of the beam would be in tension and the lower part would be in compression.

Let's compute the magnitudes of the direct tensile stress and the bending stress separately. Considering *only* the horizontal force F_h, we find resulting tensile stress to be

$$s_1 = \frac{F_h}{A}$$

Here the subscript 1 is merely indicating the first stress to be computed. The area of the aluminum $6I \times 4.03$ beam shape is 3.427 in.², as found in the Appendix. Then

$$s_1 = \frac{8660 \text{ lb}}{3.427 \text{ in.}^2} = 2527 \text{ psi}$$

Considering the beam to be a cantilever with a concentrated load of 5000 lb downward at its end, we find that the maximum bending moment would occur at the left end where it is attached to the column.

$$M = F_v(2 \text{ ft}) = 5000 \text{ lb } (2 \text{ ft})(12 \text{ in./ft})$$
$$M = 120\ 000 \text{ lb} \cdot \text{in.}$$

From the flexure formula, using a section modulus of 7.01 in.³ for the beam, the stress due to bending at the left end of the beam is

$$s_2 = \frac{M}{Z} = \frac{120\ 000 \text{ lb} \cdot \text{in.}}{7.01 \text{ in.}^3}$$
$$s_2 = 17\ 118 \text{ psi}$$

When combining the stresses, it is advisable to make a judgment as to the points in the member where the maximum tensile and compressive stresses would probably occur. In this case, the direct tensile stress s_1 is the same throughout the beam. However, the bending stress s_2 is zero at the right end, where the bending moment is zero, and reaches a maximum at the left end. Thus the left end appears to be the critical location, that is, where the maximum stresses occur.

Figure 11-4 shows a series of three sketches which illustrate the principle of superposition. Part (b) shows the uniform tensile stress $s_1 = 2527$ psi acting

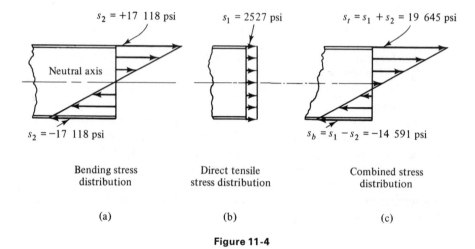

Bending stress distribution	Direct tensile stress distribution	Combined stress distribution
(a)	(b)	(c)

Figure 11-4

on the beam cross section. Part (a) shows the bending stress distribution. Establishing a sign convention which denotes tensile stresses as positive and compressive stresses as negative, we can say that at the top of the beam the bending stress is $+17\ 118$ psi. At the bottom, the stress is $-17\ 118$ psi. Now since both s_1 and s_2 are produced at the same time, the combined stress on the section has the distribution shown in Figure 11-4(c). At the top,

$$s_t = s_1 + s_2 = 2527 + 17\ 118 = 19\ 645 \text{ psi} \qquad \text{(tension)}$$

At the bottom,

$$s_b = s_1 - s_2 = 2527 - 17\ 118 = -14\ 591 \text{ psi} \qquad \text{(compression)}$$

These are the maximum tensile and compressive stresses in the beam.

The analysis of members carrying loads which produce a combination of bending stress with direct tensile and compressive stresses can be summarized by the equation

$$s = \pm \frac{M}{Z} \pm \frac{F}{A} \qquad (11\text{-}1)$$

The dual signs ($\pm$) indicate that it is necessary to determine the sense of the stresses (tensile or compressive) *at the point* where the combined stress is to be calculated.

Example Problem 11-1

A picnic table in a park is made by supporting a circular top on a pipe which is rigidly held in concerte in the ground. Figure 11-5 shows the arrangement. Compute the maximum stress in the pipe if a person

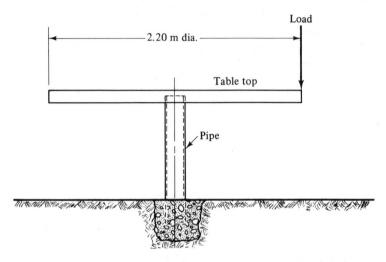

Figure 11-5

having a mass of 135 kg sits on the edge of the table. The pipe is made of an aluminum alloy and has an outside diameter of 170 mm and an inside diameter of 163 mm. If the aluminum has a yield strength of 241 MPa, what is the design factor for this loading?

Solution

The pipe is subjected to combined bending and direct compression as a result of the person sitting on the edge of the table. Therefore, it is necessary to compute the magnitude of the downward force and the bending moment on the pipe.

The force is the gravitational attraction of the 135-kg mass.

$$F = m \cdot g = 135 \text{ kg} \cdot 9.81 \text{ m/s}^2 = 1324 \text{ N}$$

Since this force acts at a distance of 1.1 m from the axis of the pipe, the moment is

$$M = (1324 \text{ N})(1.1 \text{ m}) = 1456 \text{ N} \cdot \text{m}$$

Now the direct compressive stress in the pipe is

$$s_1 = -\frac{F}{A}$$

But

$$A = \frac{\pi(D_o^2 - D_i^2)}{4} = \frac{\pi(170^2 - 163^2) \text{ mm}^2}{4}$$

$$A = 1831 \text{ mm}^2$$

Then

$$s_1 = -\frac{1324 \text{ N}}{1831 \text{ mm}^2} = -0.723 \text{ MPa}$$

This stress is a uniform, compressive stress across any cross section of the pipe.

The bending stress computation requires the application of the flexure formula,

$$s_2 = \frac{Mc}{I}$$

where $c = \dfrac{D_o}{2} = \dfrac{170 \text{ mm}}{2} = 85 \text{ mm}$

$$I = \frac{\pi}{64}(D_o^4 - D_i^4) = \frac{\pi}{64}(170^4 - 163^4) \text{ mm}^4$$

$$I = 6.347 \times 10^6 \text{ mm}^4$$

Then

$$s_2 = \frac{(1456 \text{ N} \cdot \text{m})(85 \text{ mm})}{6.347 \times 10^6 \text{ mm}^4} \times \frac{10^3 \text{ mm}}{\text{m}} = 19.5 \text{ MPa}$$

The bending stress s_2 produces compressive stress on the right side of the pipe and tension on the left side. Since the direct compression stress adds to the bending stress on the right side, that is where the maximum stress would occur. The combined stress would then be

$$s_t = -s_1 - s_2 = (-0.723 - 19.5) \text{ MPa} = -20.22 \text{ MPa}$$

Based on a yield strength of 241 MPa, the design factor would be

$$N = \frac{s_y}{s_t} = \frac{241 \text{ MPa}}{20.22 \text{ MPa}} = 11.9$$

Materials with Unequal Strengths in Tension and Compression. Earlier problems have sought to determine the maximum stress, either tensile or compressive, as an indication of the relative safety of a load-carrying member. This practice is acceptable for wrought forms of steel, aluminum, and most nonferrous alloys because the tensile and compressive strengths of these materials are very nearly the same. For cast iron and some other cast metals, however, there are quite different strengths in compression and in tension. Table A-5 lists several alloys of cast iron in the gray, ductile, and malleable grades. Comparing the ultimate strength in compression s_{uc} with the ultimate strength in tension s_{ut} shows that the compressive strengths range from 1.5 to 4 times higher than the tensile strengths. This may be an advantage or disadvantage in design, but it should be clear that the safety of *both* tensile and compressive stresses must be assured when such materials are used.

Some designs take advantage of the differing properties of the material by judiciously placing material where it will be most effective. Consider the power press frame shown in Figure 11-6. As the press operates, a force is

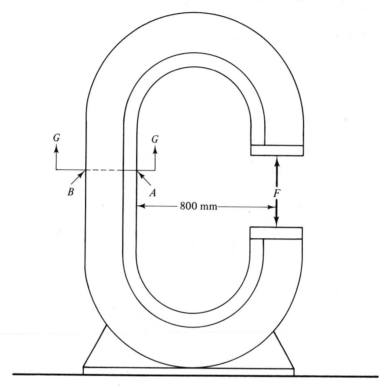

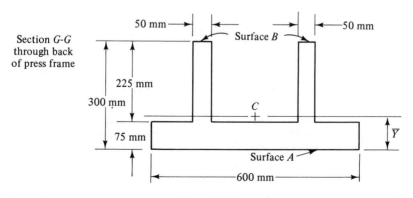

Figure 11-6

exerted between the upper and lower dies which tends to open up the frame. The back of the C-frame of the press is then subjected to a combined direct tensile stress and bending stress. Since the bending action tends to place the front of the frame at A in tension and the rear at B in compression, we will find the maximum tensile stress at A and the maximum compressive stress at B.

The cross section of the rear portion of the press frame shows a double tee shape. The centroidal axis of such a shape lies close to the front flanges of the tee. Thus, when subjected to a bending moment, the tensile stress will be less than the compressive stress. The following example problem illustrates these concepts.

Example Problem 11-2

The power press shown in Figure 11-6 can exert a force of 175 kN between the dies. Compute the maximum tensile and compressive stresses in the back part of the C-frame at the section through A and B. Then, if the frame is to be made of ductile iron, ASTM A439-D2, compute the design factor based on ultimate strength for both tension and compression. Would you judge the design of the frame satisfactory?

Solution

The back of the frame is subjected to combined direct tensile and bending stresses. Therefore, the area and the moment of inertia are needed for computing the stresses. Using the methods discussed in Chapter 8, locate the centroidal axis of the section. The cross-sectional dimensions are shown in Figure 11-6. Do this now before looking at the result below.

You should have $\bar{Y} = 87.5$ mm back from surface A. The following table shows the calculations, with part 1 referring to the flange across the front of the section and part 2 referring to the two webs toward the rear taken together as if they were one piece.

Part	A (mm^2)	y (mm)	Ay (mm^3)
1	45 000	37.5	1.688×10^6
2	22 500	187.5	4.219×10^6
	$A_T = 67\ 500$		$5.907 \times 10^6 = \sum Ay$

$$\bar{Y} = \frac{\sum Ay}{A_T} = 87.5 \text{ mm}$$

Now compute the moment of inertia with respect to the centroidal axis before looking below.

Did you get $I = 453.5 \times 10^6$ mm^4? Here are the detailed calculations.

$$I = I_1 + A_1 d_1^2 + I_2 + A_2 d_2^2$$

$$I_1 = \frac{BH^3}{12} = \frac{(600)(75)^3}{12} = 21.09 \times 10^6 \text{ mm}^4$$

$$d_1 = \bar{Y} - y_1 = 87.5 \text{ mm} - 37.5 \text{ mm} = 50 \text{ mm}$$

$$A_1 d_1^2 = (45\ 000 \text{ mm}^2)(50 \text{ mm})^2 = 112.50 \times 10^6 \text{ mm}^4$$

$$I_2 = \frac{BH^3}{12} = \frac{(100)(225)^3}{12} = 94.92 \times 10^6 \text{ mm}^4$$

$$d_2 = y_2 - \bar{Y} = 187.5 \text{ mm} - 87.5 \text{ mm} = 100 \text{ mm}$$

$$A_2 d_2^2 = (22\ 500 \text{ mm}^2)(100 \text{ mm})^2 = 225.0 \times 10^6 \text{ mm}^4$$

Then

$$I = (21.09 + 112.50 + 94.92 + 225.00)(10^6) \text{ mm}^4$$

$$I = 453.5 \times 10^6 \text{ mm}^4$$

We can now begin to compute stresses. Notice that the 175-kN force tends to place a direct tensile stress on the back of the press frame. Compute the magnitude of this stress now.

From $s = F/A$, the direct tensile stress is 2.6 MPa. The area can be found from the table used to compute the centroid.

$$s_1 = \frac{F}{A} = \frac{175\ 000 \text{ N}}{67\ 500 \text{ mm}^2} = 2.6 \text{ MPa}$$

This stress would be uniform across the section.

Bending stresses must now be computed from the flexure formula, $s = Mc/I$. The bending moment M is due to the fact that the force F acts at a large distance away from the section at the back of the frame. The magnitude of M is the product of the force times the distance from the line of action of the force to the *centroidal axis* of the section. Then

$$M = F(800 \text{ mm} + \bar{Y})$$
$$M = 175 \text{ kN} (800 \text{ mm} + 87.5 \text{ mm})$$
$$M = 0.155 \times 10^6 \text{ kN} \cdot \text{mm}$$

Now compute the tensile stress due to bending at surface A of the section.

You should have $s_{A2} = 29.9$ MPa. In the flexure formula, the distance c is from the centroidal axis out to the surface A. Then $c_A = 87.5$ mm. Now

$$s_{A2} = \frac{Mc_A}{I} = \frac{(0.155 \times 10^6 \text{ kN} \cdot \text{mm})(87.5 \text{ mm})}{453.5 \times 10^6 \text{ mm}^4}$$

$$s_{A2} = 29.9 \text{ MPa} \qquad \text{(tension)}$$

The compressive stress at the surface B will be greater than 29.9 MPa because there is a larger value of c from the centroidal axis to B. Compute the bending stress at B now.

$$c_B = 300 \text{ mm} - \bar{Y} = 212.5 \text{ mm}$$

Then

$$s_{B2} = \frac{Mc_B}{I} = \frac{(0.155 \times 10^6 \text{ kN} \cdot \text{mm})(212.5 \text{ mm})}{453.5 \times 10^6 \text{ mm}^4}$$

$$s_{B2} = 72.6 \text{ MPa} \qquad \text{(compression)}$$

The combined stresses at A and B can now be computed. Do that before going on.

At A, $s_A = 32.5$ MPa. At B, $s_B = -70.0$ MPa.
Remember, the tensile stress s_1 is uniform across the section.

$$s_A = s_1 + s_{A2} = 2.6 \text{ MPa} + 29.9 \text{ MPa} = 32.5 \text{ MPa} \qquad \text{(tension)}$$
$$s_B = s_1 - s_{B2} = 2.6 \text{ MPa} - 72.6 \text{ MPa} = -70.0 \text{ MPa} \text{ (compression)}$$

To evaluate the safety of the press under these stresses, compute the design factors based on ultimate strength at A and B for the ASTM A439-D2 ductile iron.

At A, $N_A = 12.3$. At B, $N_B = 17.7$. From Table A-5, $S_{ut} = 400$ MPa and $S_{uc} = 1240$ MPa.

Then

$$N_A = \frac{S_{ut}}{S_A} = \frac{400 \text{ MPa}}{32.5 \text{ MPa}} = 12.3$$

$$N_B = \frac{S_{uc}}{S_B} = \frac{1240 \text{ MPa}}{70.0 \text{ MPa}} = 17.7$$

Are these design factors adequate for this application?

Based on the recommended design factors shown in Table 3-1, a design factor of 12 based on ultimate strength for a ductile material is satisfactory under impact loading. In this example, the tensile stress is the most critical. Notice that, even though the compressive stress in the press frame is more than two times higher than the tensile stress, it is the tensile stress which governs the design because of the large difference in the strength of the material in tension and compression.

11-2   COMBINED BENDING AND TORSION

Rotating shafts in machines transmitting power represent good examples of members loaded in such a way as to produce combined bending and torsion. Figure 11-7 shows a shaft carrying two chain sprockets. Power is delivered to the shaft through the sprocket at C and transmitted down the

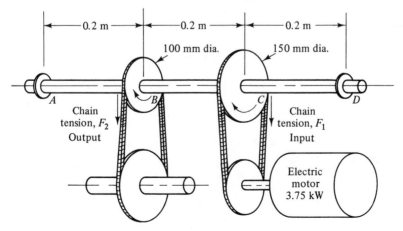

Figure 11-7

shaft to the sprocket at B, which in turn delivers it to another shaft. Because it is transmitting power, the shaft between B and C is subject to a torque and torsional shear stress, as you learned in Chapter 6. In order for the sprockets to transmit torque, they must be pulled by one side of the chain. At C, the back side of the chain must pull down with the force F_1 in order to drive the sprocket clockwise. Since the sprocket at B drives a mating sprocket, the front side of the chain would be in tension under the force F_2. The two forces, F_1 and F_2, acting downward cause bending of the shaft. Thus the shaft must be analyzed for both torsional shear stress and bending stress. Then, since both stresses act at the same place on the shaft, their combined effect must be determined. The method of analysis to be used is called the *maximum shear stress theory of failure*, which is described below. Then example problems will be shown.

Maximum Shear Stress Theory of Failure. When the tensile or compressive stress caused by bending occurs at the same place that a shearing stress occurs, the two kinds of stress combine to produce a larger shearing stress, which can be computed from

$$s_{s_{max}} = \sqrt{\left(\frac{s}{2}\right)^2 + s_s^2} \tag{11-2}$$

In Equation (11-2), s refers to the magnitude of the tensile or compressive stress at the point, and s_s is the shear stress at the same point. The result $s_{s_{max}}$ is the maximum shear stress at the point.

The maximum shear stress theory of failure states that a member fails when the maximum shear stress exceeds the yield strength of the material in shear. This failure theory shows good correlation with test results for ductile metals such as most steels.

Equation (11-2) can be expressed in a simplified form for the particular case of a circular shaft subjected to bending and torsion. Evaluating the bending stress separately, the maximum tensile or compressive stress would be

$$s = \frac{M}{Z}$$

where $Z = \dfrac{\pi D^3}{32}$

  $D =$ diameter of the shaft

The maximum stress due to bending occurs at the outside surface of the shaft, as shown in Figure 11-8.

Now consider the torsional shear stress separately. In Chapter 6 the torsional shear stress equation was derived:

$$s_s = \frac{T}{Z_p}$$

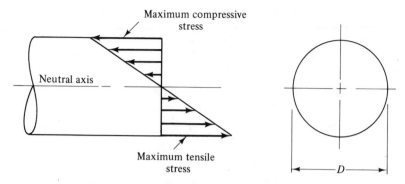

Figure 11-8   Bending Stress Distribution in a Round Shaft

where $Z_p = \dfrac{\pi D^3}{16}$

T = the torque on the section

The maximum shear stress occurs at the outer surface of the shaft all the way around the diameter, as shown in Figure 11-9.

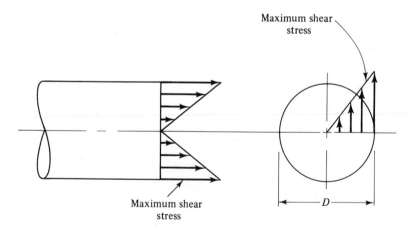

Figure 11-9   Torsional Shear Stress Distribution in a Round Shaft

Thus the maximum tensile stress and the maximum torsional shear stress both occur at the same point in the shaft. Now let's use Equation (11-2) to get an expression for the combined stress in terms of the bending moment M, the torque T, and the shaft diameter D.

$$s_{s_{max}} = \sqrt{\left(\frac{s}{2}\right)^2 + s_s^2}$$

$$s_{s_{max}} = \sqrt{\left(\frac{M}{2Z}\right)^2 + \left(\frac{T}{Z_p}\right)^2}$$

(11-3)

Notice that, from the definitions of Z and Z_p given above,

$$Z = \frac{Z_p}{2}$$

Substituting this into Equation (11-3) gives

$$s_{smax} = \sqrt{\left(\frac{M}{Z_p}\right)^2 + \left(\frac{T}{Z_p}\right)^2}$$

Factoring out Z_p,

$$s_{smax} = \frac{1}{Z_p} \sqrt{M^2 + T^2}$$

Sometimes the term $\sqrt{M^2 + T^2}$ is called the *equivalent torque* because it represents the amount of torque which would have to be applied to the shaft by itself to cause the equivalent magnitude of shear stress as the combination of bending and torsion. Calling the equivalent torque T_e,

$$T_e = \sqrt{M^2 + T^2} \tag{11-4}$$

and

$$s_{smax} = \frac{T_e}{Z_p} \tag{11-5}$$

Equations (11-4) and (11-5) greatly simplify the calculation of the maximum shear stress in a circular shaft subjected to bending and torsion.

In the design of circular shafts subjected to bending and torsion, a design stress can be specified giving the maximum allowable shear stress. This was done in Section 6-3.

$$S_{sd} = \frac{s_{ys}}{N}$$

where s_{ys} is the yield strength of the material in shear. Since s_{ys} is seldom known, the approximate value found from $s_{ys} = \frac{s_y}{2}$ can be used. Then,

$$S_{sd} = \frac{s_y}{2N} \tag{11-6}$$

where s_y is the yield strength in tension, as reported in most material property tables such as those in the Appendix. It is recommended that the value of the design factor be *no less than 4*. A rotating shaft subjected to bending is a good example of a repeated and reversed load. With each revolution of the shaft, a particular point on the surface is subjected to the maximum tensile and then the maximum compressive stress. Thus fatigue is the expected mode of failure, and $N = 4$ or greater is recommended in Table 3-1, based on yield strength.

Stress Concentrations. In shafts, stress concentrations are created by geometry factors such as key seats, shoulders, and grooves. The proper application of stress concentration factors to the equivalent torque Equations (11-4) and (11-5) should be considered carefully. If the value of K_t at a section of interest is equal for both bending and torsion, then it can be applied directly to Equation (11-5). That is,

$$s_{s_{max}} = \frac{T_e K_t}{Z_p} \qquad (11\text{-}7)$$

The form of Equation (11-7) can also be applied as a conservative calculation of $s_{s_{max}}$ by selecting K_t as the largest value for either torsion or bending.

To account for the proper K_t for both torsion and bending, Equation (11-4) can be modified as

$$T_e = \sqrt{(K_{tB}M)^2 + (K_{tT}T)^2} \qquad (11\text{-}8)$$

Then Equation (11-5) can be used directly to compute the maximum shear stress.

Example Problem 11-3

Specify a suitable material for the shaft shown in Figure 11-7. The shaft has a uniform diameter of 55 mm and rotates at 120 rpm while transmitting 3.75 kW of power. The chain sprockets at B and C are keyed to the shaft with sled-runner key seats. Sprocket C receives the power, and sprocket B delivers it to another shaft. The bearings at A and D provide simple supports for the shaft.

Solution

The several steps to be used to solve this problem are outlined below.

1. The torque in the shaft will be computed for the known power and rotational speed from $T = P/n$, as developed in Chapter 6.
2. The tensions in the chains for sprockets B and C will be computed. These are the forces which produce bending in the shaft.
3. Considering the shaft to be a beam, the shear and bending moment diagrams will be drawn for it.
4. At the section where the maximum bending moment occurs, the equivalent torque T_e will be computed from Equation (11-4).
5. The polar section modulus Z_p and the stress concentration factor K_t will be determined.
6. The maximum shear stress will be computed from Equation (11-7).

7. The required yield strength of the shaft material will be computed by letting $s_{s_{max}} = s_{sd}$ in Equation (11-6) and solving for s_y. Remember, let $N = 4$ or more.
8. A steel which has a sufficient yield strength will be selected from Table A-1.

Now let's follow the steps.

Step 1

The desirable unit for torque is N·m. Then it is most convenient to observe that the power unit of kilowatts is equivalent to the units of kN·m/s. Also, rotational speed must be expressed in rad/s.

$$n = \frac{120 \text{ rev}}{\text{min}} \times \frac{2\pi \text{ rad}}{\text{rev}} \times \frac{1 \text{ min}}{60 \text{ s}} = 12.57 \text{ rad/s}$$

We can now compute torque.

$$T = \frac{P}{n} = \frac{3.75 \text{ kN·m}}{\text{s}} \times \frac{1}{12.57 \text{ rad/s}} = 0.298 \text{ kN·m}$$

Step 2

The tensions in the chains are indicated in Figure 11-7 by the forces F_1 and F_2. In order for the shaft to be in equilibrium, the torque on both sprockets must be the same in magnitude but opposite in direction. On either sprocket the torque is the product of the chain force times the *radius* of the pulley. That is,

$$T = F_1 R_1 = F_2 R_2$$

The forces can now be computed.

$$F_1 = \frac{T}{R_1} = \frac{0.298 \text{ kN·m}}{75 \text{ mm}} \times \frac{10^3 \text{ mm}}{\text{m}} = 3.97 \text{ kN}$$

$$F_2 = \frac{T}{R_2} = \frac{0.298 \text{ kN·m}}{50 \text{ mm}} \times \frac{10^3 \text{ mm}}{\text{m}} = 5.96 \text{ kN}$$

Step 3

Figure 11-10 shows the complete shear and bending moment diagrams found by the methods of Chapter 7. The maximum bending moment is 1.06 kN·m at section B, where one of the sprockets is located.

Step 4

At section B, the torque in the shaft is 0.298 kN·m and the bending moment is 1.06 kN·m. Then

$$T_e = \sqrt{M^2 + T^2} = \sqrt{(1.06)^2 + (0.298)^2} = 1.10 \text{ kN·m}$$

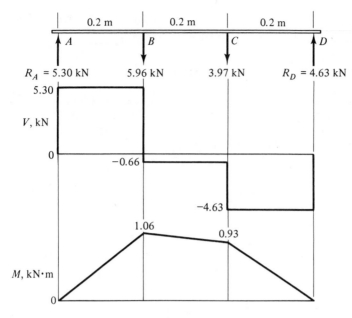

Figure 11-10

Step 5

$$Z_p = \frac{\pi D^3}{16} = \frac{\pi(55 \text{ mm})^3}{16} = 32.67 \times 10^3 \text{ mm}^3$$

For the key seat at section B securing the sprocket to the shaft, we will use $K_t = 1.6$, as reported in Figure 6-8.

Step 6

$$S_{s\text{max}} = \frac{T_e K_t}{Z_p} = \frac{1.10 \times 10^3 \text{ N}\cdot\text{m } (1.6)}{32.67 \times 10^3 \text{ mm}^3} \times \frac{10^3 \text{ mm}}{\text{m}}$$

$$S_{s\text{max}} = 53.9 \text{ MPa}$$

Step 7

Let

$$S_{s\text{max}} = S_{sd} = \frac{S_y}{2N}$$

Then

$$S_y = 2N S_{s\text{max}} = (2)(4)(53.9 \text{ MPa}) = 431 \text{ MPa}$$

Step 8

Referring to Table A-1, we find that several alloys could be used. For example, AISI 1040, cold-drawn, has a yield strength of 490 MPa. Alloy AISI 1141 OQT 1300 has a yield strength of 469 MPa and also a very good ductility, as indicated by the 28 percent elongation. Either of these would be reasonable choices.

PROBLEMS

11-1 A $2\frac{1}{2}$-inch Schedule 40 steel pipe is used as a support for a basketball backboard, as shown in Figure 11-11. It is securely fixed into the ground. Compute the stress which would be developed in the pipe if a 230-lb player hung on the base of the rim of the basket.

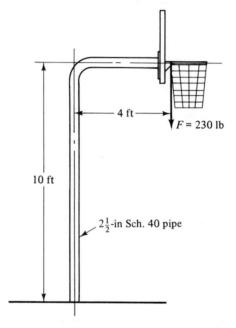

Figure 11-11

11-2 The bracket shown in Figure 11-12 has a rectangular cross section 18 mm wide by 75 mm high. It is securely attached to the wall. Compute the maximum stress in the bracket.

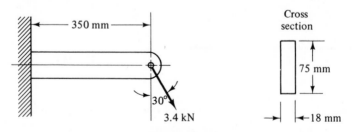

Figure 11-12

11-3 The beam shown in Figure 11-13 carries a 6000-lb load attached to a bracket below the beam. Compute the stress at points M and N, where it is attached to the column.

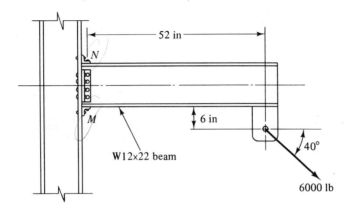

Figure 11-13

11-4 For the beam shown in Figure 11-13, compute the stress at points M and N if the 6000-lb load acts vertically downward instead of at an angle.

11-5 For the beam shown in Figure 11-13, compute the stress at points M and N if the 6000-lb load acts back toward the column at an angle of 40 deg below the horizontal instead of as shown.

11-6 Compute the maximum stress in the top portion of the coping saw frame shown in Figure 11-14 if the tension in the blade is 125 N.

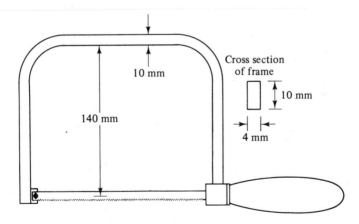

Figure 11-14

11-7 Compute the maximum stress in the crane beam shown in Figure 11-15 when a load of 12 kN is applied at the middle of the beam.

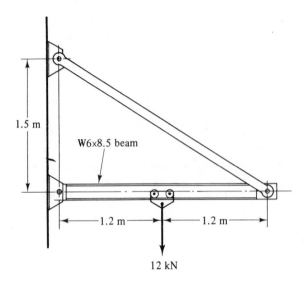

Figure 11-15

11-8 Figure 11-16 shows a metal-cutting hack saw. Its frame is made of hollow tubing having an outside diameter of 12 mm and a wall thickness of 1.0 mm. The blade is pulled taut by the wing nut so that a tensile force of 160 N is applied to the blade. Compute the maximum stress in the top section of the tubular frame.

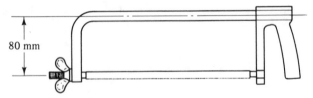

Figure 11-16

11-9 The C-clamp in Figure 11-17 is made of cast zinc, ASTM AC41A. Determine the allowable clamping force which the clamp can exert if it is desired to have a design factor of 4 based on ultimate strength in either tension or compression.

11-10 The C-clamp shown in Figure 11-18 is made of cast malleable iron, ASTM A220 Grade 40010. Determine the allowable clamping force which the clamp can exert if it is desired to have a design factor of 2 based on yield strength in either tension or compression.

11-11 A tool used for compressing a coil spring to allow its installation in a car is shown in Figure 11-19. A force of 1200 N is applied near the ends of the extended lugs, as shown. Compute the maximum tensile stress in the threaded rod. Assume a stress concentration factor of 3.0 at the thread root

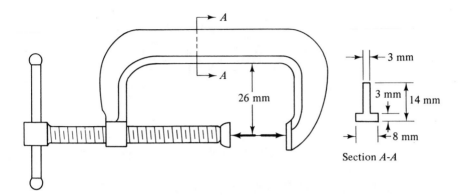

Section A-A

Figure 11-17

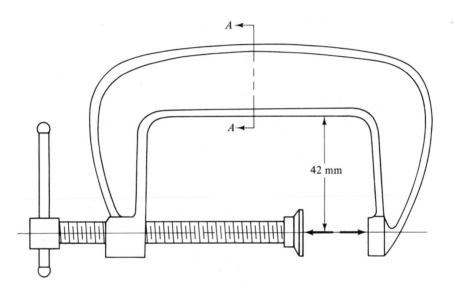

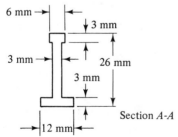

Section A-A

Figure 11-18

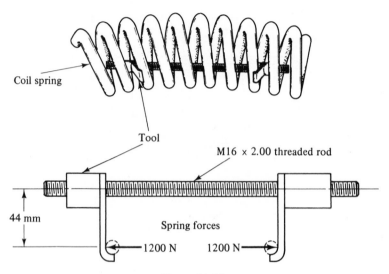

Figure 11-19

for both tensile and bending stresses. Then, using a design factor of 2 based on yield strength, specify a suitable material for the rod.

11-12 Figure 11-20 shows a portion of the steering mechanism for a car. The steering arm has the detailed design shown. Notice that the arm has a constant thickness of 10 mm so that all cross sections between the end lugs are rectangular. The 225-N force is applied to the arm at a 60-deg angle. Compute the stress in the arm at sections *A* and *B*. Then, if the arm is made of

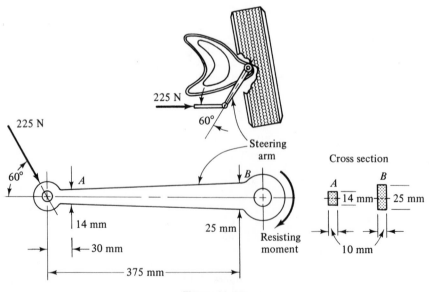

Figure 11-20

ductile iron, ASTM 80-55-6, compute the minimum design factor at these two points based on ultimate strength.

11-13 An S3 × 5.7 American Standard I-beam is subjected to the forces shown in Figure 11-21. The 4600-lb force acts directly in line with the axis of the beam. The 500-lb downward force at *A* produces the reactions shown at the supports *B* and *C*. Compute the maximum tensile and compressive stresses in the beam.

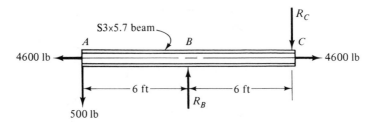

Figure 11-21

11-14 The horizontal crane boom shown in Figure 11-22 is made of a hollow rectangular steel tube. Compute the stress in the boom just to the left of point *B* when a mass of 1000 kg is supported at the end.

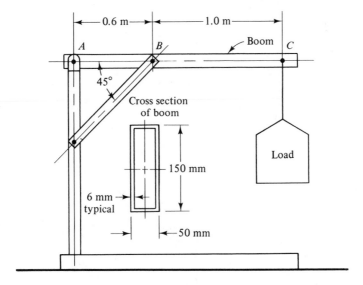

Figure 11-22

11-15 For the crane boom shown in Figure 11-22, compute the load which could be supported if a design factor of 3 based on yield strength is desired. The boom is made of AISI 1050 hot-rolled steel. Analyze only sections where the full box section carries the load, assuming that sections at connections are adequately reinforced.

11-16 A television antenna is mounted on a hollow aluminum tube as sketched in Figure 11-23. During installation, a force of 20 lb is applied to the end of the antenna as shown. Calculate the torsional shear stress in the tube and the stress due to bending. Consider the tube to be simply supported against bending at the clamps, but assume that rotation is not permitted. If the tube is made of 6061-T6 aluminum, would it be safe under this load? The tube has an outside diameter of 1.50 in. and a wall thickness of $\frac{1}{16}$ in.

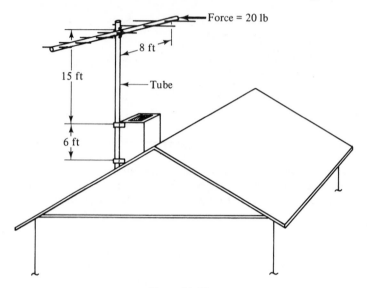

Figure 11-23

11-17 Figure 11-24 shows a crank to which a force F of 1200 N is applied. Compute the maximum stress in the circular portion of the crank.

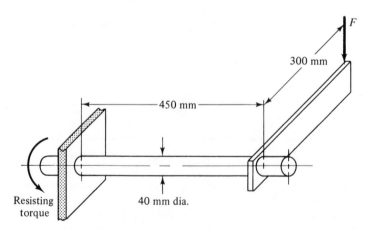

Figure 11-24

11-18   A standard steel pipe is to be used to support a bar carrying four loads, as shown in Figure 11-25. Specify a suitable pipe which would keep the maximum combined shear stress to 8000 psi.

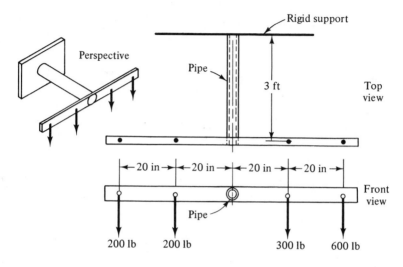

Figure 11-25

11-19   A 1.0-in. diameter solid round shaft will carry 25 hp while rotating at 1150 rpm in the arrangement shown in Figure 11-26. The total bending loads at the gears *A* and *C* are shown, along with the reactions at the bearings *B* and *D*. Use a design factor of 6 for the maximum shear stress theory of failure, and determine a suitable steel for the shaft.

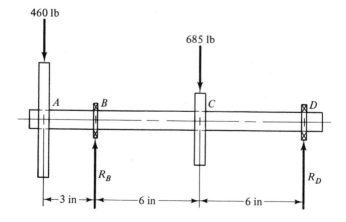

Figure 11-26

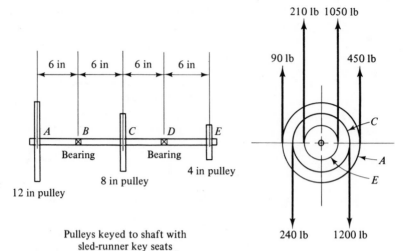

Figure 11-27

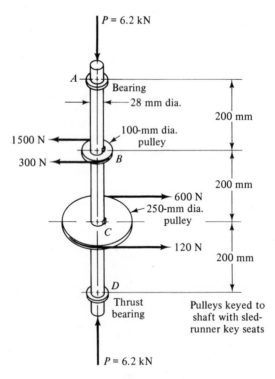

Figure 11-28

11-20   Three pulleys are mounted on a rotating shaft, as shown in Figure 11-27. Belt tensions are as shown in the end view. All power is received by the shaft through pulley *C* from below. Pulleys *A* and *B* deliver power to mating pulleys above.

   (a) Compute the torque at all points in the shaft.
   (b) Compute the bending stress and the torsional shear stress in the shaft at the point where the maximum bending moment occurs if the shaft diameter is 1.75 in.
   (c) Compute the maximum shear stress at the point used in step (b). Then specify a suitable material for the shaft.

11-21   The vertical shaft shown in Figure 11-28 carries two belt pulleys. The tensile forces in the belts under operating conditions are shown. Also, the shaft carries an axial compressive load of 6.2 kN. Considering torsion, bending, and axial compressive stresses, compute the maximum shear stress using Equation (11-2).

11-22   For the shaft in Problem 11-21, specify a suitable steel which would provide a design factor of 4 based on yield strength in shear.

11-23   The member *EF* in the truss shown in Figure 11-29 carries an axial tensile load of 54 000 lb in addition to the two 1200-lb loads shown. It is planned to use two steel angles for the member, arranged back to back. Specify a suitable size for the angles if they are to be made of A36 structural steel with an allowable stress of 0.6 times the yield strength.

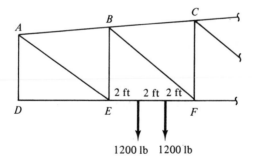

Figure 11-29

12

DEFLECTION OF BEAMS

12-1 NEED TO CONSIDER DEFLECTION IN BEAMS

The proper performance of machine parts, the structural rigidity of buildings, vehicles, and machine frames, and the tendency for a part to vibrate are all dependent on deformation in beams. Therefore, the ability to analyze beams for deflections under load is very important.

The spindle of a lathe or drill press and the arbor of a milling machine carry cutting tools for machining metals. Deflection of the spindle or arbor would have an adverse effect on the accuracy that the machine could produce. The manner of loading and support of these machine elements indicate that they are beams, and the approach to computing their deflection will be discussed in this chapter.

Precision measuring equipment must also be designed to be very rigid. Deflection caused by the application of measuring forces reduces the precision of the desired measurement.

Power-transmission shafts carrying gears must have sufficient rigidity to ensure that the gear teeth mesh properly. Excessive deflection of the shafts would tend to separate the mating gears, resulting in a movement away from the most desirable point of contact between the gear teeth. Noise generation, decreased power-transmitting capability, and increased wear would result. For straight spur gears, it is recommended that the movement between two gears not exceed 0.005 in. (0.13 mm). This limit is the *sum* of the movement of the two shafts carrying the mating gears at the location of the gears.

The floors of buildings must have sufficient rigidity to carry expected loads. Occupants of the building should not notice floor deflections. Machines and other equipment require a stable floor support for proper operation. Beams carrying plastered ceilings must not deflect excessively so as not to crack the plaster. A limit of $\frac{1}{360}$ times the span of the beam carrying a ceiling is often used for deflection.

Frames of vehicles, metal-forming machines, automation devices, and process equipment must also possess sufficient rigidity to ensure satisfactory operation of the equipment carried by the frame. The bed of a lathe, the crown of a punch press, the structure of an automatic assembly device, and the frame of a truck are examples.

Vibration is caused by the forced oscillations of parts of a structure or machine. The tendency to vibrate at a certain frequency and the severity of the vibrations are functions of the flexibility of the parts. Of course, flexibility is a term used to describe how much a part deflects under load. Vibration problems can be solved by either *increasing or decreasing* the stiffness of the part, depending on the circumstances. In either case, an understanding of how to compute deflections of beams is important.

Two basic methods of determining the deflection of beams which will be presented in this chapter are the *formula method* and the *successive integration method*. Which method to use in a specific case depends on the manner of loading and support and on what information is desired. The formula method is the simplest to apply if the proper formulas are available and if the deflection at only one or a few points is needed. For more complex loadings or support conditions, or if the shape of the entire deflected beam is desired, the other method must be used.

12-2 DEFLECTION FORMULAS

For many practical configurations of beam loading and support, formulas have been derived which allow the computation of the deflection at any point on the beam. Three examples are illustrated in Figure 12-1. The simply supported beam carrying a single concentrated load or a distributed load occurs frequently in practice. Using the mathematical derivation techniques described later in this chapter, we have derived the general equations shown in the figure. They allow the computation of the deflection at any point along the beam.

In the formulas, y is the vertical deflection of the neutral axis of the beam from its straight unloaded position due to the applied load. The term x is the

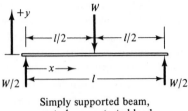

Maximum deflection at center:

$$y_{max} = \frac{-Wl^3}{48EI}$$

At any point:

$$y = \frac{-Wx}{48EI}(3l^2 - 4x^2)$$

x = distance from support to point where y is desired

Simply supported beam, central concentrated load

(a)

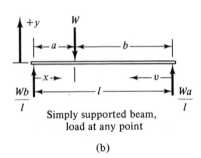

Deflection at load $= \dfrac{-Wa^2 b^2}{3EIl}$

At a point in the shorter segment a:

$$y = \frac{-Wbx}{6EIl}(l^2 - x^2 - b^2)$$

At a point in the longer segment b:

$$y = \frac{-Wav}{6EIl}(l^2 - v^2 - a^2)$$

Maximum deflection in segment b:

$$y_{max} = \frac{-Wav_1^3}{3EIl} \quad at \quad v_1 = b\sqrt{\frac{1}{3} + \frac{2a}{3b}}$$

Simply supported beam, load at any point

(b)

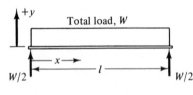

Maximum deflection at center:

$$y_{max} = \frac{-5}{384}\frac{Wl^3}{EI}$$

At any point:

$$y = \frac{-Wx(l - x)}{24EIl}[l^2 + x(l - x)]$$

Simply supported beam, uniformly distributed load

(c)

Figure 12-1   Deflection Formulas for Three Simple Cases

distance from the left end of the beam, or from a support, to the point where the deflection is to be computed.

The geometrical and material properties of the beam are included in the product EI, which appears in all the formulas. Recall that E is the modulus of elasticity of the beam material and is a measure of its stiffness. The moment of inertia I is a measure of the stiffness (resistance to deflection) of the cross-sectional shape of the beam. Thus the combined term EI is a measure of the stiffness of the beam as a whole. Deflections are inversely proportional to EI.

The deflection formulas are valid only for the cases where the cross section of the beam is uniform for its entire length. Several other formulas are included in the Appendix for other loading and support conditions. Example problems will demonstrate the application of the formulas.

Example Problem 12-1

Determine the deflection of a simply supported beam carrying a hydraulic cylinder in a machine used to press bushings into a casting, as shown in Figure 12-2. The force exerted during the pressing operation is 15 kN. The beam is rectangular, 25 mm thick and 100 mm high, and made of steel.

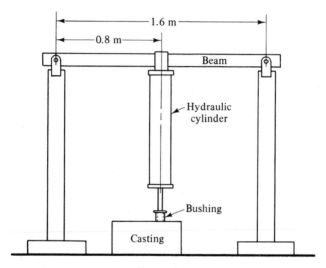

Figure 12-2

Solution

Using the formula from part (a) of Figure 12-1, we find the deflection to be

$$y = \frac{-Wl^3}{48\,EI}$$

where $W = 15\,\text{kN} = 15 \times 10^3\text{N}$
$l = 1.6\,\text{m}$

From Table A-1, $E = 207\,\text{GPa} = 205 \times 10^9\text{N/m}^2$. For the rectangular beam,

$$I = \frac{(25)(100)^3}{12} = 2.083 \times 10^6\,\text{mm}^4$$

Then

$$y = \frac{-Wl^3}{48EI} = \frac{-(15 \times 10^3 \text{ N})(1.6 \text{ m})^3}{48(207 \times 10^9 \text{ N/m}^2)(2.083 \times 10^6 \text{ mm}^4)} \times \frac{(10^3 \text{ mm})^5}{\text{m}^5}$$

$$y = -2.97 \text{ mm}$$

Example Problem 12-2

A round shaft, 45 mm in diameter, carries a 3500-N load as shown in Figure 12-3. If the shaft is steel, compute the deflection at the load and at the point C, 100 mm from the right end of the shaft.

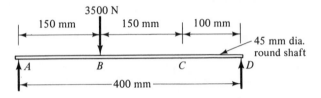

Figure 12-3

Solution

The shaft conforms to the case depicted in part (b) of Figure 12-1. Then, at the load, point B,

$$y_B = \frac{-Wa^2b^2}{3EIl}$$

But

$$I = \frac{\pi D^4}{64} = \frac{\pi(45 \text{ mm})^4}{64} = 0.201 \times 10^6 \text{ mm}^4$$

Also, in this case it would be convenient to express the modulus of elasticity E in the units of N/mm^2 instead of N/m^2 for consistency. The deflection y will then be in mm.

$$E = \frac{207 \times 10^9 \text{ N}}{\text{m}^2} \times \frac{1 \text{ m}^2}{(10^3 \text{ mm})^2} = 207 \times 10^3 \text{ N/mm}^2$$

Now

$$y_B = \frac{-Wa^2b^2}{3EIl} = \frac{-(3500)(150)^2(250)^2}{3(207 \times 10^3)(0.201 \times 10^6)(400)} = -0.0985 \text{ mm}$$

We can now compute the deflection at point C. In the set of formulas given, the one for deflections in the segment of the beam of length b applies.

$$y_C = \frac{-Wav}{6EIl}(l^2 - v^2 - a^2)$$

where $W = 3500\text{N}$
$\quad\quad a = 150\text{ mm}$
$\quad\quad l = 400\text{ mm}$
$\quad\quad v = 100\text{ mm}$ (from right support to C)

Then

$$y_C = \frac{-(3500)(150)(100)}{6(207 \times 10^3)(0.201 \times 10^6)(400)}(400^2 - 100^2 - 150^2)$$

$$y_C = -0.0670 \text{ mm}$$

Example Problem 12-3

Determine the deflection at the middle of an aluminum $10\text{I} \times 8.646$ beam when it carries a total load of 55 800 lb distributed over a length of 6.0 ft. The beam is simply supported, as that shown in Figure 12-1(c).

Solution

The deflection can be computed from

$$y = \frac{-5Wl^3}{384EI}$$

Consistency of units must be carefully observed. Express W in *pounds*, length in *inches*, E in *psi*, and I in *inches* to the fourth power. From the Appendix,

$$I = 132.09 \text{ in.}^4$$
$$E = 10 \times 10^6 \text{ psi}$$

and

$$l = 6.0 \text{ ft} \times 12 \text{ in./ft} = 72 \text{ in.}$$

Then

$$y = \frac{-5(55\ 800)(72^3)}{384(10 \times 10^6)(132.09)}\text{in.} = -0.205 \text{ in.}$$

12-3 SUPERPOSITION USING DEFLECTION FORMULAS

Formulas such as those used in the preceding section are available for a large number of cases of loading and support conditions. Table A-19 gives 15 different cases. Obviously, these cases would allow the solution of many practical beam deflection problems. An even larger number of situations can be handled by the use of the *principle of superposition*.

If a particular loading and support pattern can be broken into components such that each component is like one of the cases for which a formula is available, then the total deflection at a point on the beam is equal to the sum of the deflections caused by each component. The deflection due to one component load is *superposed* on deflections due to the other loads, thus giving the name *superposition*.

An example of where superposition can be applied is shown in Figure 12-4. Represented in the figure is a roof beam carrying a uniformly distrib-

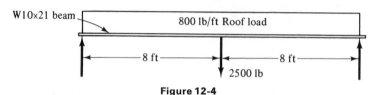

Figure 12-4

uted roof load of 800 lb/ft and also supporting a portion of a piece of process equipment which provides a concentrated load at the middle. Figure 12-5 shows how the loads are considered separately. Each component load pro-

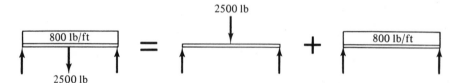

Figure 12-5   Illustration of Superposition Principle

duces a maximum deflection at the middle. Therefore, the maximum total deflection will occur there also. Let the subscript 1 refer to the concentrated load case and the subscript 2 refer to the distributed load case. Then

$$y_1 = \frac{-Wl^3}{48\ EI}$$

$$y_2 = \frac{-5}{384} \frac{Wl^3}{EI}$$

The total deflection will be

$$y_T = y_1 + y_2$$

The terms l, E, and I will be the same for both cases.

$$l = 16 \text{ ft} \times 12 \text{ in./ft} = 192 \text{ in.}$$

$$E = 30 \times 10^6 \text{ psi} \qquad \text{for steel}$$

$$I = 107 \text{ in.}^4 \qquad \text{for W10} \times 21 \text{ beam}$$

To compute y_1, let $W = 2500$ lb.

$$y_1 = \frac{-2500(192)^3}{48(30 \times 10^6)(107)}\text{in.} = -0.115\,\text{in.}$$

To compute y_2, W is the total resultant of the distributed load.

$$W = (800\,\text{lb/ft})(16\,\text{ft}) = 12\ 800\,\text{lb}$$

Then

$$y_2 = \frac{-5(12\ 800)(192)^3}{384(30 \times 10^6)(107)}\text{in.} = -0.367\,\text{in.}$$

and

$$y_T = y_1 + y_2 = -0.115\,\text{in.} - 0.367\,\text{in.} = -0.482\,\text{in.}$$

Since this is the total deflection, we could check to see if it meets the recommendation that the maximum deflection should be less than $\frac{1}{360}$ times the span of the beam. The span is l.

$$\frac{l}{360} = \frac{192\,\text{in.}}{360} = 0.533\,\text{in.}$$

The computed deflection was 0.482 in., which is satisfactory.

The superposition principle is valid for any place on the beam, not just at the loads. The following example problem illustrates this.

Example Problem 12-4

Figure 12-6 shows a shaft carrying two gears and simply supported at its ends by bearings. The mating gears above exert downward forces which tend to separate the gears. Other forces acting horizontally are

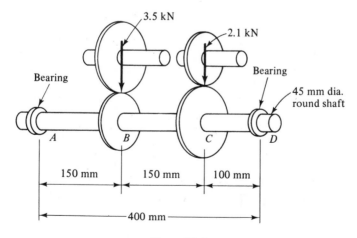

Figure 12-6

being neglected for this analysis. Determine the total deflection at the gears B and C if the shaft is steel and has a diameter of 45 mm.

Solution

The two unsymmetrically placed, concentrated loads constitute a situation for which none of the cases for beam deflection formulas is valid, either in Figure 12-1 or in the Appendix. However, it can be solved by using the formulas of Figure 12-1(b) twice. Considering each load separately, the deflections at B and C can be found for each. The total, then, would be the sum of the component deflections. Figure 12-7 shows the logic from which we can say

$$y_B = y_{B1} + y_{B2}$$
$$y_C = y_{C1} + y_{C2}$$

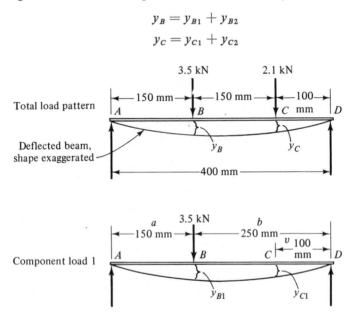

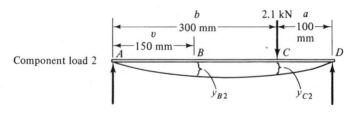

Figure 12-7 Superposition Logic for Deflection of Shaft in Fig. 12-6

where y_B = total deflection at B

$\quad\quad y_C$ = total deflection at C

$\quad\quad y_{B1}$ = deflection at B due to 3.5-kN load alone

$\quad\quad y_{C1}$ = deflection at C due to 3.5-kN load alone

$\quad\quad y_{B2}$ = deflection at B due to 2.1-kN load alone

$\quad\quad y_{C2}$ = deflection at C due to 2.1-kN load alone

For all the calculations, the values of E, I, and l will be needed. These are the same as those used in Example Problem 12-2.

$$E = 207 \times 10^3 \text{ N/mm}^2$$

$$I = 0.201 \times 10^6 \text{ mm}^4$$

$$l = 400 \text{ mm}$$

The product EIl is present in all the formulas.

$$EIl = (207 \times 10^3)(0.201 \times 10^6)(400) = 16.64 \times 10^{12}$$

Now the individual component deflections will be computed. For component 1:

$$y_{B1} = \frac{-Wa^2b^2}{3EIl} = \frac{-(3.5 \times 10^3)(150)^2(250)^2}{3(16.64 \times 10^{12})} = -0.0985 \text{ mm}$$

$$y_{C1} = \frac{-Wav}{6EIl}(l^2 - v^2 - a^2)$$

$$y_{C1} = \frac{-(3.5 \times 10^3)(150)(100)}{6(16.64 \times 10^{12})}(400^2 - 100^2 - 150^2) = -0.0670 \text{ mm}$$

For component 2, the load will be 2.1 kN at point C. Then

$$y_{B2} = \frac{-Wav}{6EIl}(l^2 - v^2 - a^2)$$

$$y_{B2} = \frac{-(2.1 \times 10^3)(100)(150)}{6(16.64 \times 10^{12})}(400^2 - 150^2 - 100^2) = -0.0402 \text{ mm}$$

$$y_{C2} = \frac{-Wa^2b^2}{3EIl} = \frac{-(2.1 \times 10^3)(100)^2(300)^2}{3(16.64 \times 10^{12})} = -0.0378 \text{ mm}$$

Now, by superposition,

$$y_B = y_{B1} + y_{B2} = -0.0985 \text{ mm} - 0.0402 \text{ mm} = -0.1387 \text{ mm}$$

$$y_C = y_{C1} + y_{C2} = -0.0670 \text{ mm} - 0.0378 \text{ mm} = -0.1048 \text{ mm}$$

In Section 12-1 it was observed that a recommended limit for the movement of one gear relative to its mating gear is 0.13 mm. Thus this shaft is too flexible since the deflection at B exceeds 0.13 mm without even considering the deflection of the mating shaft.

12-4 BEAM DEFLECTIONS— GENERAL APPROACH

While the beam deflection formulas and the application of the super-position method allow the solution of a large number of practical problems with ease, there certainly are cases for which formulas are not readily available. Most notable are beams with one or two overhanging ends carrying several loads and beams with distributed loads over only a portion of the beam length. Also, if the absolute maximum deflection is desired, superposition may have to be used many times in a trial-and-error approach to locate *where* the point of maximum deflection occurs. This is cumbersome indeed. A computerized solution of the formulas may help.

A general approach will now be presented which allows the determination of deflection at any point on the beam. The advantages of this approach are listed below.

1. The result is a set of equations for deflection at all parts of the beam. Deflection at any point can then be found by substitution of the beam "rigidity" properties of E and I, and the position on the beam.
2. Data are easily obtained from which a plot of the shape of the deflection curve may be made.
3. The equations for the *slope* of the beam at any point are generated, as are deflections. This is important in some machinery applications such as shafts at bearings and shafts carrying gears. An excessive slope of the shaft would result in poor performance and reduced life of the bearings or gears.
4. The fundamental relationships between loads, manner of support, beam rigidity properties, slope, and deflections are emphasized in the solution procedure. The designer who understads these relationships can make more efficient designs.
5. The method requires the application of only simple mathematical concepts.
6. The point of maximum deflection can be found directly from the resulting equations.

The basis for the general approach to beam deflections is illustrated in Figure 12-8, which shows the familiar shear and bending moment diagrams plus two new diagrams relating the slope of the deflected beam θ and the actual deflection y to positions on the beam. The diagrams imply that there are relationships among them, and indeed there are.

We have already used the relationships between the shear and bending moment diagrams. These were called the *laws of beam diagrams* and were stated as:

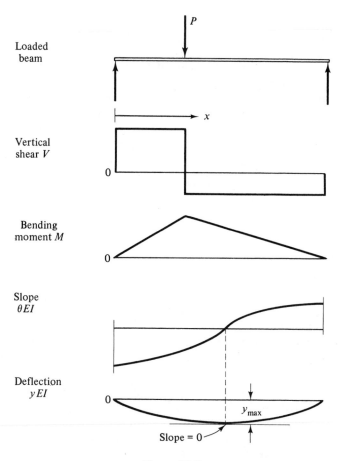

Figure 12-8

1. The *change in moment* between two points on a beam is equal to the *area* under the shear curve between those same two points.
2. The *slope* of the bending moment curve at any point is equal to the *value of the shear* at that point.

These laws could be stated mathematically as:

1. $M_B - M_A = \int_A^B V_{AB}\, dx$ (12-1)

2. $V_x = \dfrac{dM}{dx}$ (12-2)

Equation (12-1) shows that the process of integration is the mathematical equivalent to the calculation of the area under a curve. The term V_{AB}

refers to the expression for the shearing force V as a function of the position on the beam x in the segment from point A to point B. The A and B on the integral sign express the limits between which the area is to be calculated.

Equation 12-2 states that the shear at the position x on the beam is equal to the derivative of the bending moment curve at that point x. The derivative dM/dx is the mathematical method of finding the *slope* of the curve.

In using integration to determine deflections of beams, only one simple kind of integration must be performed. It can be learned and used quickly even without previous instruction in calculus. This required kind of integration process is:

$$\int ax^n \, dx = \frac{ax^{n+1}}{n+1} + C \qquad (12\text{-}3)$$

The process used to evaluate Equation (12-3) is explained in the special tinted panel accompanying this section.

EXPLANATION OF THE PROCESS: $\int ax^n \, dx = \dfrac{ax^{n+1}}{n+1} + C$

The process could be read: "The integral of the expression ax^n with respect to x equals the new expression $ax^{n+1}/(n+1)$ plus a constant of integration." The coefficient a is any constant. The term x is called the variable and will be the position on the beam for the purposes of this discussion. The exponent n is the power to which x is raised, and in our work, n will be an integer, either 0, 1, 2, 3, or 4. This greatly simplifies the process. The last term C is called the constant of integration. Its purpose is to insure that the expression resulting from the integration process fits the particular conditions of the beam. That is, the general term $ax^{n+1}/(n+1)$ gives the general form of the expression, and C ties it down to the particular beam being analyzed.

Let's look at a few examples.

Example 1

Evaluate $\int 8x^3 \, dx$.
Using Equation (12-3), we can identify $a = 8$ and $n = 3$. Then $(n + 1) = 4$. The result is

$$\int 8x^3 \, dx = \frac{8x^4}{4} + C = 2x^4 + C$$

The evaluation of the constant C will be shown in actual problems.

Example 2

Evaluate $\int 4x \, dx$.
Here $a = 4$, and it must be observed that $n = 1$. Remember that

The manner in which the four diagrams—shear, bending moment, θEI, and yEI—are related is listed below. The subscript AB indicates that the expressions hold for the particular segment of the beam from point A to point B. In general, the expressions are rewritten for each segment of the beam's length between points where abrupt changes in shear and moment occur.

Given the expression V_{AB} relating the shearing force between the points A and B to the position on the beam x,

$$M_{AB} = \int V_{AB}\,dx + C \tag{12-4}$$

$$\theta_{AB}EI = \int M_{AB}\,dx + C \tag{12-5}$$

$$y_{AB}EI = \int \theta_{AB}EI\,dx + C \tag{12-6}$$

x^1 is written as just x. Then $n + 1 = 2$. And we have

$$\int 4x\,dx = \frac{4x^2}{2} + C = 2x^2 + C$$

Example 3

Evaluate $\int 12\,dx$.
Here $a = 12$, but there is no term involving x. However, we can observe that $x^0 = 1$. Then $n = 0$, and $n + 1 = 1$. The result of the integration would then be

$$\int 12x^0\,dx = \frac{12x^1}{1} + C = 12x + C$$

Finally, several terms involving different powers of x can be handled in the same equation by repeatedly applying Equation (12-3) for each term.

Example 4

Evaluate $\int (8x^3 + 4x + 12)\,dx$.
This is the same as

$$\int 8x^3\,dx + \int 4x\,dx + \int 12\,dx$$

Therefore,

$$\int (8x^3 + 4x + 12)\,dx = \frac{8x^4}{4} + \frac{4x^2}{2} + 12x + C$$

$$= 2x^4 + 2x^2 + 12x + C$$

Only one constant of integration is required.

The constants of integration will be different for each equation and will be evaluated by identifying special *boundary conditions* for the moment, slope, and deflection curves.

The units for terms in Equations (12-4) to (12-6) are important. Combinations of force units and units of length are encountered. The following table summarizes the units for three different systems.

Term	Example Units		
Force	N	lb	lb
Length	m	in.	ft
Shear V	N	lb	lb
Bending moment M	N·m	lb·in.	lb·ft
θEI	N·m^2	lb·in.^{2}	lb·ft^2
yEI	N·m^3	lb·in.^{3}	lb·ft^3
θ	rad	rad	rad
y	m or mm	in.	in.

The step-by-step method used to find the deflection of beams using the general approach is as follows.

1. Determine the reactions at the supports for the beam.
2. Draw the shear and bending moment diagrams using the same procedures presented in Chapter 7, and identify the magnitudes at critical points.
3. Divide the beam into segments in which the shear diagram is continuous by naming points at the places where abrupt changes occur with the letters A, B, C, D, etc.
4. Write equations for the shear curve in each segment. In most cases, these will be equations of straight lines, that is, equations involving x to the first power. Sometimes, as for beams carrying concentrated loads, the equation will be simply of the form

$$V = \text{constant}$$

5. For each segment, perform the process called for in Equation (12-4).

$$M = \int V \, dx + C$$

To evaluate the constant of integration which ties the moment equation to the particular values already known for the moment diagram, insert known boundary conditions and solve for C.

6. For each segment, perform the process called for in Equation (12-5).

$$\theta EI = \int M \, dx + C$$

The constant of integration generated here cannot be evaluated directly right away. So each constant should be identified separately by a subscript such as C_1, C_2, C_3, etc. Then when they are evaluated, they can be put in their proper places.

7. For each segment, perform the process called for in Equation (12-6).

$$yEI = \int \theta EI \, dx + C$$

Here again, the constants should be labeled with subscripts.

8. Establish *boundary conditions* for the slope and deflection diagrams. As many boundary conditions must be identified as there are unknown constants from steps 6 and 7. Boundary conditions express mathematically the special values of slope and deflection at certain points and the fact that both the slope curve and the deflection curve are continuous. Typical boundary conditions are:
   a. The deflection of the beam at each support is zero.
   b. The deflection of the beam at the end of one segment is equal to the deflection of the beam at the beginning of the next segment. This follows from the fact that the deflection curve is continuous; that is, it has no abrupt changes.
   c. The slope of the beam at the end of one segment is equal to the slope at the beginning of the next segment. The slope has no abrupt changes.
   d. For the special case of a cantilever, the slope of the beam at the support is also zero.

9. Combine all the boundary conditions to evaluate all the constants of integration.

10. Substitute the constants of integration back into the slope and deflection equations, thus completing them. The value of the slope or deflection at any point can then be evaluated by simply placing the proper value of the position on the beam in the equation.

The method will now be illustrated with an example problem. Admittedly, this example is for a case where formulas are available. However, by starting with simple beams, the method is more readily understood. Other, more complex loadings will be demonstrated later.

Example Problem 12-5

Determine the equations for the shape of the deflected beam shown in Figure 12-9. If the beam is a rectangular aluminum bar having a thickness of 10 mm and a height of 30 mm, compute the deflection at the load and the maximum deflection. Also compute the value of the slope of the deflected beam at each support.

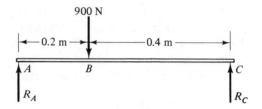

Figure 12-9

Solution

The solution will be shown using the step-by-step method outlined above. *Steps 1 and 2* are familiar from previous work, with the results shown in Figure 12-10.

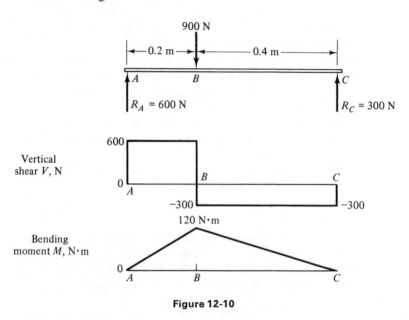

Figure 12-10

Step 3

An abrupt change in shear occurs at the left and right ends of the beam and at the load. We will label these points A, B, and C, beginning at the left end and proceeding toward the right. Now the shear curve is continuous in the segments AB and BC.

Step 4

The shear is constant in each segment. Reading right off the shear diagram, we can see that

$$V_{AB} = 600$$

$$V_{BC} = -300$$

Units are left off for these operations, but it is understood that the shear is in newtons, as indicated on the shear diagram.

Step 5

Two separate integration processes are required, one for each segment. In segment AB,

$$M_{AB} = \int V_{AB}\, dx + C$$

$$M_{AB} = \int 600\, dx + C = 600x + C$$

This is the equation of the straight line in the moment diagram from A to B. Notice that the line starts at A with a value of zero. Since $x = 0$ at A, we can say,

$$\text{At } x = 0,\ M_{AB} = 0$$

Knowing the value of M at a particular value of x allows the determination of the constant C in the moment equation. By substitution,

$$M_{AB} = 0 = 600(0) + C$$

Therefore,

$$C = 0$$

The final equation for M_{AB} is then

$$\boxed{M_{AB} = 600x}$$

We can now go on to segment BC.

$$M_{BC} = \int V_{BC}\, dx + C$$

But $V_{BC} = -300$. So,

$$M_{BC} = \int -300\, dx + C = -300x + C$$

By inspection of the moment diagram at point C, the moment is zero when $x = 0.6$ m. Remember, x is measured from the *left* end of the beam. By substitution,

$$M_{BC} = 0 = -300(0.6) + C$$

and

$$C = 180$$

The final form for M_{BC} is

$$M_{BC} = -300x + 180$$

A check can be made on these equations by noting that both should yield the result that $M = 120$ when $x = 0.2$ m. Substitute $x = 0.2$ in both the M_{AB} and M_{BC} equations. Check!

Step 6

The two equations for θEI are

$$\theta_{AB}EI = \int M_{AB}\, dx + C$$

$$= \int 600x\, dx + C = \frac{600x^2}{2} + C_1$$

$$\theta_{AB}EI = 300x^2 + C_1$$

The constant, C_1, will be evaluated later. In segment BC,

$$\theta_{BC}EI = \int M_{BC}\, dx + C$$

$$= \int (-300x + 180)\, dx + C$$

$$= \frac{-300x^2}{2} + 180x + C_2$$

$$\theta_{BC}EI = -150x^2 + 180x + C_2$$

Step 7

In segment AB,

$$y_{AB}EI = \int \theta_{AB}EI\, dx + C$$

$$= \int (300x^2 + C_1)\, dx + C$$

$$= \frac{300x^3}{3} + C_1x + C_3$$

$$y_{AB}EI = 100x^3 + C_1x + C_3$$

Notice here that, even though the constant C_1 has not yet been evaluated, it is carried along just as if it were a number. Also, the new constant of integration is called C_3, following numerical order. Now in segment BC,

$$y_{BC}EI = \int \theta_{BC}EI \, dx + C$$

$$= \int (-150x^2 + 180x + C_2) \, dx + C$$

$$= \frac{-150x^3}{3} + \frac{180x^2}{2} + C_2x + C_4$$

$$y_{BC}EI = -50x^3 + 90x^2 + C_2x + C_4$$

Step 8

Since four constants of integration are yet to be found for the θ and y equations, four boundary conditions must be identified. They are:

1. At $x = 0$, $y_{AB} = 0$ (deflection $= 0$ at support at A).
2. At $x = 0.6$, $y_{BC} = 0$ (deflection $= 0$ at support at C).
3. At $x = 0.2$, $y_{AB} = y_{BC}$ (deflection is continuous at B).
4. At $x = 0.2$, $\theta_{AB} = \theta_{BC}$ (slope is continuous at B).

All four state known relationships between x and y or between x and θ.

Step 9

In order to evaluate the four constants of integration, four independent equations involving them must be used. The four boundary conditions above allow the generation of these equations by substitution into the proper expression for y or θ.

Condition 1: At $x = 0$, $y_{AB} = 0$.

$$y_{AB}EI = 0 = 100(0) + C_1(0) + C_3$$

Therefore,

$$C_3 = 0$$

Condition 2: At $x = 0.6$, $y_{BC} = 0$.

$$y_{BC}EI = 0 = -50(0.6)^3 + 90(0.6)^2 + C_2(0.6) + C_4$$

This reduces to

$$0.6C_2 + C_4 + 21.6 = 0 \qquad\qquad (a)$$

Since this cannot be solved directly for either constant, we'll go on to condition 3.

Condition 3: At $x = 0.2$, $y_{AB} = y_{BC}$.
Of course, an equivalent expression would be

$$y_{AB}EI = y_{BC}EI$$

Then

$$100x^3 + C_1x + C_3 = -50x^3 + 90x^2 + C_2x + C_4$$

Observing that $C_3 = 0$ and letting $x = 0.2$,

$$100(0.2)^3 + C_1(0.2) = -50(0.2)^3 + 90(0.2)^2 + C_2(0.2) + C_4$$

This reduces to

$$0.2C_1 - 0.2C_2 - C_4 - 2.4 = 0 \qquad\qquad (b)$$

Condition 4: At $x = 0.2$,

$$\theta_{AB} = \theta_{BC}$$

or

$$\theta_{AB}EI = \theta_{BC}EI$$

Then

$$300x^2 + C_1 = -150x^2 + 180x + C_2$$

At $x = 0.2$,

$$300(0.2)^2 + C_1 = -150(0.2)^2 + 180(0.2) + C_2$$
$$C_1 - C_2 - 18 = 0 \qquad\qquad (c)$$

The three equations (a), (b), and (c) can be solved simultaneously for the three constants remaining, C_1, C_2, and C_4. Summarizing,

$$0.6C_2 + C_4 + 21.6 = 0 \qquad\qquad (a)$$
$$0.2C_1 - 0.2C_2 - C_4 - 2.4 = 0 \qquad\qquad (b)$$
$$C_1 - C_2 \qquad\quad - 18 = 0 \qquad\qquad (c)$$

Many ways are available for solving simultaneous equations. The one shown here seeks to combine two equations together in such a way that one or more of the unknown constants drops out. For example, multiply Equation (c) by 0.2 and subtract the result from Equation (b).

$$0.2C_1 - 0.2C_2 - C_4 - 2.4 = 0$$
$$\underline{0.2C_1 - 0.2C_2 \qquad\quad - 3.6 = 0} \quad \text{(subtract)}$$
$$-C_4 + 1.2 = 0$$

Then

$$C_4 = 1.2$$

This value can be put into Equation (a) to solve for C_2.

$$0.6C_2 + 1.2 + 21.6 = 0$$

$$C_2 = -38$$

Finally, C_2 can be put into Equation (c) to find C_1.

$$C_1 - (-38) - 18 = 0$$

$$C_1 = -20$$

We can now go to step 10.

Step 10

The final equations for θ and y can now be written by substituting the values of the constants into the equations derived earlier.

$$\theta_{AB}EI = 300x^2 - 20 \tag{d}$$
$$\theta_{BC}EI = -150x^2 + 180x - 38 \tag{e}$$
$$y_{AB}EI = 100x^3 - 20x \tag{f}$$
$$y_{BC}EI = -50x^3 + 90x^2 - 38x + 1.2 \tag{g}$$

The desired results are:

1. The deflection at the load (point B).
2. The slope at each support (points A and C).
3. The maximum deflection.

Since point B is at $x = 0.2$, we can put the value into Equation (f) and solve for yEI.

$$y_B EI = 100\,(0.2)^3 - 20\,(0.2) = -3.2 \text{ N} \cdot \text{m}^3$$

The units are needed here to ensure consistency. To get the actual deflection y, the above result must be divided by EI. That is,

$$y_B = \frac{-3.2 \text{ N} \cdot \text{m}^3}{EI}$$

For the aluminum bar, $E = 69$ GPa $= 69 \times 10^9$ N/m². The moment of inertia is

$$I = \frac{BH^3}{12} = \frac{10(30)^3 \text{ mm}^4}{12} = 22\ 500 \text{ mm}^4$$

Then at the load, point B,

$$y_B = \frac{-3.2 \text{ N} \cdot \text{m}^3}{(69 \times 10^9 \text{ N/m}^2)(22\ 500 \text{ mm}^4)} \times \frac{(10^3 \text{ mm})^5}{\text{m}^5} = -2.06 \text{ mm}$$

The slopes at A and C can be found directly from the slope equations (d) and (e). At $x = 0$, point A,

$$\theta_A EI = 300(0) - 20 = -20 \text{ N} \cdot \text{m}^2$$

Solving for θ_A,

$$\theta_A = \frac{-20 \text{ N} \cdot \text{m}^2}{EI} = \frac{-20 \text{ N} \cdot \text{m}^2}{(69 \times 10^9 \text{ N/m}^2)(22\ 500 \text{ mm}^4)} \times \frac{(10^3 \text{ mm})^4}{\text{m}^4}$$

$$\theta_A = -0.0129 \text{ rad}$$

Notice that the units cancel out completely, but the result for θ must be considered to be *radians*. Now at point C, where $x = 0.6$ m,

$$\theta_C EI = -150(0.6)^2 + 180(0.6) - 38 = 16 \text{ N} \cdot \text{m}^2$$

Notice that Equation (e) must be used since Equation (d) is not valid at point C.

$$\theta_C = \frac{16 \text{ N} \cdot \text{m}^2}{EI} = 0.0103 \text{ rad}$$

Now we will find the maximum deflection. It will help to sketch the probable slope of the deflected beam, as in Figure 12-11. At the point of maximum deflection, the slope of the curve will be zero. Therefore, we can look to the θ equations to find where $\theta = 0$. Trying Equation (d) first,

$$\theta_{AB} EI = 0 = 300x^2 - 20$$

$$x = \pm 0.258 \text{ m}$$

It must be observed that this result is *not valid*, since neither $x = +0.258$ m nor $x = -0.258$ m falls in the segment AB. The equation is only valid between $x = 0$ and $x = 0.2$ m. We can conclude then that the maximum deflection does not occur in segment AB. Let's go on to segment BC, Equation (e).

$$\theta_{BC} EI = 0 = -150x^2 + 180x - 38$$

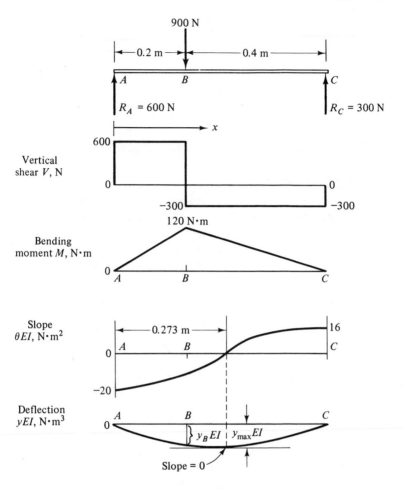

Figure 12-11

Solving by the quadratic formula,

$$x = \frac{-180 \pm \sqrt{180^2 - 4(-150)(-38)}}{2(-150)}$$

$$x = 0.273 \text{ m} \quad \text{or} \quad 0.927 \text{ m}$$

Since $x = 0.927$ m is not on the beam, the correct answer is

$$x = 0.273 \text{ m}$$

This is in the segment *BC* and is a valid result. The maximum deflection then occurs at $x = 0.273$ m. In Equation (g),

$$y_{max}EI = -50(0.273)^3 + 90(0.273)^2 - 38(0.273) + 1.2$$
$$= -3.48 \text{ N·m}^3$$

Finally,

$$y_{max} = \frac{-3.48 \text{ N·m}^3}{EI} = -2.24 \text{ mm}$$

The significant results are illustrated in Figure 12-11.

Example Problem 12-6

Figure 12-12 shows a beam used as a part of a special structure for a machine. The 20 000-lb load at A and the 30 000-lb load at C represent places where heavy equipment is supported. Between the two supports at B and D, the uniformly distributed load of 2000 lb/ft is due to stored bulk materials which are in a bin supported by the beam.

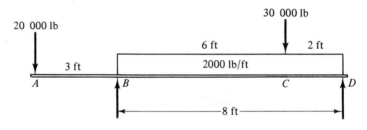

Figure 12-12

In order to maintain accuracy of the product produced by the machine, the maximum allowable deflection of the beam is 0.05 in. Specify an acceptable wide flange steel beam, and also check the stress in the beam.

Solution

The beam will be analyzed to determine where the maximum deflection will occur. Then the required moment of inertia will be determined to limit the deflection to 0.05 in. A wide flange beam which has the required moment of inertia will then be selected. The ten-step procedure discussed earlier will be used. The solution is shown in a programmed format. You should work through the problem yourself before looking at the given solution.

In order to reduce the size of numbers to a manageable number of digits, the loads should be restated in the unit of *kilopounds*, some-

times called kips. One kip equals 1000 lb. Then the 20 000-lb load
becomes 20 kips, the 30 000-lb load is 30 kips, and the distributed load
is 2 kips/ft.

Steps 1 and 2 call for drawing the shear and bending moment
diagrams. Do this now before checking the result below.

Figure 12-13 shows the results. Now do *step 3*.

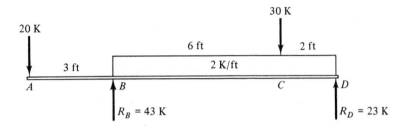

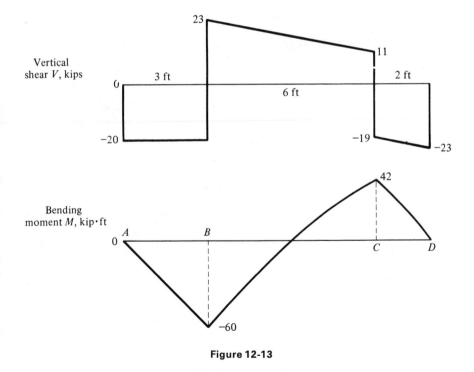

Figure 12-13

Three segments are required, *AB*, *BC*, and *CD*. These are the segments over which the shear diagram is continuous. Now do *step 4* to get the shear curve equations.

The results are:

$$V_{AB} = -20 \tag{a}$$
$$V_{BC} = -2x + 29 \tag{b}$$
$$V_{CD} = -2x - 1 \tag{c}$$

In the segments *BC* and *CD*, the shear curve is a straight line with a slope of -2 kips/ft, the same as the load. Any method of writing the equation of a straight line can be used to derive these equations.

Now do *step 5* of the procedure.

You should have the following for the moment equations. First

$$M_{AB} = \int V_{AB}\, dx + C = \int -20\, dx + C$$
$$= -20x + C$$

At $x = 0$, $M_{AB} = 0$. Therefore, $C = 0$, and,

$$M_{AB} = -20x \tag{d}$$

Next,

$$M_{BC} = \int V_{BC}\, dx + C = \int (-2x + 29)\, dx + C = -x^2 + 29x + C$$

At $x = 3$, $M_{BC} = -60$. Therefore, $C = -138$, and

$$M_{BC} = -x^2 + 29x - 138 \tag{e}$$

Finally,

$$M_{CD} = \int V_{CD}\, dx + C = \int (-2x - 1)\, dx + C$$
$$= -x^2 - x + C$$

At $x = 9$, $M_{CD} = 42$. Therefore, $C = 132$, and

$$M_{CD} = -x^2 - x + 132 \tag{f}$$

Now do *step 6* to get equations for θEI.

By integrating the moment equations,

$$\theta_{AB}EI = \int M_{AB}\,dx + C = \int -20x\,dx + C$$

$$\theta_{AB}EI = -10x^2 + C_1 \tag{g}$$

$$\theta_{BC}EI = \int M_{BC}\,dx + C = \int (-x^2 + 29x - 138)\,dx + C$$

$$\theta_{BC}EI = -0.333x^3 + 14.5x^2 - 138x + C_2 \tag{h}$$

$$\theta_{CD}EI = \int M_{CD}\,dx + C = \int (-x^2 - x + 132)\,dx + C$$

$$\theta_{CD}EI = -0.333x^3 - 0.5x^2 + 132x + C_3 \tag{i}$$

Now in *step 7*, integrate Equations (g), (h), and (i) to get the *yEI* equations.

You should have

$$y_{AB}EI = \int \theta_{AB}EI\,dx + C$$

$$y_{AB}EI = -3.33x^3 + C_1 x + C_4 \tag{j}$$

$$y_{BC}EI = \int \theta_{BC}EI\,dx + C$$

$$y_{BC}EI = -0.0833x^4 + 4.83x^3 - 69x^2 + C_2 x + C_5 \tag{k}$$

$$y_{CD}EI = \int \theta_{CD}EI\,dx + C$$

$$y_{CD}EI = -0.0833x^4 - 0.167x^3 + 66x^2 + C_3 x + C_6 \tag{l}$$

Step 8 calls for identifying boundary conditions. Six are required since there are six unkown constants of integration in Equations (g) through (l). Write them now.

Considering zero deflection points and the continuity of the slope and deflection curves, we can say

1. At $x = 3$, $y_{AB}EI = 0$ ⎫
2. At $x = 3$, $y_{BC}EI = 0$ ⎬(zero deflection at supports).
3. At $x = 11$, $y_{CD}EI = 0$ ⎭

4. At $x = 9$, $y_{BC}EI = y_{CD}EI$ (continuous deflection curve at C).
5. At $x = 3$, $\theta_{AB}EI = \theta_{BC}EI$ (continuous slope curve at B and
6. At $x = 9$, $\theta_{BC}EI = \theta_{CD}EI$ C).

We can now substitute the values of x above into the proper equations and solve for C_1 through C_6. First make the substitutions and reduce the resulting equations to the form involving the constants.

For the six conditions listed above, the following equations result.

1. $3C_1 + C_4$ $= 90$
2. $3C_2 + C_5$ $= 497.25$
3. $11C_3 + C_6$ $= -6544$
4. $9C_2 - 9C_3 + C_5 - C_6 = 7290$
5. $C_1 - C_2$ $= -202.5$
6. $C_2 - C_3$ $= 1215$

Now solve the six equations simultaneously for the values of C_1 through C_6.

The results are:

$$C_1 = 132.5, \qquad C_4 = -307.5$$
$$C_2 = 335, \qquad C_5 = -507$$
$$C_3 = -880, \qquad C_6 = 3138$$

The final equations for θ and y can now be written by substituting the constants into Equations (g) through (l).

$$\theta_{AB}EI = -10x^2 + 132.5$$

$$\theta_{BC}EI = -0.333x^3 + 14.5x^2 - 138x + 335$$

$$\theta_{CD}EI = -0.333x^3 - 0.5x^2 + 132x - 880$$

$$y_{AB}EI = -3.33x^3 + 132.5x - 307.5$$

$$y_{BC}EI = -0.0833x^4 + 4.83x^3 - 69x^2 + 335x - 507$$

$$y_{CD}EI = -0.0833x^4 - 0.167x^3 + 66x^2 - 880x + 3138$$

Having the completed equations, we can now determine the point of maximum deflection, which is the primary objective of the analysis. Based on the loading, the probable shape of the deflected beam would be like that shown in Figure 12-14. Therefore, the maximum deflection could occur at point A at the end of the overhang, at a point to the right of B (upward), or at a point near the load at C (downward).

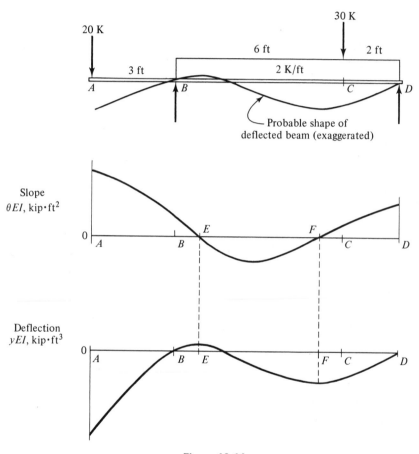

Figure 12-14

It is probable that there are two points of zero slope at the points E and F, as shown in Figure 12-14. We would need to know where the slope equation $\theta_{BC}EI$ equals zero to determine where the maximum deflections occur. Notice that this equation is a third degree equation and that it would be difficult to solve for the value of x where θEI

equals zero. A graphical approach is easier. We can substitute a few values of x into the equation at points near where θEI is likely to be zero. Then plotting the resulting values on a graph would reveal where θEI equals zero. Figure 12-15 shows such a graph, which locates point E at $x = 3.84$ ft and point F at $x = 8.36$ ft.

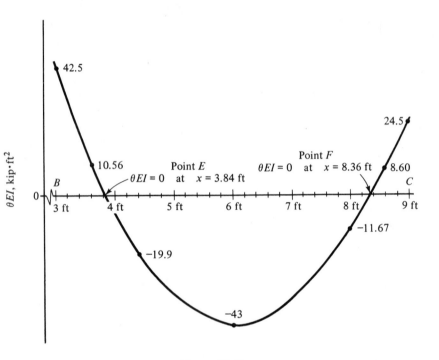

Figure 12-15

We can now determine the values for yEI at point A, E, and F to find out which is larger.

Point A: At $x = 0$,

$$y_A EI = -307.5 \text{ kip} \cdot \text{ft}^3$$

Point E: At $x = 3.84$ ft in segment BC,

$$y_E EI = -0.0833(3.84)^4 + 4.83(3.84)^3 - 69(3.84)^2$$
$$+ 335(3.84) - 507$$

$$y_E EI = +17.5 \text{ kip} \cdot \text{ft}^3$$

Point F: At $x = 8.36$ ft in segment BC,

$$y_F EI = -0.0833(8.36)^4 + 4.83(8.36)^3 - 69(8.36)^2$$
$$+ 335(8.36) - 507$$

$$y_F EI = -111.8 \text{ kip} \cdot \text{ft}^3$$

The largest value occurs at point A; so that is the critical point. We must choose a beam which limits the deflection at A to 0.05 in. or less.

$$y_A EI = -307 \text{ kip} \cdot \text{ft}^3$$

Let $y_A = -0.05$ in. Then the required I is

$$I = \frac{-307 \text{ kip} \cdot \text{ft}^3}{E y_A} \times \frac{1000 \text{ lb}}{\text{kip}} \times \frac{(12 \text{ in.})^3}{\text{ft}^3}$$

$$I = \frac{(-307)(1000)(1728) \text{ lb} \cdot \text{in.}^3}{(30 \times 10^6 \text{ lb/in.}^2)(-0.05 \text{ in.})} = 354 \text{ in.}^4.$$

Consult the table of wide flange beams and select a suitable beam.

A W16×31 is the best choice since it is the lightest beam which has a large enough value for I. For this beam, $I = 374$ in.^{4}, and the section modulus is $Z = 47.2$ in.^{3}. Now compute the maximum bending stess in the beam.

In Figure 12-13 we find the maximum bending moment to be 60 kip·ft. Then

$$s = \frac{M}{Z} = \frac{60 \text{ kip} \cdot \text{ft}}{47.2 \text{ in.}^3} \times \frac{1000 \text{ lb}}{\text{kip}} \times \frac{12 \text{ in.}}{\text{ft}} = 15\ 250 \text{ psi}$$

Since the allowable stress for structural steel is 22 000 psi, the selected beam is safe.

The next example problem deals with a cantilever beam.

Example Problem 12-7

A cover over a loading dock extends out from a building a distance of 2.8 m as shown in Figure 12-16. At one place, a cantilever beam carries the roof load plus an electric hoist rated at 20 kN. The hoist can travel out a total of 4.2 m. The expected load on the beam due to the cover is 4 kN/m. The plan calls for a standard wide flange shape for the beam with a limit of 150 MPa on stress due to bending. Select

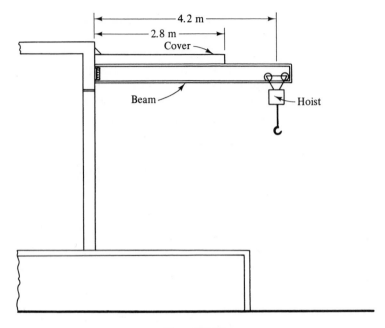

Figure 12-16

a beam based on strength, and then determine the deflection at the right end and at the end of the cover.

Solution

Since the data is given in SI units, all of the analysis will be done in those units. However, since the geometrical properties of W-beams are listed in conventional English units, some conversions will have to be made.

Let's follow the ten-step procedure for finding beam deflections as used in the preceding example problems. The results are shown in steps. Work the steps yourself before checking the solution.

Start with the reactions and the shear and bending moment diagrams.

Figure 12-17 shows the results. As with all cantilevers, the maximum bending moment occurs at the support. The beam can now be selected since selection is to be based on strength. The deflection analysis will follow.

The required section modulus for the beam can first be computed in mm³ units to limit the stress to 150 MPa. Do that now.

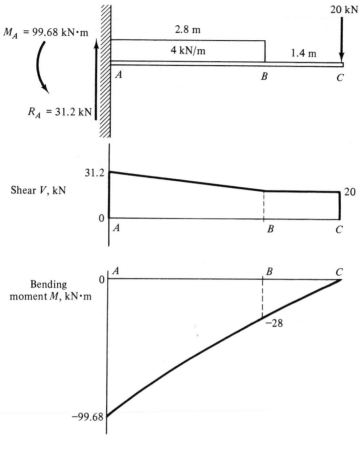

Figure 12-17

You should have $Z = 0.665 \times 10^6$ mm³.

$$Z = \frac{M}{s_d} = \frac{99.68 \times 10^3 \text{ N} \cdot \text{m}}{150 \text{ N/mm}^2} \times \frac{10^3 \text{ mm}}{1 \text{ m}} = 0.665 \times 10^6 \text{ mm}^3$$

Now using 1 in. = 25.4 mm,

$$Z = 0.665 \times 10^6 \text{ mm}^3 \times \frac{1 \text{ in.}^3}{(25.4 \text{ mm})^3} = 40.5 \text{ in.}^3$$

Now consult the table of properties for W-beams and select the lightest one having this required value of Z.

The proper selection is W14×30 with $Z = 41.9$ in.^{3} and $I = 290$ in.^{4}. Now we can continue our analysis for deflection. The equations for the deflected shape of the beam must be derived. Next we need the equations for the shear curve for the segments AB and BC.

You should have

$$V_{AB} = -4x + 31.2$$
$$V_{BC} = 20$$

Now derive the moment equations.

By integration, for segment AB,

$$M_{AB} = \int (-4x + 31.2)\, dx + C$$
$$= -2x^2 + 31.2x + C$$

At $x = 0$, $M_{AB} = -99.68$. Therefore, $C = -99.68$.

$$M_{AB} = -2x^2 + 31.2x - 99.68$$

For segment BC,

$$M_{BC} = \int 20\, dx + C$$
$$= 20\,x + C$$

At $x = 4.2$ m, $M_{BC} = 0$.

$$0 = 20(4.2) + C$$
$$C = -84$$

Finally,

$$M_{BC} = 20x - 84$$

Using the moment equations, derive the θEI and yEI equations.

For both segments,

$$\theta_{AB}EI = -0.667x^3 + 15.6x^2 - 99.68x + C_1$$
$$\theta_{BC}EI = 10x^2 - 84x + C_2$$
$$y_{AB}EI = -0.1667x^4 + 5.2x^3 - 49.87x^2 + C_1x + C_3$$
$$y_{BC}EI = 3.33x^3 - 42x^2 + C_2x + C_4$$

With four unknown constants, four boundary conditions are required. Remember the special condition for a cantilever beam: The slope at the support is zero.

Here's the list.

1. At $x = 0$, $\theta_{AB}EI = 0$ } (zero slope and deflection at A).
2. At $x = 0$, $y_{AB}EI = 0$
3. At $x = 2.8$, $y_{AB}EI = y_{BC}EI$ } (continuity at B).
4. At $x = 2.8$, $\theta_{AB}EI = \theta_{BC}EI$

After substituting these conditions into the proper equations, you should solve for C_1 through C_4.

The constants are

$$C_1 = 0, \qquad C_3 = 0$$
$$C_2 = -14.64, \qquad C_4 = 10.56$$

Putting these into the proper equations gives

$$\theta_{AB}EI = -0.667x^2 + 15.6x^2 - 99.68x$$
$$\theta_{BC}EI = 10x^2 - 84x - 14.64$$
$$y_{AB}EI = -0.1667x^4 + 5.2x^3 - 49.84x^2$$
$$y_{BC}EI = 3.33x^3 - 42x^2 - 14.64x + 10.56$$

The deflections at points B and C are desired. Use $E = 207$ GPa for the steel beam. Let's convert the moment of inertia to the units of mm⁴.

$$I = 290 \text{ in.}^4 \times \frac{(25.4 \text{ mm})^4}{\text{in.}^4} = 120.7 \times 10^6 \text{ mm}^4$$

Since points B and C are both in segment BC, use the equation for $y_{BC}EI$ and evaluate it at $x = 2.8$ m and $x = 4.2$ m.

$$y_B EI = -286.5 \text{ kN} \cdot \text{m}^3$$
$$y_C EI = -544.8 \text{ kN} \cdot \text{m}^3$$

Now solve for y_B and y_C.

The deflections are

$$y_B = -11.5 \text{ mm}$$
$$y_C = -21.8 \text{ mm}$$

PROBLEMS

12-1 A round shaft having a diameter of 32 mm is 700 mm long and carries a 3.0-kN load at its center. If the shaft is steel and simply supported at its ends, compute the deflection at the center.

12-2 For the shaft in Problem 12-1, compute the deflection if the shaft is 6061-T6 aluminum instead of steel.

12-3 For the shaft in Problem 12-1, compute the deflection if the ends are fixed against rotation instead of simply supported.

12-4 For the shaft in Problem 12-1, compute the deflection if the shaft is 350 mm long rather than 700 mm.

12-5 For the shaft in Problem 12-1, compute the deflection if the diameter is 25 mm instead of 32 mm.

12-6 For the shaft in Problem 12-1, compute the deflection if the load is placed 175 mm from the left support rather than in the center. Compute the deflection both at the load and at the center of the shaft.

12-7 A wide flange steel beam, W10×21, carries the load shown in Figure 7-35(d). Compute the deflection at the loads and at the center of the beam.

12-8 A standard steel construction pipe, TS1½×0.145, carries a 650-lb load at the center of a 28-in. span, simply supported. Compute the deflection of the pipe at the load.

12-9 An Aluminum Association Standard I-beam, 8I×6.181, carries a uniformly distributed load of 1125 lb/ft over a 10-ft span. Compute the deflection at the center of the span.

12-10 For the beam in Problem 12-9, compute the deflection at a point 3.5 ft from the left support of the beam.

12-11 A wide flange steel beam, W12×36, carries the load shown in Figure 7-37(d). Compute the deflection at the load.

12-12 For the beam in Problem 12-11, compute the deflection at a point 4.0 ft from the right reaction.

12-13 For the beam in Problem 12-11, compute the maximum upward deflection and determine its location.

12-14 The load shown in Figure 7-36(b) is being carried by an extruded aluminum (6061-T6) beam having a moment of inertia of 0.186×10^6 mm⁴ for its cross section. Compute the deflection of the beam at each load. The beam shape is shown in Figure 8-20(b).

12-15 The loads shown in Figure 7-36(a) represent the feet of a motor on a machine frame. The frame member has the cross section shown in Figure 8-20(c), which has a moment of inertia of 16 956 mm^4. Compute the deflection at each load. The aluminum alloy 2014-T5 is used for the frame.

12-16   The built-up section shown in Figure 8-17(d) is made of 6061-T6 aluminum channels and plates and has a moment of inertia of 710.4 in.^{4}. Compute the maximum deflection of the beam when it carries the load shown in Figure 7-36(c).

12-17   A 1-in. Schedule 40 steel pipe is used as a cantilever beam 8 in. long to support a load of 120 lb at its end. Compute the deflection of the pipe at the end.

12-18   A 1-in. Schedule 40 steel pipe carries the two loads shown in Figure 7-39(b). Compute the deflection of the pipe at each load.

12-19   A cantilever beam carries two loads as shown in Figure 7-40(a). If a rectangular steel bar, 20 mm wide by 80 mm high, is used for the beam, compute the deflection at the end of the beam.

12-20   For the beam in Problem 12-19, compute the deflection if the bar is aluminum 2014-T4 rather than steel.

12-21   For the beam in Problem 12-19, compute the deflection if the bar is magnesium, ASTM AZ 63A-T6, instead of steel.

12-22   The load shown in Figure 7-48(a) is carried by a round steel bar having a diameter of 0.800 in. Compute the deflection of the bar at the right end.

12-23   A round steel bar is to be used to carry a single concentrated load of 3.0 kN at the center of a 700-mm long span on simple supports. Determine the required diameter of the bar if its deflection must not exceed 0.12 mm.

12-24   For the bar designed in Problem 12-23, compute the stress in the bar and specify a suitable steel to provide a design factor of 8 based on ultimate strength.

12-25   Specify a standard wide flange steel beam which would carry the loading shown in Figure 7-36(c) with a maximum deflection less than $\frac{1}{360}$ times the length of the beam.

12-26   It is planned to use an Aluminum Association channel with its legs down to carry the loads shown in Figure 7-45(a) so that the flat face can contact the load. The maximum allowable deflection is 0.080 in. Specify a suitable channel.

12-27   A flat strip of steel, 0.100 in. wide and 1.200 in. long, is clamped at one end and loaded at the other like a cantilever beam (as in Case 11 in. Table A-19). What should be the thickness of the strip if it is to deflect 0.15 in. under a load of 0.52 lb?

12-28   A wood joist in a commercial building is 14 ft 6 in. long and carries a uniformly distributed load of 50 lb/ft. It is 1.50 in. wide and 9.25 in. high. If it is made of southern pine, compute the maximum deflection of the joist. Also compute the stress in the joist due to bending and horizontal shear, and compare them with the allowable stresses for No. 2 stress rated southern pine.

For the following problems, use the general approach outlined in Section 12-4 to determine the equations for the shape of the deflection curves of the beams. Unless otherwise noted, compute the maximum deflection of the beam from the equation and tell where it occurs.

12-29 The loading is shown in Figure 12-18(a). The beam is a rectangular steel bar, 1.0 in. wide and 2.0 in. high.

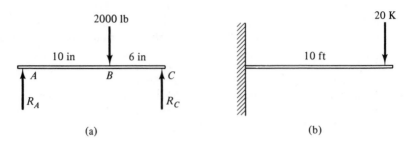

(a)   (b)

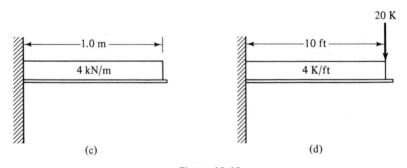

(c)   (d)

Figure 12-18

12-30 The loading is shown in Figure 12-18(b). The beam is steel, wide flange shape W14×78.

12-31 The loading is shown in Figure 12-18(c). The beam is a 2½-in. Schedule 40 steel pipe.

12-32 The loading is shown in Figure 12-18(d). The beam is steel, wide flange shape W24×100.

12-33 The load is shown in Figure 12-19(a). Design a round steel bar which will limit the deflection at the end of the beam to 5.0 mm.

12-34 The load is shown in Figure 12-19(b). Design a steel beam which will limit the maximum deflection to 10.0 mm. Use any shape, including those listed in the Appendix tables.

12-35 The load is shown in Figure 12-19(c). Select an aluminum I-beam which will limit the stress to 120 MPa; then compute the maximum deflection in the beam.

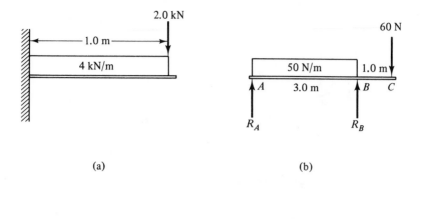

(a)   (b)

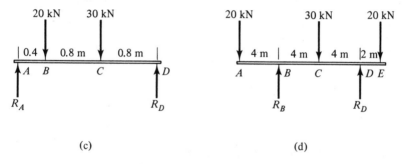

(c)   (d)

Figure 12-19

12-36   A steel, wide flange beam W16×26 carries the loading shown in Figure 12-19(d). Compute the maximum deflection between the supports and at each end.

12-37   The loading shown in Figure 12-20 represents a steel shaft for a large machine. The loads are due to gears mounted on the shaft. Assuming that the entire shaft will be the same diameter, determine the required diameter to limit the deflection at any gear to 0.13 mm.

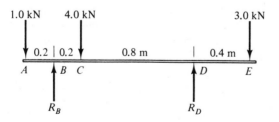

Figure 12-20

13

STATICALLY
INDETERMINATE BEAMS

13-1 EXAMPLES OF STATICALLY INDETERMINATE BEAMS

Beams with more than two simple supports, cantilevers with a second support, or beams with two fixed ends are important examples of statically indeterminate beams. Figure 13-1 shows the conventional method of representing these kinds of beams. The typical shapes for the deflection curves of the beams are also shown. Significant differences should be noted between these and the curves for beams in the preceding chapter.

Part (a) of Figure 13-1 is called a *continuous beam*, the name coming from the fact that the beam is continuous over several supports. It is important to note that the shape of the deflection curve of the beam is also continuous through the supports. This fact will be useful in analyzing such beams. Continuous beams occur frequently in building structures and highway bridges. Many ranch-type homes with basements contain such a beam running the length of the house to carry loads from floor joists and partitions. Expressway bridges carrying local traffic over through lanes are frequently supported at the ends on each side of the roadway and also at the middle in the median strip. Notice that the beams of such bridges are usually one piece or are connected to form a rigid *continuous* beam.

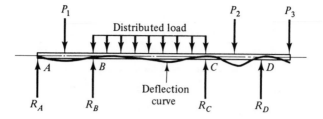

(a) Continuous beam

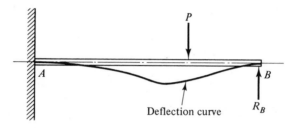

(b) Supported cantilever

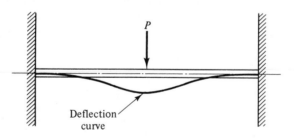

(c) Beam with two fixed ends

Figure 13-1

The *fixed-fixed* beam shown in Figure 13-1(c) is used in building structures and also in machine structures because of the high degree of rigidity
provided. The creation of the fixed-fixed type of end condition requires that
the connections at the ends restrain the beam from rotation as well as provide
support for vertical loads. Figure 13-2 shows one way in which the fixed end
condition can be achieved. Welding of the cross beam to the supporting
columns would achieve the same result. Care must be used in evaluating
fixed-ended beams to ensure that the connections can indeed restrain the

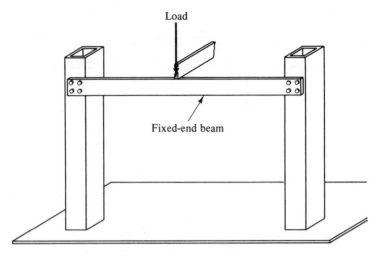

Figure 13-2   Fixed-end Beam

beam from rotating at the support and resist the moments created because of the restraint. Without due care, a condition *between* that of fixed ends and simple supports could result, making analysis difficult and subject to error.

The supported cantilever could be constructed as shown in Figure 13-3. The load on the flat roof is carried by a beam rigidly fixed to a column at one end and resting simply on another column at the other end.

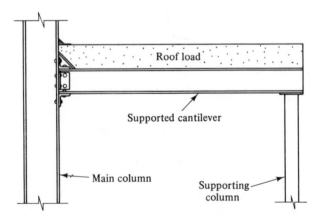

Figure 13-3   Supported Cantilever

Continuous beams and beams with fixed ends have certain advantages over simply supported beams: A greater rigidity is provided and generally lower stresses would be produced in beams of equivalent size; alternatively, smaller and lighter beams can be used to give the same degree of strength and rigidity. The disadvantages are that the analyses are more difficult and the effect of small variations in support conditions can be significant.

The reasons why beams such as those described above are called statically indeterminate are discussed in the next section.

13-2 WHEN IS A BEAM STATICALLY INDETERMINATE?

Beams considered in earlier chapters could be analyzed to determine support reactions by the principles of static equilibrium. By applying the familiar equations,

$$\sum M = 0$$
$$\sum F = 0$$

the unknown reactions could be computed directly. Usually, only two unknown reactions, either forces or moments at supports, needed to be found. Equations were easily developed to determine the unknowns. These beams are called *statically determinate*.

Continuous beams, supported cantilevers, and fixed-ended beams all have at least three unknown reaction forces or moments which *cannot* be found from the conditions of static equilibrium. Thus they are called *statically indeterminate*.

Two mothods available to allow the complete analysis of statically indeterminate beams are the *superposition method* and the *three-moment method*. The superposition method is useful for continuous beams having three supports (two spans) and for the supported cantilever. The process is similar to that used in Chapter 12 to evaluate the deflection of beams carrying a variety of loads. In that process, the effect of each load was considered separately, and then the combined effect of all loads was found by superposition.

The three-moment method can also be applied to a continuous beam on three supports. But its major advantage comes in analyzing beams on more than three supports. In fact, any number of supports can be considered by successive applications of the process.

13-3 SUPERPOSITION METHOD

Consider first the supported cantilever shown in Figure 13-4. Because of the restraint at A and the simple support at B, the unknown reactions include:

1. The vertical force R_B.
2. The vertical force R_A.
3. The restraining moment M_A.

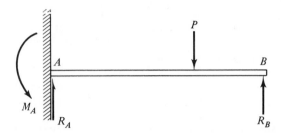

Figure 13-4

The conditions assumed for this beam are that the supports at A and B are absolutely rigid and at the same level, and that the connection at A does not allow any rotation of the beam at that point. Conversely, the support at B will allow rotation and cannot resist moments.

If the support at B is removed, the beam would deflect downward, as shown in Figure 13-5(a), by an amount y_{B1} due to the load P. Now if the

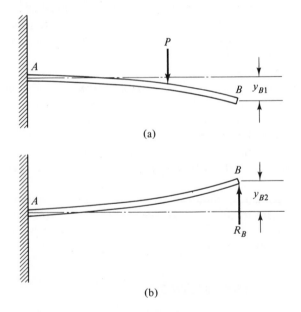

(a)

(b)

Figure 13-5

load is removed and the reaction force R_B is applied upward at B, the beam would deflect upward by an amount y_{B2}, as shown in Figure 13-5(b). In reality, of course, both forces are applied, and the deflection at B is *zero*. The principle of superposition would then provide the conclusion that

$$y_{B1} + y_{B2} = 0 \qquad (13\text{-}1)$$

This equation, along with the normal equations of static equilibrium, will permit the evaluation of all three unknowns, as demonstrated in the example problem which follows. It must be recognized that the principles of static equilibrium are still valid for statically indeterminate beams. However, they are not sufficient to allow a direct solution.

Example Problem 13-1

Determine the support reactions at A and B for the supported cantilever shown in Figure 13-4 if the load P is 2600 N and placed 1.20 m out from A. The total length of the beam is 1.80 m. Then draw the complete shear and bending moment diagrams, and design the beam by specifying a configuration, a material, and the required dimensions of the beam. Use a design factor of 8 based on ultimate strength since the load will be repeated often.

Solution

We must first determine the reaction at B using superposition. Equation (13-1) states that, for this situation,

$$y_{B1} + y_{B2} = 0 \qquad (13\text{-}1)$$

The equation for y_{B1} can be found from the beam deflection formulas in the Appendix. As suggested in Figure 13-5(a), the deflection at the end of a cantilever carrying an intermediate load is required. Then

$$y_{B1} = \frac{-Pl^2}{6EI}(2l + 3b)$$

where $p = 2600$ N
$\quad\quad\ l = 1.20$ m
$\quad\quad\ b = 0.60$ m

The values of E and I are still unknown, but we can express the deflection in terms of EI.

$$y_{B1} = \frac{(-2600 \text{ N})(1.20 \text{ m})^2}{6EI}[2(1.20 \text{ m}) + 3(0.6 \text{ m})]$$

$$y_{B1} = \frac{-2621 \text{ N} \cdot \text{m}^3}{EI}$$

Now looking at Figure 13-5(b), we need the deflection at the end of the cantilever due to a concentrated load there. Then

$$y_{B2} = \frac{Pl^3}{3EI} = \frac{R_B(1.8 \text{ m})^3}{3EI} = \frac{R_B(1.944 \text{ m}^3)}{EI}$$

Putting these values in Equation (13-1) gives

$$\frac{-2621 \text{ N} \cdot \text{m}^3}{EI} + \frac{R_B(1.944 \text{ m}^3)}{EI} = 0$$

The term EI can be cancelled out, allowing the solution for R_B.

$$R_B = \frac{2621 \text{ N} \cdot \text{m}^3}{1.944 \text{ m}^3} = 1348 \text{ N}$$

The values of R_A and M_A can now be found using the equations of static equilibrium.

$$\sum F = 0 \qquad \text{(in the vertical direction)}$$
$$R_A + R_B - P = 0$$
$$R_A = P - R_B = 2600 \text{ N} - 1348 \text{ N} = 1252 \text{ N}$$

Summing moments about point A gives

$$0 = M_A - 2600 \text{ N } (1.2 \text{ m}) + 1348 \text{ N } (1.8 \text{ m})$$
$$M_A = 693 \text{ N} \cdot \text{m}$$

The positive sign for the result indicates that the assumed sense of the reaction moment in Figure 13-4 is correct. However, this is a negative moment because it causes the beam to bend concave downward near the support A.

The shear and bending moment diagrams can now be drawn, as shown in Figure 13-6, using conventional techniques. The maximum bending moment occurs at the load where $M = 809 \text{ N} \cdot \text{m}$.

The beam can now be designed. Let's assume that the actual installation is similar to that sketched in Figure 13-7, with the beam welded at its left end and resting on another beam at the right end. A rectangular bar would work well in this arrangement, and a ratio of $h = 3t$ will be assumed. A carbon steel such as AISI 1040, hot-rolled, provides an ultimate strength of 524 MPa. Its percent elongation—18 percent—suggests good ductility, which will help it resist the repeated loads. The design should be based on bending stress.

$$s = \frac{M}{Z}$$

But let

$$s = s_d = \frac{s_u}{N} = \frac{524 \text{ MPa}}{8} = 65.5 \text{ MPa}$$

Then

$$Z = \frac{M}{s_d} = \frac{809 \text{ N} \cdot \text{m}}{65.5 \text{ N/mm}^2} \times \frac{10^3 \text{ mm}}{\text{m}}$$
$$Z = 12 \ 350 \text{ mm}^3$$

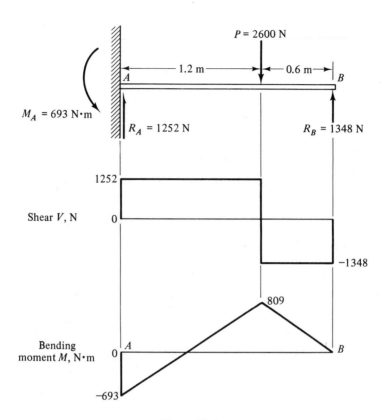

Figure 13-6

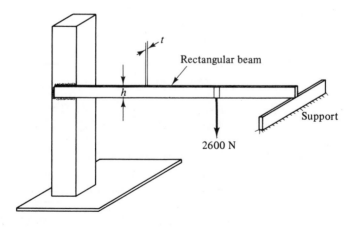

Figure 13-7

For a rectangular bar,

$$Z = \frac{th^2}{6} = \frac{t(3t)^2}{6} = \frac{9t^3}{6} = 1.5t^3$$

Then

$$1.5t^3 = 12\ 350\ \text{mm}^3$$

$$t = 20.2\ \text{mm}$$

Let's use the preferred size of 22 mm for t. Then

$$h = 3t = 3(22\ \text{mm}) = 66\ \text{mm}$$

The final design can be summarized as a rectangular steel bar of AISI 1040, hot-rolled, 22 mm thick and 66 mm high, welded to a rigid support at its left end and resting on a simple support at its right end. The maximum stress in the bar would be less than 65.5 MPa, providing a design factor of at least 8 based on ultimate strength.

The superposition method can be applied to any supported cantilever beam analysis for which the equation for the deflection due to the applied load can be found. Either the beam deflection formulas like those in the Appendix or the mathematical analysis method developed in Chapter 12 can be used.

Continuous beams can also be analyzed using superposition. Consider the beam on three supports shown in Figure 13-8. The three unknown

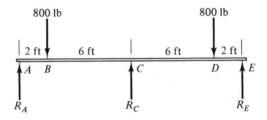

Figure 13-8

support reactions make the beam statically indeterminate. The "extra" reaction R_C can be found using the technique suggested in Figure 13-9. Removing the support at C would cause the deflection y_{C1} downward due to the two 800-lb loads. Case 4 of Table A-19 can be used to find y_{C1}. Then if the loads are imagined to be removed and the reaction R_C is replaced, the upward deflection y_{C2} would result. The formulas of Case 2 in Table A-19 can be used.

Here again, of course, the actual deflection at C is zero because of the unyielding support. Therefore,

$$y_{C1} + y_{C2} = 0$$

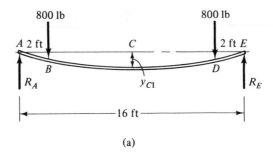

(a)

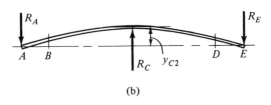

(b)

Figure 13-9

From this relationship, the value of R_C can be computed. The remaining reactions R_A and R_E can then be found in the conventional manner, allowing the creation of the shear and bending moment diagrams.

13-4 CONTINUOUS BEAMS—THEOREM OF THREE MOMENTS

A continuous beam on any number of supports can be analyzed by using the *theorem of three moments*. The theorem actually relates the bending moments at three successive supports to each other and to the loads on the beam. For a beam with only three supports, the theorem allows the direct computation of the moment at the middle support. Known end conditions provide data for computing moments at the ends. Then the principles of statics can be used to find reactions.

For beams on more than three supports, the theorem is applied successively to sets of three adjacent supports (two spans), yielding a set of equations which can be solved simultaneously for the unknown moments.

The theorem of three moments can be used for any combination of loads. Special forms of the theorem have been developed for uniformly distributed loads and concentrated loads. These forms will be used in this chapter.

Uniformly Distributed Loads on Adjacent Spans. Figure 13-10 shows the arrangement of loads and the definition of terms applicable to Equation (13-2).

$$M_A L_1 + 2M_B(L_1 + L_2) + M_C L_2 = \frac{-w_1 L_1^3}{4} - \frac{w_2 L_2^3}{4} \qquad (13\text{-}2)$$

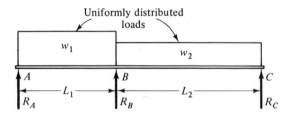

Figure 13-10

The values of w_1 and w_2 are expressed in units of force per unit length such as N/m, lb/ft, etc. The bending moments at the supports A, B, and C are M_A, M_B, and M_C. If M_A and M_C at the ends of the beam are known, M_B can be found from Equation (13-2) directly. Example problems will demonstrate the application of this equation.

The special case in which two equal spans carry equal uniform loads allows the simplification of Equation (13-2). If $L_1 = L_2 = L$ and $w_1 = w_2 = w$, then

$$M_A + 4M_B + M_C = \frac{-wL^2}{2} \qquad (13\text{-}3)$$

Concentrated Loads on Adjacent Spans. If adjacent spans carry only one concentrated load each, as shown in Figure 13-11, then Equation (13-4) applies.

$$M_A L_1 + 2M_B(L_1 + L_2) + M_C L_2 = \frac{-P_1 a}{L_1}(L_1^2 - a^2) - \frac{P_2 b}{L_2}(L_2^2 - b^2)$$
$$(13\text{-}4)$$

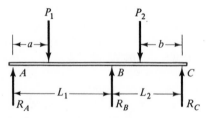

Figure 13-11

Combinations of Uniformly Distributed Loads and Several Concentrated Loads. This is a somewhat general case, allowing each span to carry a uniformly distributed load and any number of concentrated loads, as suggested in Figure 13-12. The general equation for such a loading is a combination of Equations (13-2) and (13-4).

$$M_A L_1 + 2M_B(L_1 + L_2) + M_C L_2 = \frac{-\sum P_1 a}{L_1}(L_1^2 - a^2)$$

$$- \sum \frac{P_2 b}{L_2}(L_2^2 - b^2) - \frac{w_1 L_1^3}{4} - \frac{w_2 L_2^3}{4} \qquad (13\text{-}5)$$

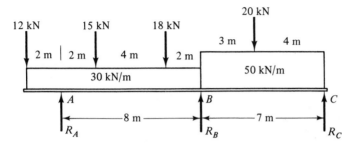

Figure 13-12

The term involving P_1 is to be evaluated for *each* concentrated load on span 1 and then summed together. Similarly, the term having P_2 is repeatedly applied for all loads on span 2. Notice that the distance a is measured from the reaction at A for each load on span 1, and the distance b is measured from the reaction at C for each load on span 2. Any of the terms in Equation (13-5) may be left out of a problem solution if there is no appropriate load or moment existing at a particular section for which the equation is being written.

Example problems illustrating the use of Equations (13-2) to (13-5) will now be shown. Many calculations are required; so be sure you understand how each number was obtained.

Example Problem 13-2

A beam of two 3-m spans carries a uniformly distributed load of 20 kN/m, as shown in Figure 13-13. Determine the reactions at all three supports, and draw the complete shear and bending moment diagrams. Tell where the maximum positive and maximum negative bending moments occur.

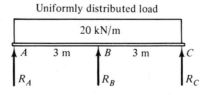

Figure 13-13

Solution

In general, for a problem of this type we would have to find the moments and reactions at all three supports. In this case, we can observe that $M_A = 0$ and $M_C = 0$. Thus only M_B needs to be found. Also, because of the symmetry of the loading, $R_A = R_C$. Equation (13-3) can be used to find M_B since this beam conforms to the special condition of symmetry. In that equation,

$$M_A = M_C = 0$$
$$w = 20 \text{ kN/m}$$
$$L = 3 \text{ m}$$

Then

$$0 + 4M_B + 0 = \frac{-(20 \text{ kN/m})(3 \text{ m})^2}{2} = -90 \text{ kN} \cdot \text{m}$$

$$M_B = -22.5 \text{ kN} \cdot \text{m}$$

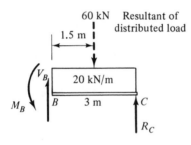

Figure 13-14

The general procedure used to compute reactions for any continuous beam will be demonstrated here. A segment of the beam can be isolated as a free body, as shown in Figure 13-14 for segment BC. All external loads on the segment are shown, plus the shear force and bending moment at B. Now simple equations of statics can be used.

Summing moments about point B,

$$60 \text{ kN} (1.5 \text{ m}) - R_C(3 \text{ m}) - 22.5 \text{ kN} \cdot \text{m} = 0$$

$$R_C = 22.5 \text{ kN}$$

It is important to place the moment M_B on the diagram in the proper direction, noting that it is a *negative* moment. Figure 13-15 shows the directions of positive and negative moments at each end of free-body diagrams of beam segments.

Because of symmetry,

$$R_A = R_C = 22.5 \text{ kN}$$

Now we can sum forces in the vertical direction to find R_B.

$$\Sigma F = 0$$
$$0 = 22.5 \text{ kN} + R_B + 22.5 \text{ kN} - (20 \text{ kN/m})(6 \text{ m})$$
$$R_B = 120 \text{ kN} - 45 \text{ kN} = 75 \text{ kN}$$

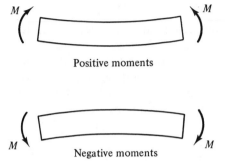

Figure 13-15

Having R_A, R_B, and R_C allows the preparation of the complete shear and bending moment diagrams using conventional methods. The result is shown in Figure 13-16. The beam could now be designed and analyzed for stress and deflection.

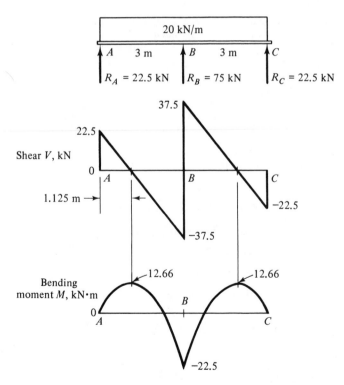

Figure 13-16

Example Problem 13-3

The continuous beam carrying three concentrated loads, shown in Figure 13-17, is to be analyzed to determine the maximum positive and negative bending moments. The beam is a large shaft in a piece of mining equipment. The bending moments are required so they can be combined with the torque in the shaft to determine the diameter. Only the bending moments are to be considered in the problem.

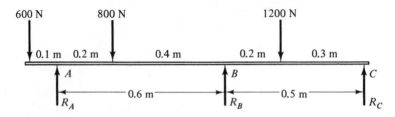

Figure 13-17

Solution

Because the beam has three supports and carries all concentrated loads, Equation (13-4) can be used.

$$M_A L_1 + 2M_B(L_1 + L_2) + M_C L_2 = \frac{-P_1 a}{L_1}(L_1^2 - a^2) - \frac{P_2 b}{L_2}(L_2^2 - b^2)$$

We can observe:

$$M_A = -600 \text{ N } (0.1 \text{ m}) = -60 \text{ N} \cdot \text{m}$$
$$\text{(due to the overhanging load)}$$
$$M_C = 0$$
$$L_1 = 0.6 \text{ m}$$
$$L_2 = 0.5 \text{ m}$$
$$P_1 = 800 \text{ N on span 1}$$
$$P_2 = 1200 \text{ N on span 2}$$
$$a = 0.2 \text{ m}$$
$$b = 0.3 \text{ m}$$

Then the only unknown in the equation is M_B.

$$-60(0.6) + 2M_B(1.1) + 0 = \frac{-800(0.2)}{0.6}(0.6^2 - 0.2^2)$$
$$- \frac{1200(0.3)}{0.5}(0.5^2 - 0.3^2)$$

Solving for M_B gives $M_B = -74.8$ N·m.

The reactions can now be determined. Free-body diagrams are shown in Figure 13-18 for the two segments BC and AB. In segment BC, summing moments about point B,

$$0 = 1200(0.2) - R_C(0.5) - 74.8$$

$$R_C = 330 \text{ N}$$

Summing moments about point B for segment AB,

$$0 = 800(0.4) + 600(0.7) - R_A(0.6) - 74.8$$

$$R_A = 1109 \text{ N}$$

Now by summing forces in the vertical direction, R_B can be found.

$$0 = 1109 + 330 + R_B - 600 - 800 - 1200$$

$$R_B = 1161 \text{ N}$$

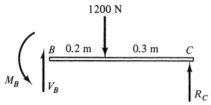

(a) Free-body diagram, segment BC

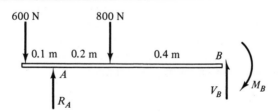

(b) Free-body diagram, segment AB

Figure 13-18

The reactions can now be shown on the beam to permit the construction of the shear and bending moment diagrams, as shown in Figure 13-19. The maximum positive moment is 99 N·m at the location of the 1200-N load, and the maximum negative moment is −74.8 N·m at B.

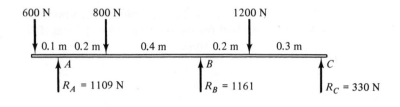

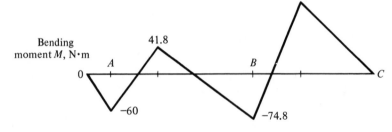

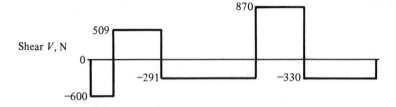

Figure 13-19

Example Problem 13-4

The loading composed of a combination of distributed loads and concentrated loads, shown in Figure 13-12, is to be analyzed to determine the reactions at the three supports and the complete shear and bending moment diagrams. The 17-m beam is to be used as a floor beam in an industrial building.

Solution

Because of the combined loading, Equation (13-5) must be used. Remember, the subscript 1 refers to span 1 between A and B, and subscript 2 refers to span 2 between B and C. From the loading, we can say

$$M_C = 0$$

$$M_A = -12 \text{ kN } (2 \text{ m}) - 60 \text{ kN } (1 \text{ m}) = -84 \text{ kN·m}$$

Each remaining term in Equation (13-5) will be evaluated.

$$M_A L_1 = -84 \text{ kN} \cdot \text{m}(8 \text{ m}) = -672 \text{ kN} \cdot \text{m}^2$$

$$2M_B(L_1 + L_2) = 2M_B(8 + 7) = M_B(30 \text{ m})$$

$$-\sum \frac{P_1 a}{L_1}(L_1^2 - a^2) = -\frac{15(2)}{8}(8^2 - 2^2) - \frac{18(6)}{8}(8^2 - 6^2)$$

$$= -603 \text{ kN} \cdot \text{m}^2$$

$$-\sum \frac{P_2 b}{L_2}(L_2^2 - b^2) = -\frac{20(4)}{7}(7^2 - 4^2) = -377 \text{ kN} \cdot \text{m}^2$$

$$-\frac{w_1 L_1^3}{4} = -\frac{30(8)^3}{4} = -3840 \text{ kN} \cdot \text{m}^2$$

$$-\frac{w_2 L_2^3}{4} = -\frac{50(7)^3}{4} = -4288 \text{ kN} \cdot \text{m}^2$$

Now putting these values into Equation (13-5) gives

$$-672 \text{ kN} \cdot \text{m}^2 + M_B(30 \text{ m}) + 0$$
$$= (-603 - 377 - 3840 - 4288) \text{ kN} \cdot \text{m}^2$$

Solving for M_B,

$$M_B = 281 \text{ kN} \cdot \text{m}$$

Having the three moments at the supports allows the computation of the reactions using the procedure shown in earlier example problems. From this we get

$$R_A = 183 \text{ kN}$$
$$R_B = 389 \text{ kN}$$
$$R_C = 143 \text{ kN}$$

The maximum positive bending moment is 204 kN·m at a point 2.86 m from C. The maximum negative moment is -281 kN·m at the support B. (See Figure 13-20.)

The following example problem shows the analysis of a beam on four supports, and thus having three spans. The solution requires successive applications of the theorem of three moments, in this case twice. The procedure can be expanded to any number of spans by continued application of the theorem. A set of equations will be obtained containing moments at all interior supports. Simultaneous solution of the equations can then be used to determine the moments.

Example Problem 13-5

Determine the reactions at all supports and the shear and bending moment diagrams for the continuous beam with three spans shown in Figure 13-21.

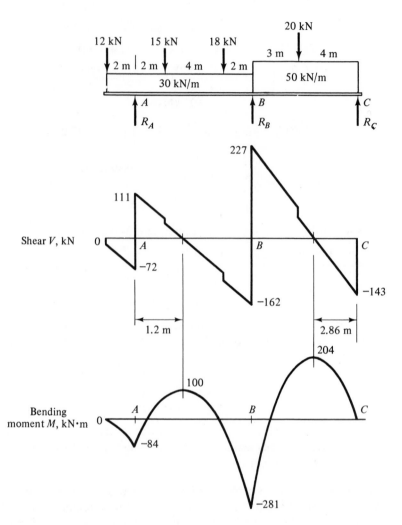

Figure 13-20

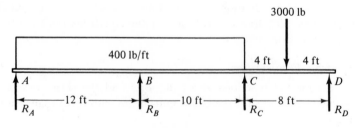

Figure 13-21

Solution

The three-moment equation is applicable to any two adjacent spans of a continuous beam. In this problem, we can first look at the spans *AB* and *BC*. The moment at *A* is zero, so an equation involving the moments at *B* and *C* will be obtained. Then the spans *BC* and *CD* will be analyzed, yielding a second equation involving the moments at *B* and *C*. Simultaneous solution of the two equations will result in the values of M_B and M_C. Spans *AB* and *BC* are shown in Figure 13-22(a). Equation (13-2) applies here. Since $M_A = 0$, we get

$$0 + 2M_B(22) + 10M_C = -\frac{400(12)^3}{4} - \frac{400(10)^3}{4}$$

$$44M_B + 10M_C = -272\ 800 \tag{13-6}$$

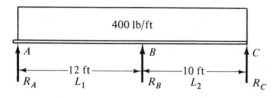

(a) Spans *AB* and *BC*

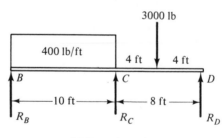

(b) Spans *BC* and *CD*

Figure 13-22

Spans *BC* and *CD* are shown in Figure 13-22(b). We know $M_D = 0$. A simplified form of Equation (13-5) can be used to write the three-moment equation. Be careful of the subscripts on the terms.

$$M_B(10) + 2M_C(18) + 0 = -\frac{400(10)^3}{4} - \frac{3000(4)}{8}(8^2 - 4^2)$$

$$10M_B + 36M_C = -172\ 000 \tag{13-7}$$

Equations (13-6) and (13-7) can be solved simultaneously. To simplify the process, divide Equation (13-6) by 10 and divide Equation
(13-7) by 36. Then subtract one from the other.

$$4.4M_B + M_C = -27\ 280 \qquad\qquad (13\text{-}6)$$
$$0.278M_B + M_C = -\ 4\ 778 \qquad \text{(subtract)} \qquad (13\text{-}7)$$
$$\overline{4.122M_B + \ 0\ \ = -22\ 502}$$

Solving for M_B,

$$M_B = -5459\ \text{lb} \cdot \text{ft}$$

Substituting back into Equation (13-6) gives

$$4.4(\ -5459) + M_C = -27\ 280$$
$$M_C = -3260\ \text{lb} \cdot \text{ft}$$

The reactions, shears, and moments at all points on the beam can now
be computed. An important difference occurs in the analysis for the
reactions using the free-body diagrams. Consider the segments AB and
BC, for which the free-body diagrams are shown in Figure 13-23. The
beam is cut at B in such a way that the shear force just to the *left* of
B, called V_{BL}, is acting on the segment AB. On the segment BC, the
shear force to the *right* of B, called V_{BR}, acts at B. The total reaction

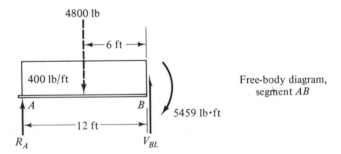

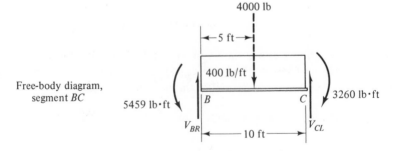

Figure 13-23

at *B*, then, is the *change* in shear which occurs at *B*, just as we have always observed in shear diagrams.

Understanding the signs for shear forces will help to interpret the results in the following analyses. Figure 13-24 shows the proper directions for positive and negative shears on segments of beams.

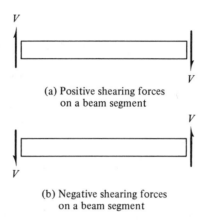

(a) Positive shearing forces
on a beam segment

(b) Negative shearing forces
on a beam segment

Figure 13-24

For the segment *AB*, summing moments about *B*,

$$0 = 4800(6) - R_A(12) - 5459$$

$$R_A = 1945 \text{ lb}$$

Summing forces in the vertical direction,

$$0 = R_A - 4800 + V_{BL}$$

$$V_{BL} = 4800 - R_A = 4800 - 1945 = 2855 \text{ lb}$$

The positive sign shows that the assumed direction for V_{BL} was correct. However, this is by definition a *negative shear*.

Now for the segment *BC*, summing moments about *C*,

$$0 = 4000(5) - V_{BR}(10) - 3260 + 5459$$

$$V_{BR} = 2220 \text{ lb}$$

This is a *positive shear*.

Notice that the moments at both points *B* and *C* are included in the equation with their proper directions. The value of V_{CL} can be found by summing forces in the vertical direction.

$$0 = 2220 - 4000 + V_{CL}$$

$$V_{CL} = 1780 \text{ lb} \qquad \text{(negative shear)}$$

For segment *CD*, see Figure 13-25. Summing moments about *C* gives

$$0 = 3000(4) - R_D(8) - 3260$$

$$R_D = 1093 \text{ lb}$$

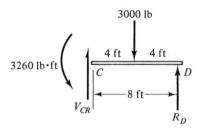

Figure 13-25 Free Body Diagram Segment *CD*

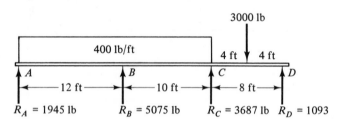

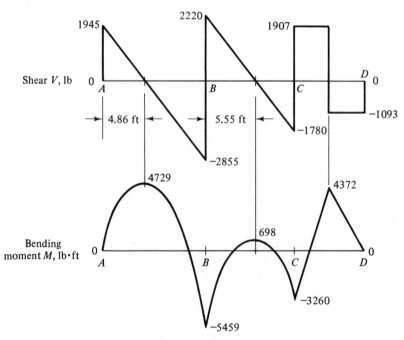

Figure 13-26

and

$$0 = V_{CR} - 3000 + R_D$$

$$V_{CR} = 1907 \text{ lb} \quad \text{(positive shear)}$$

The complete shear and bending moment diagrams are shown in Figure 13-26. Notice that

$$R_B = V_{BR} - V_{BL} = 2220 \text{ lb} - (-2855 \text{ lb}) = 5075 \text{ lb}$$

$$R_C = V_{CR} - V_{CL} = 1907 \text{ lb} - (-1780 \text{ lb}) = 3687 \text{ lb}$$

The maximum positive moment is 4729 lb·ft in span 1, and the maximum negative moment is −5459 lb·ft at *B*.

PROBLEMS

For Problems 13-1 and 13-2, use the superposition method to determine the reactions at all supports and draw the complete shear and bending moment diagrams. Indicate the maximum shear and bending moment for each beam.

13-1 Use Figure 13-27(a), (b), (c), and (d).

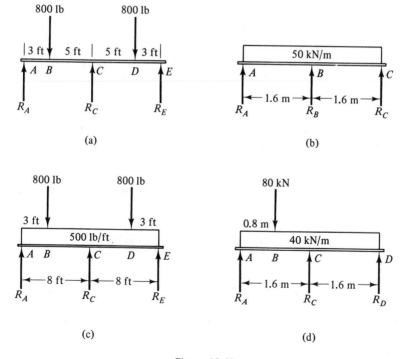

(a)

(b)

(c)

(d)

Figure 13-27

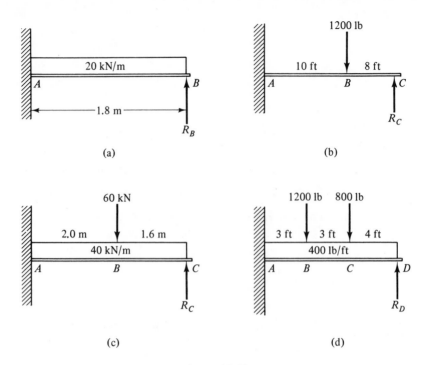

Figure 13-28

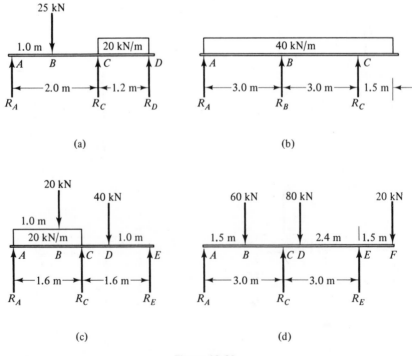

Figure 13-29

13-2   Use Figure 13-28(a), (b), (c), and (d).

For Problems 13-3, 13-4, and 13-5, use the theorem of three moments to determine the reactions at all supports and draw the complete shear and bending moment diagrams. Indicate the maximum shear and bending moment for each beam.

13-3   Use Figure 13-27(a), (b), (c), and (d).
13-4   Use Figure 13-29(a), (b), (c), and (d).
13-5   Use Figure 13-30(a), (b), (c), and (d).

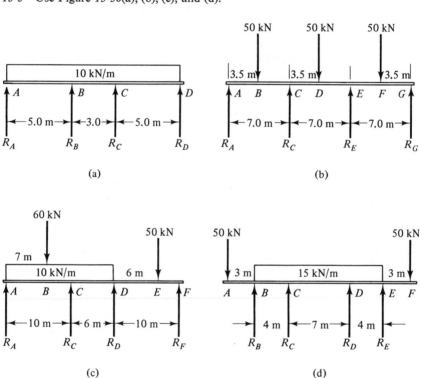

Figure 13-30

14

COLUMNS

14-1 BUCKLING IN COLUMNS

A column is a relatively long member loaded in compression. The analysis of columns is different from what has been studied before because the mode of failure is different. In Chapter 3, when members loaded in compression was discussed, it was assumed that the member failed by yielding of the material. A stress qreater than the yield strength of the material would have to be applied. This is true for short members.

A long, slender column fails by *buckling*, a common name given to *elastic instability*. Instead of crushing or tearing the material, the column deflects drastically at a certain critical load and then collapses suddenly. Any thin member can be used to illustrate the buckling phenomenon. Try it with a wood or plastic ruler, a thin rod or strip of metal, or a drinking straw. By gradually increasing the force, applied directly downward, the critical load is reached when the column begins to bend. Normally, the load can be removed without causing permanent damage since yielding does not occur. Thus a column fails by buckling at a stress lower than the yield strength of the material in the column. The objective of column analysis methods is to predict the load or stress level at which a column would become unstable and buckle.

The tests of simple columns called for in the previous paragraph should also indicate one other important behavior of columns. A column will always buckle about the axis with the *least radius of gyration*. In the case of a thin ruler, the axis through the thin part serves as the buckling axis because of its low radius of gyration. Called *r*, the radius of gyration is defined as

$$r = \sqrt{\frac{I}{A}} \qquad (14\text{-}1)$$

Since *I* and *A* are geometrical properties of the cross section of the column, so also is *r*. Table A-8 lists the formulas for *r* for a variety of section shapes. For a rectangular section which is used for a common ruler, *r* is directly proportional to the thickness of the rectangle measured in a direction perpendicular to the bending axis. Then obviously the least radius of gyration will be that with respect to the axis through the thinnest part, as indicated in Figure 14-1.

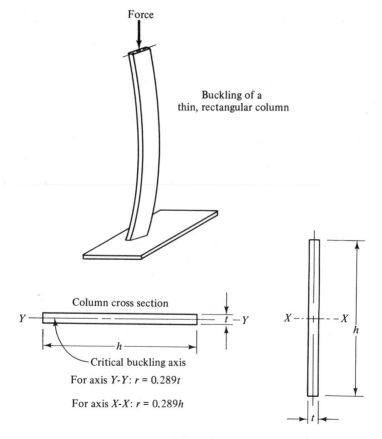

Force

Buckling of a
thin, rectangular column

Column cross section

For axis *Y-Y*: *r* = 0.289*t*

For axis *X-X*: *r* = 0.289*h*

Figure 14-1

Likewise, for other shapes, buckling will occur with respect to the axis having the least radius of gyration.

The value of the radius of gyration is also important in the calculations in order to predict the critical load on a column, as will be discussed in the following sections.

The analysis of the critical load on a column depends also on how "long" it is. The methods discussed in this chapter are the Euler formula for long columns, the Johnson formula for columns of intermediate length, and special formulas specified by codes and standards. All of these formulas depend on the evaluation of the *slenderness ratio*, as defined in the next section.

14-2 SLENDERNESS RATIO AND EFFECTIVE LENGTH

The ratio of the length of a column to the least radius of gyration of its cross section is called the *slenderness ratio*. All of the column formulas involve this ratio, as it is an indication of how "long" the column is. The equation for slenderness ratio is simply L/r. However, the manner in which the column is installed also affects its behavior. Therefore, the length term in the slenderness ratio should be adjusted to determine an *effective length* L_e. Then the effective slenderness ratio is

$$\frac{KL}{r} = \frac{L_e}{r} = \text{effective slenderness ratio} \qquad (14\text{-}2)$$

The value of K depends on how the ends of the column are secured, as shown in Figure 14-2. It should be noted that the values listed for K are theoretical,

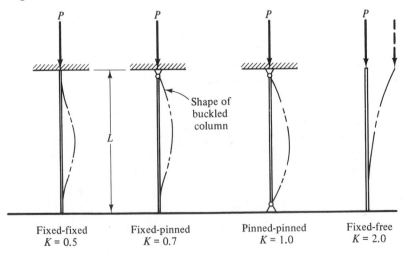

Figure 14-2 Values of K for Effective Length, $L_e = KL$, for Different Connections

based on the expected shape of the deflected column when buckling takes place. Care should be taken when assuming fixed ends to ensure that sufficient rigidity does indeed exist to restrain the end of the column from both rotation and lateral motion.

14-3   TRANSITION SLENDERNESS RATIO

When is a column considered long? The answer to this question requires the determination of the *transition slenderness ratio*, or column constant C_c.

$$C_c = \sqrt{\frac{2\pi^2 E}{s_y}} \qquad (14\text{-}3)$$

If the actual effective slenderness ratio L_e/r is greater than C_c, then the column is long, and the Euler formula, defined in the next section, should be used to analyze the column. If the actual ratio L_e/r is less than C_c, then the column is intermediate or short. In these cases, either the J. B. Johnson formula, special codes, or the direct compressive stress formula should be used, as discussed in later sections.

Where a given column is being analyzed to determine the load it will carry, the value of C_c and the actual ratio L_e/r should be computed first to determine which method of analysis should be used. Notice that C_c depends on the material properties of yield strength s_y and modulus of elasticity E. In working with structural steel, E is usually taken to be 29×10^6 psi (200 GPa). Using this value and assuming a range of values for yield strength, we obtain the values for C_c shown in Figure 14-3. Using $E = 30 \times 10^6$ psi (207 GPa), typical for carbon or alloy steels, would result in C_c being less than 2 percent greater. Thus it is recommended that Figure 14-3 also be used to determine the transition slenderness ratio C_c for any carbon or alloy steel.

For aluminum, E is approximately 69 GPa (10×10^6 psi). The corresponding values for C_c are shown in Figure 14-4.

14-4   THE EULER FORMULA
FOR LONG COLUMNS

For long columns having an effective slenderness ratio greater than the transition value C_c, the Euler formula can be used to predict the critical load at which the column would be expected to buckle. The formula is

$$P_{cr} = \frac{\pi^2 E A}{(L_e/r)^2} \qquad (14\text{-}4)$$

where A is the cross-sectional area of the column. An alternative form can be expressed in terms of the moment of inertia by noting that $r^2 = I/A$.

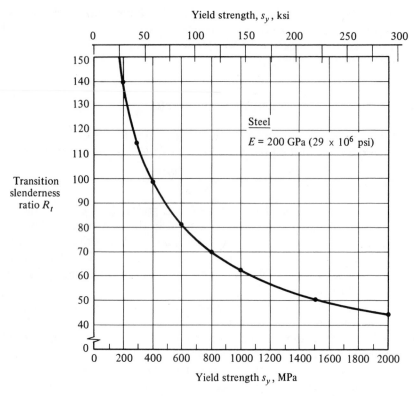

Figure 14-3   Transition Slenderness Ratio C_c Vs. Yield Strength for Steel

Then the formula becomes

$$P_{cr} = \frac{\pi^2 EI}{L_e^2} \tag{14-5}$$

Example Problem 14-1

A round compression member with both ends pinned and made of AISI 1020 cold-drawn steel is to be used in a machine. Its diameter is 25 mm, and its length is 950 mm. What maximum load can the member take before buckling would be expected?

Solution

Let's first evaluate C_c and the actual ratio L_e/r for this column.

$$C_c = \sqrt{\frac{2\pi^2 E}{s_y}} = \sqrt{\frac{2\pi^2(200 \times 10^9 \text{ N/m}^2)}{(352 \times 10^6 \text{ N/m}^2)}} = 106$$

For a round bar,

$$r = \frac{D}{4} = \frac{25 \text{ mm}}{4} = 6.25 \text{ mm}$$

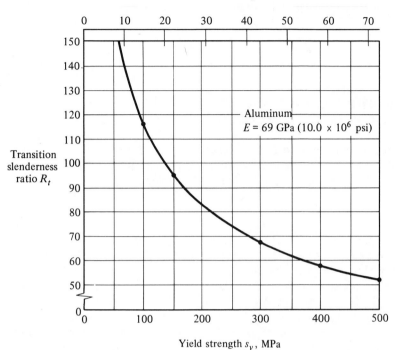

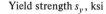

Figure 14-4   Transition Slenderness Ratio C_c Vs. Yield Strength for Aluminum

For a pinned-end column, $K = 1.0$, and $L_e = L$. Then

$$\frac{L_e}{r} = \frac{950 \text{ mm}}{6.25 \text{ mm}} = 152$$

Since L_e/r is greater than C_c, Euler's formula applies.

$$P_{cr} = \frac{\pi^2 EA}{(L_e/r)^2}$$

The area is

$$A = \frac{\pi D^2}{4} = \frac{\pi (25 \text{ mm})^2}{4} = 491 \text{ mm}^2$$

Then

$$P_{cr} = \frac{\pi^2 (200 \times 10^9 \text{ N/m}^2)(491 \text{ mm}^2)}{(152)^2} \times \frac{1 \text{ m}^2}{(10^3 \text{ mm})^2} = 41.9 \text{ kN}$$

Design factors to be applied to columns depend on the degree of confidence in the accuracy of material properties, the geometry of the cross section, and whether the load will be applied directly along the axis of the column. A minimum design factor N of 1.25 is recommended only for well-controlled conditions. A factor of 1.92 is used in general construction work.

For very long columns and for conditions of greater uncertainty, a factor of 3 is recommended. The design factor is applied to the critical buckling load to obtain an allowable load P_a. That is,

$$P_a = \frac{P_{cr}}{N} \qquad (14\text{-}6)$$

14-5   THE J. B. JOHNSON FORMULA
FOR INTERMEDIATE COLUMNS

If a column has an actual effective slenderness ratio L_e/r less than the transition value C_c, the Euler formula predicts an unreasonably high critical load. One formula recommended for machine design applications in the range of L_e/r less than C_c is the J. B. Johnson formula.

$$P_{cr} = As_y \left[1 - \frac{s_y(L_e/r)^2}{4\pi^2 E} \right] \qquad (14\text{-}7)$$

This is one form of a set of equations called parabolic formulas, and it agrees well with the performance of steel columns in typical machinery.

The Johnson formula gives the same result for the critical load at the transition slenderness ratio C_c. Then for very short columns the critical load approaches that which would be predicted from the direct compressive stress equation, $s = P/A$. Therefore, it could be said that the Johnson formula applies best to columns of intermediate length.

Example Problem 14-2

Determine the critical load on a steel column having a square cross section 12 mm on a side with a length of 300 mm. The column is to be made of AISI 1040, hot-rolled. It will be rigidly welded to a firm support at one end and connected by a pin joint at the other. Also, compute the allowable load on the column if a design factor of 3 is desired.

Solution

The values of the actual ratio L_e/r for the column and C_c are needed to determine which method of analysis should be used. For the fixed-pinned ends, $K = 0.7$, and

$$r = \frac{b}{\sqrt{12}} = \frac{12 \text{ mm}}{\sqrt{12}} = 3.46 \text{ mm}$$

$$\frac{L_e}{r} = \frac{KL}{r} = \frac{(0.7)(300 \text{ mm})}{3.46 \text{ mm}}$$

$$\frac{L_e}{r} = 60.6$$

From Figure 14-3, and using $s_y = 290$ MPa, the value of C_c is about 117. Since L_e/r is less than C_c, the Johnson formula, Equation (14-7), should be used.

$$A = b^2 = (12 \text{ mm})^2 = 144 \text{ mm}^2$$

$$P_{cr} = (144 \text{ mm}^2)\left(\frac{290 \text{ N}}{\text{mm}^2}\right)\left[1 - \frac{290 \times 10^6 \text{ N/m}^2 (60.6)^2}{4\pi^2(200 \times 10^9 \text{ N/m}^2)}\right]$$

$$P_{cr} = 36.1 \text{ kN}$$

To get a design factor of 3,

$$P_{cr} = NP_a$$

$$P_a = \frac{P_{cr}}{N} = \frac{36.1 \text{ kN}}{3} = 12.0 \text{ kN}$$

14-6 DESIGN OF COLUMNS USING EULER'S AND JOHNSON'S FORMULAS

The design decisions facing the designer of a column are:

1. What material to use.
2. What basic configuration to use (round, square, pipe, rolled shape, built-up section, etc.).
3. What design factor to use.
4. What method of analysis to use.
5. What final dimensions to specify.

Material selection is subject to the same considerations as used in the design of beams, shafts, and other load-carrying members. However, for long columns, for which Euler's formula applies, *the strength of the material is not important!* According to Equation (14-4), the critical load is dependent only on the modulus of elasticity of the material and on the geometry of the cross section. This means that for a long column made of steel, a plain carbon steel such as AISI 1020, hot-rolled ($s_y = 207$ MPa or 30 000 psi), would carry just as much load as a high alloy steel such as AISI 4140 OQT 400 ($s_y = 1730$ MPa or 251 000 psi).

The selection of a configuration is somewhat of an art and dependent on the application. Where possible, a shape should be chosen which has much of its material far away from the centroid of the section. Pipes, such as the construction pipe shown in the Appendix, are very efficient to use for columns, as are hollow square tubes, which are commercially available. Wide flange shapes or special H shapes are frequently used for columns. The radius of gyration with respect to both major axes is reported in the tables of data such as those shown in the Appendix. Remember to choose the least radius of gyration. Built-up column sections composed of channels, angles, and plates are also used frequently.

Design factors range from a low of about 1.25 for steady loads on short columns under well-controlled conditions up to 3.0 or higher for long columns under more uncertain conditions. Some codes specify the design factor to be used or specify a formula to be used which has a design factor already included.

The method of analysis depends most heavily on whether the column is long, intermediate, or short, as measured by the slenderness ratio. Since this is dependent on the geometry, which is unknown at the start of a design, an assumption must be made. In most machine design applications, the column will be of intermediate length, and the Johnson formula should be used. However, after the design is completed, the actual ratio L_e/r should be computed and compared with the transition ratio C_c. If L_e/r is less than C_c, then the method used is correct. If L_e/r is greater than C_c, the design must be redone using the Euler formula. In construction applications, the code may specify the method of analysis. Otherwise, a similar approach to that just described could be used.

The final dimensions to be specified are found by algebraically solving the particular formula to be used for the geometrical properties of the cross section of the column. For example, when the Euler formula is to be used, Equation (14-5) is a convenient form.

$$P_{cr} = \frac{\pi^2 EI}{L_e^2} \qquad (14\text{-}5)$$

The moment of inertia I is the geometrical property, so the equation can be solved for I. Also, to design a column for a given load, that load should be the allowable load. The design factor is thus introduced into the equation, as follows:

$$P_{cr} = NP_a = \frac{\pi^2 EI}{L_e^2}$$

Then

$$I = \frac{NP_a L_e^2}{\pi^2 E} \qquad (14\text{-}8)$$

Having I, we can find the specific dimensions.

Similar solutions for geometrical variables can be made with the Johnson formula, as illustrated in the example problem to follow.

Example Problem 14-3

Determine the required diameter of a round bar of AISI 1020 cold-drawn steel with a length of 625 mm if it must carry a compressive load of 44 kN. The ends are to be pinned. Use a design factor of 3. The bar will be used as a link in a transfer mechanism on an automated machining system.

Solution

Let's assume that the J. B. Johnson formula will apply here. Looking at Equation (14-7), we see that the diameter of the bar is involved in the area A and in the radius of gyration r.

$$A = \frac{\pi D^2}{4}$$

$$r = \frac{D}{4}$$

Then, letting $P_{cr} = NP$,

$$NP = \frac{\pi D^2}{4} s_y \left[1 - \frac{s_y L_e^2}{4\pi^2 E(D/4)^2} \right]$$

$$\frac{4NP}{\pi s_y} = D^2 \left(1 - \frac{s_y L_e^2 16}{4\pi^2 E D^2} \right)$$

Solving for D,

$$D = \left(\frac{4NP}{\pi s_y} + \frac{4 s_y L_e^2}{\pi^2 E} \right)^{1/2} \tag{14-9}$$

Putting in the known values and letting $L_e = L$,

$$D = \left[\frac{(4)(3)(44 \times 10^3 \text{ N})}{\pi(352 \text{ N/mm}^2)} + \frac{4(352 \text{ N/mm}^2)(625 \text{ mm})^2}{\pi^2(200 \times 10^3 \text{ N/mm}^2)} \right]^{1/2}$$

$$D = (477 \text{ mm}^2 + 279 \text{ mm}^2)^{1/2} = 27.5 \text{ mm}$$

Now the actual ratio L_e/r must be computed and compared to C_c. Let's specify the preferred size of $D = 28$ mm.

$$r = \frac{D}{4} = \frac{28 \text{ mm}}{4} = 7.0 \text{ mm}$$

$$\frac{L_e}{r} = \frac{625 \text{ mm}}{7 \text{ mm}} = 89.3$$

$$C_c = \sqrt{\frac{2\pi^2 E}{s_y}} = \sqrt{\frac{2\pi^2(200 \times 10^9 \text{ N/m}^2)}{352 \times 10^6 \text{ N/m}^2}} = 106$$

Since L_e/r is less than C_c, the use of the Johnson formula was valid.

14-7 SPECIFICATIONS OF THE AISC

Columns are essential elements of most structures. The design and analysis of steel columns in construction applications are governed by the specifications of the AISC, the American Institute of Steel Construction. The specification defines an allowable unit load or stress for columns which is the allowable total load P_a divided by the area of the column cross section. The design formulas are expressed in terms of the transition slenderness ratio C_c, already defined in Equation (14-3), the yield strength of the column material, and the effective slenderness ratio L_e/r. For $L_e/r < C_c$,

$$\frac{P_a}{A} = \left[1 - \frac{(L_e/r)^2}{2C_c^2}\right]\frac{s_y}{FS} \qquad (14\text{-}10)$$

where P_a = allowable or design load

$$C_c = \sqrt{\frac{2\pi^2 E}{s_y}}$$

FS = factor of safety

Equation (14-10) was developed by the Column Research Council and is used for intermediate length columns. The factor of safety FS is a function of the effective slenderness ratio and C_c in order to include the effect of accidental crookedness, a small eccentricity of the load, residual stresses, and any uncertainties in the evaluation of the effective length factor K. The equation for FS is,

$$FS = \frac{5}{3} + \frac{3(L_e/r)}{8C_c} - \frac{(L_e/r)^3}{8C_c^3} \qquad (14\text{-}11)$$

For long columns, $L_e/r > C_c$, Euler's equation is used as defined before but with a factor of safety of 1.92.

$$\frac{P_a}{A} = \frac{\pi^2 E}{(L_e/r)^2(1.92)} \qquad (14\text{-}12)$$

For structural steel with $E = 29 \times 10^6$ psi,

$$\frac{P_a}{A} = \frac{149 \times 10^6}{(L_e/r)^2}\text{ psi} \qquad (14\text{-}13)$$

In the SI system, using $E = 200$ GPa for structural steel,

$$\frac{P_a}{A} = \frac{1028}{(L_e/r)^2}\text{ GPa} \qquad (14\text{-}14)$$

14-8   SPECIFICATIONS OF THE ALUMINUM ASSOCIATION

The Aluminum Association publication, *Specifications for Aluminum Structures*, defines allowable stresses for columns for each of several aluminum alloys and their heat treatments. Three different equations are given for short, intermediate, and long columns defined in relation to slenderness limits. The equations are of the form,

$$\frac{P_a}{A} = \frac{s_y}{FS} \qquad \text{(short columns)} \qquad (14\text{-}15)$$

$$\frac{P_a}{A} = \frac{[B_c - D_c(L/r)]}{FS} \qquad \text{(intermediate columns)} \qquad (14\text{-}16)$$

$$\frac{P_a}{A} = \frac{\pi^2 E}{FS(L/r)^2} \qquad \text{(long columns)} \qquad (14\text{-}17)$$

In all three cases, it is recommended that $FS = 1.95$ for buildings and similar structures. The short column analysis assumes that buckling will not occur and that safety is dependent on the yield strength of the material. Equation (14-17) for long columns is the Euler formula with a factor of safety applied. The intermediate column formula, Equation (14-16), depends on buckling constants B_c and D_c, which are functions of the yield strength of the aluminum alloy and the modulus of elasticity. The division between intermediate and long columns is similar to the C_c used previously in this chapter.

Following are the specific equations for the alloy 6061-T6 used in building structures in the forms of sheet, plate, extrusions, structural shapes, rod, bar, tube, and pipe. The slenderness ratio L/r should be evaluated using the actual L (pinned ends). Any end restraint is assumed to be allowed for in the factor of safety.

Short columns: $L/r < 9.5$

$$\frac{P_a}{A} = 19 \text{ ksi (131 MPa)} \tag{14-18}$$

Intermediate columns: $9.5 < L/r < 66$

$$\frac{P_a}{A} = (20.2 - 0.126 \, L/r) \text{ ksi} \tag{14-19a}$$

$$\frac{P_a}{A} = (139 - 0.869 \, L/r) \text{ MPa} \tag{14-19b}$$

Long columns: $L/r > 66$

$$\frac{P_a}{A} = \frac{51\ 000}{(L/r)^2} \text{ ksi} \tag{14-20a}$$

$$\frac{P_a}{A} = \frac{352\ 000}{(L/r)^2} \text{ MPa} \tag{14-20b}$$

REFERENCES

1. *Manual of Steel Construction*, 7th ed., American Institute of Steel Construction, New York, 1970.
2. *Specifications for Aluminum Structures*, 2nd ed., The Aluminum Association, New York, 1971.
3. Johnston, B. G., and F. Lin, *Basic Steel Design*, Prentice-Hall, Inc., Englewood Cliffs, N.J., 1974.

PROBLEMS

14-1   Determine the critical load for a pinned-end column made of a circular bar of AISI 1020 hot-rolled steel. The diameter of the bar is 20 mm, and its length is 800 mm.

14-2   Repeat Problem 14-1 with the length of 350 mm.

14-3   Repeat Problem 14-1 with the bar made of 6061-T6 aluminum instead of steel.

14-4   Repeat Problem 14-1 with the column ends fixed instead of pinned.

14-5   Repeat Problem 14-1 with a square steel bar with the same cross-sectional area as the circular bar.

14-6   For a 1-in. Schedule 40 steel pipe, used as a column, determine the critical load if it is 2.05 m long. The material is similar to AISI 1020 hot-rolled steel. Compute the critical load for each of the four end conditions described in Figure 14-2.

14-7   A rectangular steel bar has cross-sectional dimensions of 12 mm by 25 mm and is 210 mm long. Assuming that the bar has pinned ends and is made from AISI 1141 OQT 1300 steel, compute the critical load when the bar is subjected to an axial compressive load.

14-8   Compute the allowable load on a column with fixed ends if it is 5.45 m long and made from an S6×12.5 beam. The material is ASTM A36 steel. Use the AISC formula.

14-9   A raised platform is 20 ft by 40 ft in area and is being designed for 75 pounds per square foot uniform loading. It is proposed that standard construction pipe, TS3×0.156, be used as columns to support the platform 8 ft above the ground with the base fixed and the top free. How many columns would be required if a design factor of 3.0 is desired? Use $s_y = 30\ 000$ psi.

14-10   An aluminum I-beam, I10×8.646, is used as a column with two pinned ends. It is 2.80 m long and made of 6061-T6 aluminum. Using Equations (14-18) through (14-20b), compute the allowable load on the column.

14-11   Compute the allowable load for the column described in Problem 14-10 if the length is only 1.40 m.

14-12   A column is a W8×31 steel beam, 12.50 ft long, and made of A36 steel. Its ends are attached in such a way that L_e is approximately 0.80L. Using the AISC formulas, determine the allowable load on the column.

14-13   A built-up column is made of four angles as shown in Figure 14-5. The

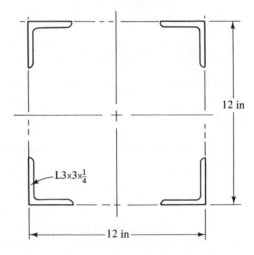

Figure 14-5

angles are held together with lacing bars, which can be neglected in the analysis of geometrical properties. Using the standard Johnson or Euler equations with $L_e = L$ and a design factor of 3.0, compute the allowable load on the column if it is 18.4 ft long. The angles are of ASTM A36 steel.

14-14 Compute the allowable load on a built-up column having the cross section shown in Figure 14-6. Use $L_e = L$ and 6061-T6 aluminum. The column is 3.20 m long. Use the Aluminum Association formulas.

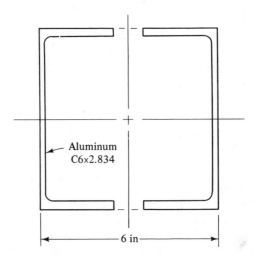

Aluminum
C6×2.834

◄———— 6 in ————►

Figure 14-6

14-15 Determine the required diameter of a round rod to be used as a column if it carries 37.8 kN of axial compressive load. It is to be 1250 mm long, and its ends will be pinned. Use AISI 4140 OQT 1300 steel for the rod. Use $N = 3$.

14-16 A toggle mechanism is used to apply a load as shown in Figure 14-7. For the position shown, determine the required diameter of the circular rods for the two arms of the toggle if a 20-kN load is applied. That is, $P = 20$ kN. Use $N = 3$ and AISI 1020 cold drawn steel.

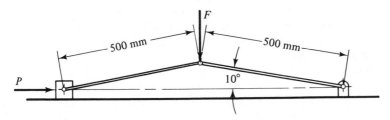

F

500 mm   500 mm

P 10°

Figure 14-7

14-17 Figure 14-8 shows a beam supported at its ends by pin joints. The inclined bar at the top supports the right end of the beam, but also places an axial compressive force in the beam. Would a standard S6×12.5 beam be satis-

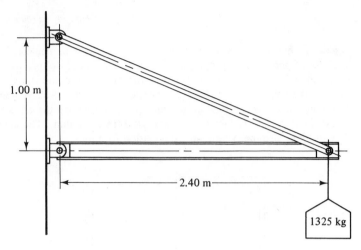

1.00 m

2.40 m

1325 kg

Figure 14-8

factory in this application if it carries 1320 kg at its end? The beam is made from A36 steel.

14-18 Problem 4-9 described a link in a mechanism which was 8.40 in. long, had a rectangular cross section $\frac{1}{4}$ in. $\times \frac{1}{8}$ in., and was subjected to a compressive load of 50 lb. If the link has pinned ends, is it safe from buckling? Cold-drawn AISI 1050 steel is used in the link.

14-19 A piston rod on a shock absorber is 12 mm in diameter and has a maximum length of 190 mm outside the shock absorber body. The rod is made of AISI 1141 OQT 1300 steel. Consider one end to be pinned and the other to be fixed. What axial compressive load on the rod would be one-third of the critical buckling load?

14-20 A stabilizing rod in an automobile suspension system is a round bar loaded in compression. It is subjected to 1375 lb of axial load and supported at its ends by pin-type connections, 28.5 in. apart. Would a 0.800-in. diameter bar of AISI 1020 hot-rolled steel be satisfactory for this application?

14-21 A structure is being designed to support a large hopper over a plastic extruding machine, as sketched in Figure 14-9. The hopper is to be carried by four columns which share the load equally. The structure is cross-braced by rods. It is proposed that the columns be made from standard construction pipe, TS2×0.218. They will be fixed at the floor. Because of the cross-bracing, the top of each column is guided so that it behaves as if it was rounded or pinned. The material for the pipe is AISI 1020 steel, hot-rolled. The hopper is designed to hold 20 000 lb of plastic powder. Are the proposed columns adequate for this load?

14-22 Discuss how the column design in Problem 14-21 would be affected if a careless fork-lift driver runs into the cross braces and breaks them.

14-23 The assembly shown in Figure 14-10 is used to test parts by pulling on them repeatedly with the hydraulic cylinder. A maximum force of 3000 lb can be exerted by the cylinder. The parts of the assembly of concern here are the columns. It is proposed that the two columns be made of $1\frac{1}{4}$-in. square bars of aluminum alloy 6061-T6. The columns are fixed at the bottom and free at the top. Determine the acceptability of the proposal.

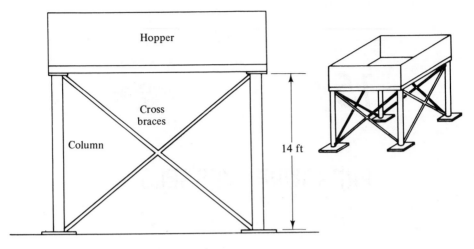

Figure 14-9

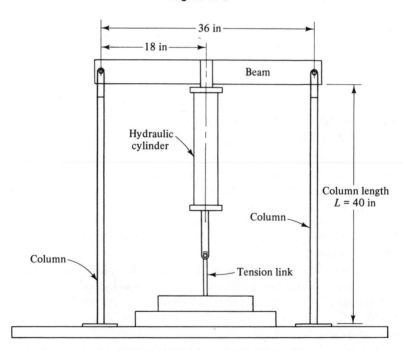

Note: Cylinder pulls up on tension link
and down on beam with a force of 3000 lb.

Figure 14-10

15

PRESSURE VESSELS

15-1 STRESSES DUE TO INTERNAL PRESSURE

Pressure vessels, such as cylinders and spheres, having an internal pressure are subjected to stresses in their walls which tend to burst the walls of the vessel. The stresses are caused by the forces which must be developed to maintain all parts of the vessel in equilibrium. The free-body diagram approach will be used to show the magnitude of the forces in the walls of the vessels.

The methods used to calculate the stresses in thin-walled spheres and cylinders are developed in this chapter. Also, stresses in thick-walled cylinders are discussed. The assumption of thin walls for a sphere or cylinder is reasonable for most typical pressure vessels. If the ratio of the mean radius of the vessel to its wall thickness is ten or greater, the stress distribution across the wall is very nearly uniform, and it can be assumed that all the material of the wall shares equally to resist the applied forces. Conversely, for thick-walled cylinders there is a significant variation in the magnitude of the stress at different points in the wall of the vessel. The method of computing stresses, therefore, must be different.

15-2 THIN-WALLED SPHERES

In analyzing a spherical pressure vessel, the objective is to determine the stress in the wall of the vessel in order to ensure safety. Because of the symmetry of a sphere, a convenient free body for use in the analysis is one-half of the sphere, as shown in Figure 15-1. The internal pressure of the liquid or

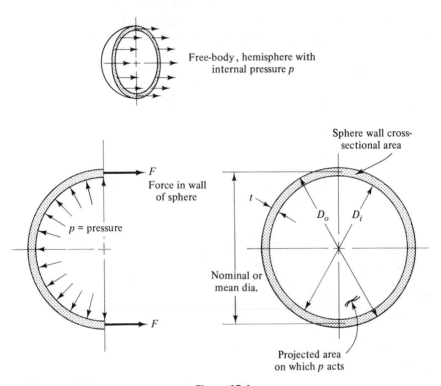

Figure 15-1

gas contained in the sphere acts perpendicular to the walls, uniformly over all the interior surface. Because the sphere was cut through a diameter, the forces in the walls all act horizontally. Therefore, only the horizontal component of the forces due to the fluid pressure needs to be considered in determining the magnitude of the force in the walls. If a pressure P acts on an area A, the force exerted on the area is

$$F = pA \tag{15-1}$$

Taking all of the force acting on the entire inside of the sphere and finding the horizontal component, we find the resultant force in the horizontal direction to be

$$F_R = pA_p \tag{15-2}$$

where A_p is the projected area of the sphere on the plane through the diameter. Therefore,

$$A_p = \frac{\pi D^2}{4} \tag{15-3}$$

Because of equilibrium of the horizontal forces on the free body, the forces in the walls must also equal F_R, as computed in Equation (15-2). These tensile forces acting on the cross-sectional area of the walls of the sphere cause tensile stresses to be developed. That is,

$$s = \frac{F_R}{A_w} \tag{15-4}$$

where A_w is the area of the annular ring cut to create the free body, as shown in Figure 15-1. The actual area is

$$A_w = \frac{\pi}{4}(D_o^2 - D_i^2) \tag{15-5}$$

However, for thin-walled spheres with a wall thickness t, less than about $\frac{1}{10}$ of the radius of the sphere, the wall area can be closely approximated as

$$A_w = \pi D t \tag{15-6}$$

This is the area of a rectangular strip having a thickness t and a length equal to the circumference of the sphere, πD.

Equations (15-2) and (15-4) can be combined to yield an equation for stress,

$$s = \frac{F_R}{A_w} = \frac{pA_p}{A_w} \tag{15-7}$$

Expressing A_p and A_w in terms of D and t from Equations (15-3) and (15-6) gives

$$s = \frac{p(\pi D^2/4)}{\pi D t} = \frac{pD}{4t} \tag{15-8}$$

This is the expression for the stress in the wall of a thin-walled sphere subjected to internal pressure.

Example Problem 15-1

A spherical container is being designed to carry nitrogen gas at a pressure of 3500 kPa. The required volume of the container dictates that it be 300 mm in diameter. Compute the required thickness of the sphere if it is to be made of AISI 5160 OQT 1300 steel. Use a design factor of 4 based on yield strength to determine the design stress.

Solution

Equation (15-8) can be used by solving for t.

$$s = \frac{pD}{4t}$$

Letting $s = s_d$, the required thickness is

$$t = \frac{pD}{4s_d} \tag{15-9}$$

But, using $s_y = 731$ MPa,

$$s_d = \frac{s_y}{4} = \frac{731 \text{ MPa}}{4} = 182.8 \text{ MPa}$$

Then

$$t = \frac{(3500 \times 10^3 \text{ Pa})(300 \text{ mm})}{(4)(182.8 \times 10^6 \text{ Pa})} = 1.44 \text{ mm}$$

15-3 THIN-WALLED CYLINDERS

Cylinders are frequently used for pressure vessels, for example, as storage tanks, hydraulic and pneumatic actuators, and for piping of fluids under pressure. The stresses in the walls of cylinders are similar to those found for spheres, although the maximum value is greater.

Two separate analyses are shown here. In one case, the tendency for the internal pressure to pull the cylinder apart in a direction parallel to its axis is found. This is called *longitudinal stress*. Next, a ring around the cylinder is analyzed to determine the stress tending to pull the ring apart. This is called *hoop stress*, or *tangential stress*.

Longitudinal Stress. Figure 15-2 shows a part of a cylinder, which is subjected to an internal pressure, cut perpendicular to its axis to create a free body. Assuming that the end of the cylinder is closed, the pressure acting on the circular area of the end would produce a resultant force of

$$F_R = pA = p\left(\frac{\pi D^2}{4}\right) \tag{15-10}$$

This force must be resisted by the force in the walls of the cylinder, which, in turn, creates a tensile stress in the walls. The stress is

$$s = \frac{F_R}{A_w} \tag{15-11}$$

Assuming that the walls are thin, as we did for spheres,

$$A_w = \pi D t$$

where t is the wall thickness.

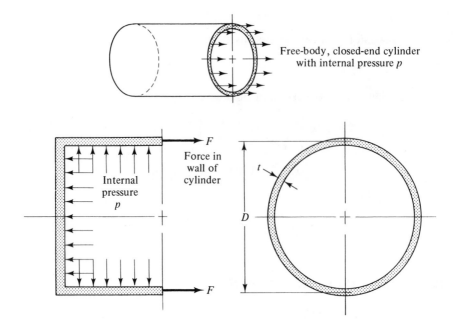

Figure 15-2

Now combining Equations (15-10) and (15-11),

$$s = \frac{F_R}{A_w} = \frac{p(\pi D^2/4)}{\pi D t} = \frac{pD}{4t} \tag{15-12}$$

This is the stress in the wall of the cylinder in a direction parallel to the axis, called the longitudinal stress. Notice that it is of the same magnitude as that found for the wall of a sphere.

Hoop Stress. The presence of the tangential or hoop stress can be visualized by isolating a ring from the cylinder, as shown in Figure 15-3. The internal pressure pushes outward evenly all around the ring. The ring must develop a tensile stress in a direction tangential to the circumference of the ring to resist the tendency of the pressure to burst the ring. The magnitude of the stress can be determined by using half of the ring as a free body, as shown in Figure 15-3(b).

The resultant of the forces due to the internal pressure must be determined in the horizontal direction and balanced with the forces in the walls of the ring. Using the same reasoning as we did for the analysis of the sphere, we find that the resultant force is the product of the pressure and the *projected area* of the ring. For a ring with a diameter D and a length L,

$$F_R = pA_p = p(DL) \tag{15-13}$$

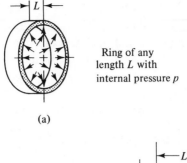

Ring of any
length L with
internal pressure p

(a)

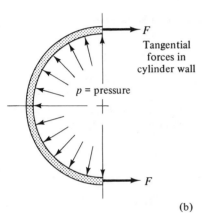

F

Tangential
forces in
cylinder wall

p = pressure

F

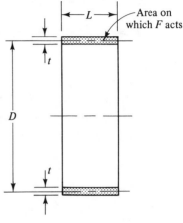

Area on
which F acts

(b)

Figure 15-3

The tensile stress in the wall of the cylinder is equal to the product of the resisting force and the cross-sectional area of the wall. Again assuming that the wall is thin, the wall area is

$$A_w = 2tL \tag{15-14}$$

Then the stress is

$$s = \frac{F_R}{A_w} = \frac{F_R}{2tL} \tag{15-15}$$

Combining Equations (15-13) and (15-15) gives

$$s = \frac{F_R}{A_w} = \frac{pDL}{2tL} = \frac{pD}{2t} \tag{15-16}$$

This is the equation for the hoop stress in a thin cylinder subjected to internal pressure. Notice that the magnitude of the hoop stress is *twice* that of the longitudinal stress. Also, the hoop stress is twice that of the stress in a spherical container of the same diameter carrying the same pressure.

Example Problem 15-2

A cylindrical tank holding oxygen at 2000 kPa pressure has a diameter of 450 mm and a wall thickness of 10 mm. Compute the hoop stress and the longitudinal stress in the wall of the cylinder.

Solution

The hoop stress can be found from Equation (15-16).

$$s = \frac{pD}{2t} = \frac{(2000 \times 10^3 \text{ Pa})(450 \text{ mm})}{2(10 \text{ mm})} = 45.0 \text{ MPa}$$

The longitudinal stress, from Equation (15-12), is

$$s = \frac{pD}{4t} = 22.5 \text{ MPa}$$

Example Problem 15-3

Determine the pressure required to burst a standard 8-in., Schedule 40 steel pipe if the ultimate tensile strength of the steel is 40 000 psi.

Solution

The dimensions of the pipe are found in the Appendix to be

$$\text{Outside diameter} = 8.625 \text{ in.} = D_o$$

$$\text{Inside diameter} \ = 7.981 \text{ in.} = D_i$$

$$\text{Wall thickness} \ = 0.322 \text{ in.} = t$$

We should first check to determine if the pipe should be called a thin-walled cylinder by computing the ratio of the mean diameter to the wall thickness.

$$D_m = \text{mean diameter} = \frac{(D_o + D_i)}{2} = 8.303 \text{ in.}$$

$$\frac{D_m}{t} = \frac{8.303 \text{ in.}}{0.322 \text{ in.}} = 25.8$$

Since this ratio is greater than 20, the thin-wall equations can be used. The hoop stress is the maximum stress and should be used to compute the bursting pressure.

$$s = \frac{pD}{2t} \tag{15-16}$$

Letting $s = 40\ 000$ psi and using the mean diameter,

$$p = \frac{2ts}{D} = \frac{(2)(0.322 \text{ in.})(40\ 000 \text{ lb/in.}^2)}{8.303 \text{ in.}}$$

$$p = 3102 \text{ psi}$$

A design factor of 6 or greater is usually applied to the bursting pressure to get an allowable operating pressure.

15-4   THICK-WALLED CYLINDERS AND SPHERES

The formulas in the preceding sections for thin-walled cylinders and spheres were derived under the assumption that the stress is uniform throughout the wall of the container. As stated, if the ratio of the diameter of the container to the wall thickness is greater than 20, this assumption is reasonably correct. Conversely, if the ratio is less than 20, the walls are considered to be thick, and a different analysis technique is required.

The detailed derivation of the thick-wall formulas will not be given here because of their complexity. But the application of the formulas will be shown.

For a thick-walled cylinder, Figure 15-4 shows the notation to be used.

s_2 = longitudinal stress s_3 = radial stress
s_1 = hoop stress

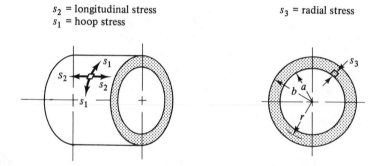

Figure 15-4   Notation for Stresses in Thick-walled Cylinders

The geometry is characterized by the inner radius a, the outer radius b, and any radial position between a and b, called r. The *longitudinal stress* is called s_1; the *hoop stress* is s_2. These have the same meaning as they did for thin-walled vessels, except now they will have varying magnitudes at different positions in the wall. In addition to hoop and longitudinal stresses, a *radial stress* s_3 is created in a thick-walled vessel. As the name implies, the radial stress acts along a radius of the cyinder or sphere. It is a compressive stress and varies from a magnitude of zero at the outer surface to a maximum at the inner surface, where it is equal to the internal pressure. Table 15-1 shows a summary of the formulas needed to compute the three stresses in the walls of thick-walled cylinders and spheres subjected to internal pressure.

TABLE 15-1   Stresses in Thick-Walled Cylinders and Spheres*

Thick-Walled Cylinder	
Stress at Position r	**Maximum Stress**
$s_1 = \dfrac{pa^2}{b^2 - a^2}$	$s_1 = \dfrac{pa^2}{b^2 - a^2}$ (uniform throughout wall)
$s_2 = \dfrac{pa^2(b^2 + r^2)}{r^2(b^2 - a^2)}$	$s_2 = \dfrac{p(b^2 + a^2)}{b^2 - a^2}$ (at inner surface)
$s_3 = \dfrac{-pa^2(b^2 - r^2)}{r^2(b^2 - a^2)}$	$s_3 = -p$ (at inner surface)
Thick-Walled Sphere	
Stress at Position r	**Maximum Stress**
$s_1 = s_2 = \dfrac{pa^3(b^3 + 2r^3)}{2r^3(b^3 - a^3)}$	$s_1 = s_2 = \dfrac{p(b^3 + 2a^3)}{2(b^3 - a^3)}$ (at inner surface)
$s_3 = \dfrac{-pa^3(b^3 - r^3)}{r^3(b^3 - a^3)}$	$s_3 = -p$ (at inner surface)

*Symbols used here are: a = inner radius; b = outer radius; r = any radius between a and b; p = internal pressure, uniform in all directions. Stresses are tensile when positive, compressive when negative. s_1 = longitudinal stress; s_2 = hoop stress; s_3 = radial stress.

Example Problem 15-4

Compute the magnitude of the maximum stresses s_1, s_2, and s_3 in a cylinder carrying helium at a pressure of 10 000 psi. The outside diameter is 8.00 in. and the inside diameter is 6.40 in.

Solution

Referring to Table 15-1, with $a = 3.20$ in. and $b = 4.00$ in.,

$$s_1 = \frac{pa^2}{b^2 - a^2} = \frac{(10\ 000 \text{ psi})(3.20 \text{ in.})^2}{(4.00^2 - 3.20^2) \text{ in.}^2} = 17\ 780 \text{ psi}$$

$$s_2 = \frac{p(b^2 + a^2)}{b^2 - a^2} = \frac{(10\ 000 \text{ psi})(4.00^2 + 3.20^2) \text{ in.}^2}{(4.00^2 - 3.20^2) \text{ in.}^2} = 45\ 560 \text{ psi}$$

$$s_3 = -p = -10\ 000 \text{ psi}$$

All three stresses are a maximum at the inner surface of the cylinder.

Example Problem 15-5

Compute the maximum stresses for a sphere having the same internal and outside diameters as did the cylinder in the preceding example problem for the same internal pressure, 10 000 psi.

Solution

From Table 15-1,

$$s_1 = s_2 = \frac{p(b^3 + 2a^3)}{2(b^3 - a^3)} = \frac{(10\ 000\ \text{psi})[4.00^3 + 2(3.20)^3]\ \text{in.}^3}{2(4.00^3 - 3.20^3)\ \text{in.}^3}$$

$$s_1 = s_2 = 20\ 740\ \text{psi}$$

$$s_3 = -p = -10\ 000\ \text{psi}$$

Each of these stresses is a maximum at the inner surface.

Example Problem 15-6

A cylindrical vessel has an outside diameter of 400 mm and an inside diameter of 300 mm. For an internal pressure of 20.1 MPa, compute the hoop stress s_2 at the inner and outer surfaces and at points within the wall at intervals of 10 mm. Plot a graph of s_2 versus the radial position in the wall.

Solution

At the inner surface, the maximum value of s_2 occurs.

$$s_2 = \frac{p(b^2 + a^2)}{b^2 - a^2} = \frac{(20.1\ \text{MPa})(200^2 + 150^2)}{200^2 - 150^2} = 71.8\ \text{MPa}$$

At the outer surface, $r = b = 200$ mm, and

$$s_2 = \frac{pa^2(b^2 + r^2)}{r^2(b^2 - a^2)} = \frac{2pa^2}{b^2 - a^2} = \frac{(2)(20.1\ \text{MPa})(150^2)}{200^2 - 150^2} = 51.7\ \text{MPa}$$

At $r = 160$ mm,

$$s_2 = \frac{(20.1\ \text{MPa})(150^2)(200^2 + 160^2)}{160^2(200^2 - 150^2)} = 66.2\ \text{MPa}$$

Using similar calculations, the stresses s_2 for other radial positions are:

r, mm	s_2, MPa
170	61.6
180	57.7
190	54.5

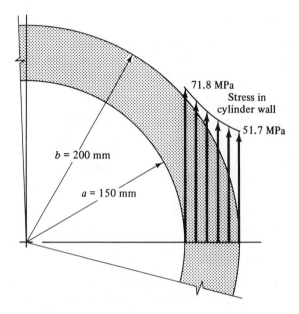

Figure 15-5

Figure 15-5 shows the graph of s_2 versus position in the wall. The graph illustrates clearly that the assumption of uniform stress in the wall of a thick-walled cylinder would *not* be valid.

PROBLEMS

15-1 Compute the stress in a sphere having an outside diameter of 200 mm and an inside diameter of 184 mm if an internal pressure of 19.2 MPa is applied.

15-2 A large, spherical storage tank for a compressed gas in a chemical plant is 10.5 m in diameter and is made of AISI 1050 hot-rolled steel plate, 12 mm thick. What internal pressure could the tank withstand if a design factor of 3.0 based on yield strength is desired?

15-3 Titanium 6Al-4V is to be used to make a spherical tank having an outside diameter of 1200 mm. The working pressure in the tank is to be 4.20 MPa. Determine the required thickness of the tank wall if a design factor of 2.5 based on yield strength is desired.

15-4 If the tank of Problem 15-3 was made of aluminum 2014-T6 sheet instead of titanium, compute the required wall thickness. Which design would weigh less?

15-5 Compute the hoop stress in the walls of a 10-in. Schedule 40 steel pipe if it carries water at 150 psi.

15-6 A pneumatic cylinder has a bore of 80 mm and a wall thickness of 3.5 mm. Compute the hoop stress in the cylinder wall if an internal pressure of 2.85 MPa is applied.

15-7   A cylinder for carrying acetylene has a diameter of 300 mm and will hold the acetylene at 1.7 MPa. If a design factor of 3 is desired, compute the required wall thickness for the tank. Use AISI 1050 cold-drawn steel.

15-8   The companion oxygen cylinder for the acetylene discussed in the preceding problem carries oxygen at 15.2 MPa. Its diameter is 250 mm. Compute the required wall thickness using the same design criteria.

15-9   A propane tank for a recreation vehicle is made of AISI 1040 hot-rolled steel, 2.20 mm thick. The tank diameter is 450 mm. Determine what design factor would result based on yield strength if propane at 750 kPa is put into the tank.

15-10   The supply tank for propane at the distributor is a cylinder having a diameter of 1800 mm. If it is desired to have a design factor of 3 based on yield strength using AISI 1040 hot-rolled steel, compute the required thickness of the tank walls when the internal pressure is 750 kPa.

15-11   Oxygen on a spacecraft is carried at a pressure of 70.0 MPa in order to minimize the volume required. The spherical vessel has an outside diameter of 250 mm and a wall thickness of 18 mm. Compute the maximum hoop and radial stresses in the sphere.

15-12   Compute the maximum longitudinal, hoop, and radial stresses in the wall of a standard $\frac{1}{2}$-in. Schedule 40 steel pipe when carrying an internal pressure of 1.72 MPa (250 psi).

15-13   The barrel of a large field-artillery piece has a bore of 220 mm and an outside diameter of 300 mm. Compute the magnitude of the hoop stress in the barrel at points 10 mm apart from the inside to the outside surfaces. The internal pressure is 50 MPa.

15-14   A $1\frac{1}{2}$-in. Schedule 40 steel pipe has a mean radius less than 10 times the wall thickness and thus should be classified as a thick-walled cylinder. Compute what maximum stresses would result from both the thin-wall and the thick-wall formulas due to an internal pressure of 10.0 MPa.

16

CONNECTIONS

16-1 TYPES OF CONNECTIONS

Structures and mechanical devices rely on the connections between load-carrying elements to maintain the integrity of the assemblies. The connections provide the path by which loads are transferred from one element to another.

Three common types of connections are riveting, welding, and bolting. Figure 16-1 shows a bulk storage hopper supported by rectangular straps from a tee beam. During fabrication of the hopper, the support tabs were welded to the outside of the side walls. The tabs contain a pattern of holes, allowing the straps to be bolted on at the assembly site. Prior to installation of the tee beam, the straps were riveted to the web.

The load due to the weight of the hopper and the materials in it must be transferred from the hopper walls into the tabs through the welds. Then the four bolts transfer the load to the straps, which act as tension members. Finally, the six rivets transfer the load into the tee.

16-2 MODES OF FAILURE

For riveted and bolted connections, four possible modes of failure exist related to four different types of stresses present in the vicinity of the joint. Figure 16-2 illustrates these kinds of stresses for a simple lap joint in which

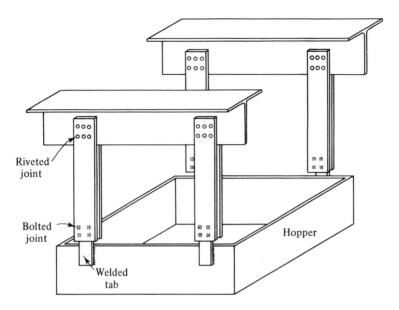

Figure 16-1

two flat plates are joined together by two rivets. The joint is prepared by drilling or punching matching holes in each plate. Then the rivets, originally having one of the shapes shown in Figure 16-3, are inserted into the holes, and the heads are upset, gripping the two plates together. In a good riveted joint, the body of the rivet is also upset somewhat, causing the rivet to completely fill the hole. This makes a tight joint which does not allow relative motion of the members joined.

As the joint is loaded with a tensile force, a shearing force is transmitted across the cross section of the rivets between the two plates. Thus *shear failure* is a mode of joint failure. The body of the rivet must bear against the material of the plates being joined, with the possibility of *failure in bearing*. This would result in a crushing of the material, normally in the plates. *Tensile failure* of the plates being joined must be investigated because the presence of the holes for the rivets causes the cross section of material at the joint to be less than in the main part of the tension member. The fourth possible failure mode is *end tear out*, in which the rivet causes the material between the edge of the plate and the hole to tear out.

Properly designed riveted and bolted joints should have an edge distance from the center of the rivet or bolt to the edge of the plate being joined of at least two times the diameter of the bolt or rivet. The edge distance is measured in the direction toward which the bearing pressure is directed. If this recommendation is heeded, then end tear out should not occur. This will be assumed in example problems for this chapter. Thus shear, bearing, and tensile failure modes only will be considered in evaluating joint strength.

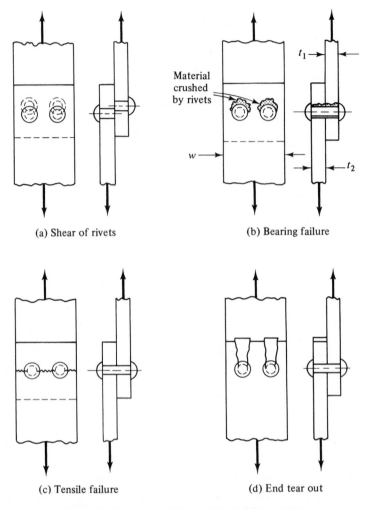

(a) Shear of rivets

(b) Bearing failure

(c) Tensile failure

(d) End tear out

Figure 16-2   Types of Failure of Riveted Connections

Welded joints fail by shear in the weld material or by fracture of the base metal of the parts joined by the welds. A properly designed and fabricated welded joint will always fail in the base metal. Thus the design of welded connections has the objective of determining the required size of weld and the length of weld in the joint.

16-3   RIVETED CONNECTIONS

In riveted connections, it is assumed that the joined plates are *not* clamped together tightly enough to cause frictional forces between the plates to transmit loads. Therefore, the rivets do bear on the holes, and bearing failure

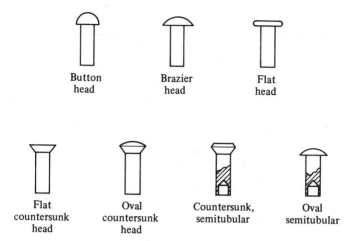

Button
head

Brazier
head

Flat
head

Flat
countersunk
head

Oval
countersunk
head

Countersunk,
semitubular

Oval
semitubular

Figure 16-3 Examples of Rivet Styles

must be investigated. Both shear failure and tensile failure could also occur. The method of analysis of these three modes of failure is outlined below.

Shear Failure. The rivet body is assumed to be in direct shear when a tensile load is applied to a joint, provided that the line of action of the load passes through the centroid of the pattern of rivets. It is also assumed that the total applied load is shared equally among all the rivets. The capacity of a joint with regard to shear of the rivets is

$$F_s = s_{sa} A_s \qquad (16\text{-}1)$$

where F_s = capacity of the joint in shear
s_{sa} = allowable shear stress in rivets
A_s = area in shear

The area in shear is dependent on the number of cross sections of rivets available to resist shear. Calling this number N_s,

$$A_s = \frac{N_s \pi D^2}{4} \qquad (16\text{-}2)$$

where D is the rivet diameter. In some cases, particularly for rivets driven hot, the body expands to fill the hole, and a larger area is available for resisting shear. However, the increase is small, and only the nominal diameter will be used here.

To determine N_s, it must be observed whether *single shear* or *double shear* exists in the joint. Figure 16-2 shows an example of single shear. Only one cross section of each rivet resists the applied load. Then N_s is equal to the number of rivets in the joint. The straps used to support the bin in Figure 16-1 place the rivets and bolts in double shear. Two cross sections of each rivet resist the applied load. Then N_s is twice the number of rivets in the joint.

Bearing Failure. When a cylindrical rivet bears against the wall of a hole in the plate, a nonuniform pressure exists between them. As a simplification of the actual stress distribution, it is assumed that the area in bearing, A_b, is the rectangular area found by multiplying the plate thickness by the diameter of the rivet. This can be considered to be the *projected area* of the rivet hole. Then the bearing capacity of a joint is

$$F_b = s_{ba} A_b \qquad (16\text{-}3)$$

where F_b = capacity of the joint in bearing
$\qquad s_{ba}$ = allowable bearing stress
$\qquad A_b$ = bearing area = $N_b D t$ $\qquad\qquad\qquad\qquad (16\text{-}4)$
$\qquad N_b$ = number of bearing surfaces

Tensile Failure. A direct tensile force applied through the centroid of the rivet pattern would produce a tensile stress. Then the capacity of the joint in tension would be

$$F_t = s_{ta} A_t \qquad (16\text{-}5)$$

where F_t = capacity of the joint in tension
$\qquad s_{ta}$ = allowable stress in tension
$\qquad A_t$ = net tensile area

The evaluation of A_t requires the subtraction of the diameter of all the holes from the width of the plates being joined. Then

$$A_t = (w - N D_H)t \qquad (16\text{-}6)$$

where $\quad w$ = width of plate
$\qquad D_H$ = hole diameter (in structures use $D_H = D + \frac{1}{8}$ in.)
$\qquad N$ = number of holes at the section of interest
$\qquad t$ = thickness of plate

16-4 ALLOWABLE STRESSES

For members not covered by codes and specifications, the allowable stresses can be determined using the design factors presented in Table 3-1. For the design of steel building structures, the specifications of the American Institute of Steel Construction (AISC) are usually used. For aluminum structures, the Aluminum Association has published its *Specifications for Aluminum Structures*. Table 16-1 shows allowable stresses for steel structures. Table 16-2 summarizes the allowable stresses for aluminum.

TABLE 16-1 Allowable Stresses for Steel Structual Connections*

Rivets	Allowable Shear Stress		Allowable Tensile Stress	
	ksi	MPa	ksi	MPa
ASTM A502				
Grade 1	15	103	20	138
Grade 2	20	138	27	186
Bolts	Allowable Shear Stress†		Allowable Tensile Stress	
	ksi	MPa	ksi	MPa
ASTM A325 and A449	15	103	40	276
ASTM A490	20	138	54	372
Connected Members	Allowable Bearing Stress‡§		Allowable Tensile Stress‡	
All alloys	$1.35s_y$		$0.6s_y$	

*AISC specifications.
†For friction-type connection. For bearing-type connection with no threads in the shear plane, use 22 ksi (152 MPa) for A325, and 32 ksi (221 MPa) for A490.
‡See Table A-3 for s_y for structural steels.
§Bearing stress not considered in friction-type bolted joint.

16-5 BOLTED CONNECTIONS

The analysis of bolted connections is the same as for riveted connections if the bolt is allowed to bear on the hole, a bearing-type connection. This would occur in joints where the clamping force provided by the bolts is small. However, most bolted connections are made with high strength bolts, such as A325 and A490, and tightened to a high tension level. The resulting large clamping forces make a friction-type joint in which the friction forces between the two mating surfaces transmit much of the joint load. The bolts are still designed for shear, using the strengths listed in Table 16-1. But bearing stress is not considered for a friction-type joint.

16-6 EXAMPLE PROBLEMS—RIVETED AND BOLTED JOINTS

Example Problem 16-1

For the single lap joint shown in Figure 16-2, determine the allowable load on the joint if the two plates are $\frac{1}{4}$ in. thick by 2 in. wide and joined by two steel rivets, $\frac{1}{4}$ in. diameter, ASTM A502, Grade 1. The plates are A36 structural steel.

TABLE 16-2 Allowable Stresses for Aluminum Structural Connections*

Rivets			
Alloy and Temper		**Allowable Shear Stress**	
Before Driving	**After Driving†**	**ksi**	**MPa**
1100-H14	1100-F	4	27
2017-T4	2017-T3	14.5	100
6053-T61	6053-T61	8.5	58
6061-T6	6061-T6	11	76

Bolts				
	Allowable Shear Stress‡		**Allowable Tensile Stress‡**	
Alloy and Temper	**ksi**	**MPa**	**ksi**	**MPa**
2024-T4	16	110	26	179
6061-T6	12	83	18	124
7075-T73	17	117	28	193

Connected Members			
		Allowable Bearing Stress	
Alloy and Temper	**Allowable Tensile Stress**	**ksi**	**MPa**
All alloys	$0.6s_y$ (See Table A-4 for s_y)		
1100-H12		11.0	76
2014-T6		49	338
3003-H12		11.5	79
6061-T6		34	234
6063-T6		24	165

*Specifications for Aluminum Structures, The Aluminum Association.
†All cold-driven.
‡Stresses are based on the area corresponding to the nominal diameter of the bolt unless the threads are in the shear plane. Then the shear area is based on the root diameter.

Solution

Shear, bearing, and tensile failure will be investigated to determine the capacity of the joint relative to all three modes. The lowest of the three values is then the limiting load for the joint.

Shear failure

$$F_s = s_{sa}A_s \qquad (16\text{-}1)$$

From Table 16-1, $s_{sa} = 15$ ksi. For the two rivets in single shear, there are two shear planes. Then

$$A_s = \frac{N_s\pi D^2}{4} \qquad (16\text{-}2)$$

$$= \frac{2\pi(0.25 \text{ in.})^2}{4} = 0.098 \text{ in.}^2$$

The capacity of the joint in shear is

$$F_s = (15\ 000\ \text{lb/in.}^2)(0.098\ \text{in.}^2) = 1470\ \text{lb}$$

Bearing failure

$$F_b = s_{ba}A_b \tag{16-3}$$

In bearing, $s_{ba} = 1.35 s_y$. For A36 steel, $s_y = 36$ ksi. Then

$$s_{ba} = 1.35(36\ 000\ \text{psi}) = 48\ 600\ \text{psi}$$

The bearing area is

$$A_b = N_b D t \tag{16-4}$$

The force on either plate is resisted by two bearing surfaces. Then

$$A_b = 2(0.25\ \text{in.})(0.25\ \text{in.}) = 0.125\ \text{in.}^2$$

The capacity of the joint in bearing is

$$F_b = (48\ 600\ \text{lb/in.}^2)(0.125\ \text{in.}^2) = 6075\ \text{lb}$$

Tensile failure

The plates would fail in tension across a section through the rivet holes, as indicated in Figure 16-2(c).

$$F_t = s_{ta}A_t$$

From Table 16-1,

$$s_{ta} = 0.6 s_y = 0.6(36\ 000\ \text{psi}) = 21\ 600\ \text{psi}$$

The net tensile area, assuming $D_H = D + \frac{1}{8}$ in. is

$$A_t = (w - ND_H)t = [2.0\ \text{in.} - 2(0.25 + 0.125)\ \text{in.}](0.25\ \text{in.})$$
$$A_t = 0.313\ \text{in.}^2$$

The capacity of the joint in tension is

$$F_t = (21\ 600\ \text{lb/in.}^2)(0.313\ \text{in.}^2) = 6760\ \text{lb}$$

Since shear failure would occur at a load of 1470 lb, that is the capacity of the joint.

Example Problem 16-2

Determine the allowable load for a joint of the same dimensions as used in Example Problem 16-1, except use two $\frac{3}{8}$-in. diameter A490 bolts in a bearing-type connection with no threads in the shear plane.

Solution

Shear failure

$$F_s = s_{sa}A_s$$
$$s_{sa} = 32\ 000\ \text{psi} \qquad \text{(Table 16-1)}$$
$$A_s = \frac{2\pi(0.375\ \text{in.})^2}{4} = 0.221\ \text{in.}^2$$

Then

$$F_s = (32\ 000\ \text{lb/in.}^2)(0.221\ \text{in.}^2) = 7070\ \text{lb}$$

Bearing failure

$$F_b = s_{ba}A_b$$
$$s_{ba} = 1.35(36\ 000\ \text{psi}) = 48\ 600\ \text{psi}$$
$$A_b = N_bDt = (2)(0.375\ \text{in.})(0.25\ \text{in.}) = 0.188\ \text{in.}^2$$

Then

$$F_b = (48\ 600\ \text{lb/in.}^2)(0.188\ \text{in.}^2) = 9140\ \text{lb}$$

Tensile failure

$$F_t = s_{ta}A_t$$
$$s_{ta} = 0.6(36\ 000\ \text{psi}) = 21\ 600\ \text{psi}$$
$$A_t = [2.0\ \text{in.} - 2(0.375 + 0.125)\ \text{in.}]0.25\ \text{in.} = 0.25\ \text{in.}^2$$

Then

$$F_t = (21\ 600\ \text{lb/in.}^2)/0.25\ \text{in.}^2) = 5400\ \text{lb}$$

Now the capacity in tension is the lowest, so the capacity of the joint is 5400 lb.

Example Problem 16-3

Figure 16-4 shows a joint called a butt splice, in which two $\frac{1}{2}$-in. plates of A242 steel are joined by using two $\frac{3}{8}$-in. cover plates of A36 steel.

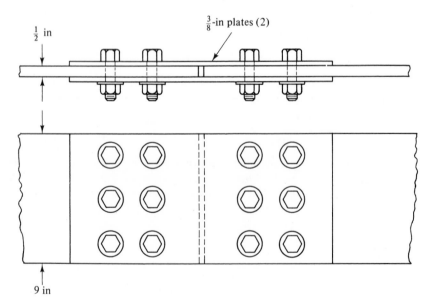

$\frac{1}{2}$ in

$\frac{3}{8}$-in plates (2)

9 in

Figure 16-4

The assembly is bolted tightly with a friction-type connection by $\frac{3}{4}$-in. diameter A449 bolts. Determine the capacity of the joint.

Solution

Because of the friction-type connection, only shear and tensile failure will have to be investigated.

Shear failure

The joint load is assumed to be shared equally by six bolts, each in double shear. Then

$$A_s = \frac{(6)(2)(\pi)(0.75 \text{ in.})^2}{4} = 5.30 \text{ in.}^2$$

Using $s_{sa} = 15$ ksi from Table 16-1,

$$F_s = (15\ 000 \text{ lb/in.}^2)(5.30 \text{ in.}^2) = 79\ 500 \text{ lb}$$

Tensile failure

Either the main $\frac{1}{2}$-in. plate or the two cover plates could fail in tension. For the $\frac{1}{2}$-in. plate (A242),

$$F_t = s_{ta}At \cdot$$
$$s_{ta} = 0.6s_y = 0.6(46 \text{ ksi}) = 27.6 \text{ ksi}$$
$$A_t = [9.0 \text{ in.} - 3(0.75 + 0.125) \text{ in.}]0.5 \text{ in} = 3.188 \text{ in.}^2$$

Then

$$F_t = (27\ 600 \text{ lb/in.}^2)(3.188 \text{ in.}^2) = 87\ 990 \text{ lb}$$

For the two $\frac{3}{8}$-in. thick cover plates (A36),

$$F_t = s_{ta}A_t$$
$$s_{ta} = (0.6)(36 \text{ ksi}) = 21.6 \text{ ksi}$$
$$A_t = [9.0 \text{ in.} - 3(0.75 + 0.125) \text{ in.}](2)(0.375 \text{ in.}) = 4.78 \text{ in.}^2$$
$$F_t = (21\ 600 \text{ lb/in.}^2)(4.78 \text{ in.}^2) = 103\ 200 \text{ lb}$$

The shear strength governs, so the capacity of the joint is 79 500 lb.

16-7 ECCENTRICALLY LOADED RIVETED AND BOLTED JOINTS

Previously considered joints were restricted to cases in which the line of action of the load on the joint passed through the centroid of the pattern of rivets or bolts. In such cases, the applied load is divided equally among all the fasteners. When the load does not pass through the centroid of the fastener pattern, it is called an *eccentrically loaded joint,* and a nonuniform distribution of forces occurs in the fasteners.

In eccentrically loaded joints, the effect of the moment or couple on the fastener must be considered. Figure 16-5 shows a bracket attached to the

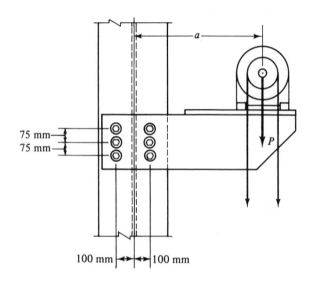

Figure 16-5

side of a column and used to support an electric motor. The net downward force exerted by the weight of the motor and the belt tension acts at a distance *a* from the center of the column flange. Then the total force system acting on the bolts of the bracket consists of the direct shearing force *P* plus the moment $P \times a$. Each of these components can be considered separately and then added together by using the principle of superposition.

Figure 16-6(a) shows that for the direct shearing force *P*, each bolt is assumed to carry an equal share of the load, just as in concentrically loaded joints. But in part (b) of the figure, each bolt is subjected to a force acting perpendicular to a radial line from the centroid of the bolt pattern. It is assumed that the magnitude of the force in a bolt due to the moment load is proportional to its distance *r* from the centroid. This magnitude is

$$R_i = \frac{Mr_i}{\sum r^2} \tag{16-7}$$

where R_i = shearing force in bolt *i* due to the moment *M*

r_i = radial distance to bolt *i* from the centroid of the bolt pattern

$\sum r^2$ = sum of the radial distances to *all* bolts in the pattern squared

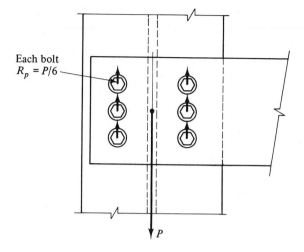

(a) Forces resisting P

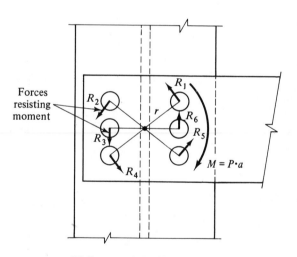

(b) Forces resisting the moment

Figure 16-6

If it is more convenient to work with horizontal and vertical components of forces, they can be found from

$$R_{ix} = \frac{My_i}{\sum r^2} = \frac{My_i}{\sum (x^2 + y^2)} \tag{16-8}$$

$$R_{iy} = \frac{Mx_i}{\sum r^2} = \frac{Mx_i}{\sum (x^2 + y^2)} \tag{16-9}$$

where $\qquad\qquad y_i =$ vertical distance to bolt i from centroid

$x_i =$ horizontal distance to bolt i from centroid

$\sum (x^2 + y^2) =$ sum of horizontal and vertical distances squared for all bolts in the pattern

Finally, all horizontal forces are summed and all vertical forces are summed for any particular bolt. Then the resultant of the horizontal and vertical forces is determined.

Example Problem 16-4

In Figure 16-5, the net downward force P is 26.4 kN on each side plate of the bracket. The distance a is 0.75 m. Determine the required size of ASTM A325 bolts to secure the bracket.

Solution

The force to be resisted by the most highly stressed bolt must be determined. Then the required area and diameter will be computed.

One compotent of the force on each bolt is the share of the 26.4 kN shearing force. Let's call this force R_p.

$$R_p = \frac{P}{6} = \frac{26.4 \text{ kN}}{6} = 4.4 \text{ kN}$$

This is an upward reaction force produced by each bolt, as shown in Figure 16-6.

Now, to find the horizontal and vertical components of the reaction forces due to the eccentric moment, we need the value of

$$\sum (x^2 + y^2) = 6(100 \text{ mm})^2 + 4(75 \text{ mm})^2$$
$$= 82\ 500 \text{ mm}^2$$

The moment on the joint is

$$P \times a = 26.4 \text{ kN} (0.75 \text{ m}) = 19.8 \text{ kN} \cdot \text{m}$$

Starting first with bolt 1 at the upper right (see Figure 16-7),

$$R_{1x} = \frac{My_1}{\sum (x^2 + y^2)} = \frac{19.8 \text{ kN} \cdot \text{m}(75 \text{ mm})}{82\ 500 \text{ mm}^2} \times \frac{10^3 \text{ mm}}{\text{m}}$$

$$R_{ix} = 18.0 \text{ kN} \leftarrow \text{(acts toward the left)}$$

$$R_{1y} = \frac{Mx_1}{\sum (x^2 + y^2)} = \frac{(19.8 \text{ kN} \cdot \text{m})(100 \text{ mm})}{82\ 500 \text{ mm}^2} \times \frac{10^3 \text{ mm}}{\text{m}}$$

$$R_{1y} = 24.0 \text{ kN} \uparrow \text{ (acts upward)}$$

Now the resultant of these forces can be found. In the vertical direction, R_p and R_{1y} both act upward.

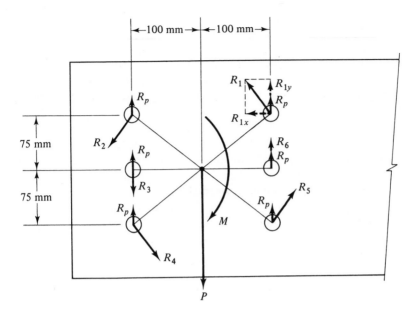

Figure 16-7

$$R_p + R_{1y} = 4.4 \text{ kN} + 24.0 \text{ kN} = 28.4 \text{ kN}$$

Only R_{1x} acts in the horizontal direction. Calling the resultant force on bolt 1, R_{R1},

$$R_{R1} = \sqrt{28.4^2 + 18.0^2} = 33.6 \text{ kN}$$

Investigating the other five bolts in a similar manner would show that bolt 1 is the most highly stressed. Then its diameter will be determined to limit the shear stress to 103 MPa (15 ksi) for the A325 bolts.

$$s_s = \frac{R_{R1}}{A}$$

$$A = \frac{R_{R1}}{s_{sa}} = \frac{33.6 \text{ kN}}{103 \text{ N/mm}^2} = 326 \text{ mm}^2$$

$$D = \sqrt{\frac{4A}{\pi}} = \sqrt{\frac{4(326) \text{ mm}^2}{\pi}} = 20.4 \text{ mm}$$

The nearest preferred metric size is 22 mm. If conventional inch units are to be specified,

$$D = 20.4 \text{ mm} \times \frac{1 \text{ in.}}{25.4 \text{ mm}} = 0.80 \text{ in.}$$

Use $D = \frac{7}{8}$ in. $= 0.875$ in.

16-8 WELDED JOINTS WITH CONCENTRIC LOADS

Welding is a joining process in which heat is applied to cause two pieces of metal to become metallurgically bonded. The heat may be applied by a gas flame, an electric arc, or by a combination of electric resistance heating and pressure.

Types of welds include groove, fillet, and spot welds (as shown in Figure 16-8), and others. Groove and fillet welds are most frequently used in structural connections since they are readily adaptable to the shapes and plates which make up the structures. Spot welds are used for joining relatively light gauge steel sheets and cold-formed shapes.

The variables involved in designing welded joints are the shape and size of the weld, the choice of filler metal, the length of weld, and the position of the weld relative to the applied load.

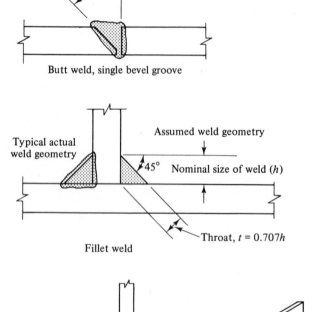

Figure 16-8

Fillet welds are assumed to have a slope of 45 deg between the two surfaces joined, as shown in Figure 16-8. The size of the weld is denoted as the height of one side of the triangular-shaped fillet. Typical sizes range from $\frac{1}{8}$ in. to $\frac{1}{2}$ in. in steps of $\frac{1}{16}$ in. The stress developed in fillet welds is assumed to be *shear stress* regardless of the direction of application of the load. The maximum shear stress would occur at the throat of the fillet (see Figure 16-8), where the thickness is 0.707 times the nominal size of the weld. Then the shearing stress in the weld due to a load P is

$$s_s = \frac{P}{Lt} \qquad (16\text{-}10)$$

where L is the length of weld and t is the thickness at the throat. Equation (16-10) is used *only* for concentrically loaded members. This requires that the line of action of the force on the welds passes through the centroid of the weld pattern. Eccentricity of the load produces a moment, in addition to the direct shearing force, which must be resisted by the weld metal. References 1 and 2 at the end of this chapter contain pertinent information with regard to eccentrically loaded welded joints.

In electric arc welding, used mostly for structural connections, a filler rod is normally used to add metal to the welded zone. As the two parts to be joined are heated to a molten state, filler metal is added, which combines with the base metal. On cooling, the resulting weld metal is normally stronger than the original base metal. Therefore, a properly designed and made welded joint should fail in the base metal rather than in the weld. In structural welding, the electrodes are given a code beginning with an E and followed by two or three digits, such as E60, E80, or E100. The number denotes the ultimate tensile strength in ksi of the weld metal in the rod. Thus an E80 rod would have a tensile strength of 80 000 psi. Other digits may be added to the code number to denote special properties. Complete specifications can be found in the standards of ASTM A233 and A316. The allowable stress for fillet welds using electrodes is 0.3 times the tensile strength of the electrode according to AISC. Table 16-3 lists some common electrodes and their allowable stresses.

TABLE 16-3   Properties of Welding Electrodes

Electrode Type	Minimum Tensile Strength		Allowable Shear Stress		Typical Metals Joined
	ksi	MPa	ksi	MPa	
E60	60	414	18	124	A36, A500
E70	70	483	21	145	A242, A441
E80	80	552	24	165	A572, Grade 65
E90	90	621	27	186	—
E100	100	690	30	207	—
E110	110	758	33	228	A514

Aluminum products are welded using either the inert gas–shielded arc process or the resistance welding process. For the inert gas–shielded arc process, filler alloys are specified by the Aluminum Association for joining particular base metal alloys, as indicated in Table 16-4. The allowable shear stress for such welds is also listed. It should be noted that the heat of welding lowers the properties of most aluminum alloys within 1.0 in. of the weld, and allowance for this must be made in the design of welded assemblies.

TABLE 16-4 Allowable Shear Stresses in Fillet Welds in Aluminum Building-Type Structures

Base Metal	Filler Alloy							
	1100		4043		5356		5556	
	ksi	MPa	ksi	MPa	ksi	MPa	ksi	MPa
1100	3.2	22	4.8	33	—	—	—	—
3003	3.2	22	5.0	34	—	—	—	—
6061	—	—	5.0	34	7.0	48	8.5	59
6063	—	—	5.0	34	6.5	45	6.5	45

Example Problem 16-5

A lap joint is made by placing two $\frac{3}{8}$-in. fillet welds across the full width of two $\frac{1}{2}$-in. A36 steel plates, as shown in Figure 16-9. The shielded metal–arc method is used, using an E60 electrode. Compute the allowable load P which can be applied to the joint.

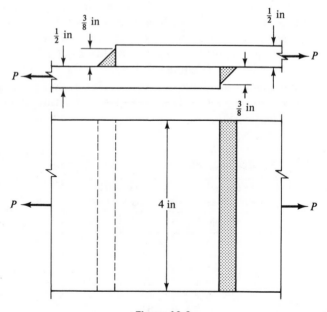

Figure 16-9

Solution

The load is assumed to be equally distributed on all parts of the weld, so that Equation (16-10) can be used with $L = 8.0$ in.

$$s_s = \frac{P}{Lt}$$

Let s_s equal the allowable stress of 18 ksi, listed in Table 16-3. The thickness t is

$$t = 0.707 \, (\tfrac{3}{8} \text{ in.}) = 0.265 \text{ in.}$$

Now we can solve for P.

$$P = s_{sa} Lt$$
$$P = (18\ 000 \text{ lb/in.}^2)(8.0 \text{ in.})(0.265 \text{ in.}) = 38\ 200 \text{ lb}$$

Example Problem 16-6

The hopper shown in Figure 16-1 at the beginning of this chapter is designed to be supported at four places with straps attached to the lugs welded to the sides of the hopper. The total load of the hopper and its contents is expected to be 120 000 lb. A design for the lugs is shown in Figure 16-10, with $\frac{5}{16}$-in. fillet welds all around on the $\frac{3}{4}$-in.

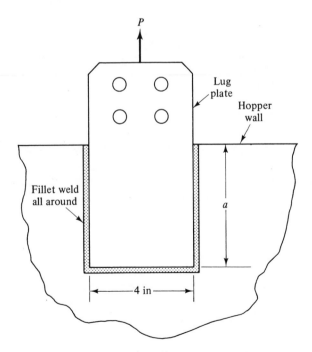

Figure 16-10

thick lug plate. It is desired to determine the required length a to ensure a safe weld. Because of the probability of some mild shock as the bin is loaded, use only one-half of the allowable shear stress for the weld as that listed in Table 16-3. Use an E70 electrode.

Solution

Assuming that the total load of the hopper is shared equally by the four hangers, each carries 30 000 lb. This is P in Equation (16-10).

$$s_s = \frac{P}{Lt}$$

Solving for L,

$$L = \frac{P}{s_{sa}t}$$

For t and s_{sa},

$$t = 0.707 \left(\tfrac{5}{16} \text{ in.}\right) = 0.221 \text{ in.}$$

$$s_{sa} = \frac{21\ 000 \text{ psi}}{2} = 10\ 500 \text{ psi}$$

Then

$$L = \frac{30\ 000 \text{ lb}}{(10\ 500 \text{ lb/in.}^2)\,(0.221 \text{ in.})} = 12.9 \text{ in.}$$

But

$$L = 2a + 4.0 \text{ in.}$$

Then

$$a = \frac{L - 4.0}{2} = \frac{12.9 - 4.0}{2} = 4.45 \text{ in.}$$

Let $a = 4.50$ in.

Example Problem 16-7

A rectangular aluminum bar is welded between two posts and loaded in tension as shown in Figure 16-11. Fillet welds are to be run on both sides of the bar, both top and bottom, making the total length of weld 200 mm. The bar is 6061-T6, and the filler alloy is to be 5556. Determine the required size of the welds if a tensile load of 65 kN is applied to the bar.

Solution

Equation 16-10 can be solved for t,

$$t = \frac{P}{Ls_{sa}}$$

From Table 16-4, $s_{sa} = 59$ MPa. Then

$$t = \frac{65\ 000 \text{ N}}{(200 \text{ mm})(59 \text{ N/mm}^2)} = 5.51 \text{ mm}$$

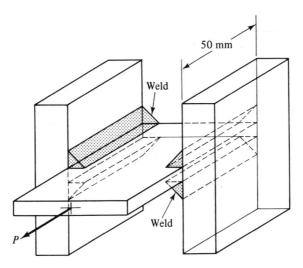

Figure 16-11

This is the dimension of the throat of the weld. Calling the height of the weld h,

$$t = 0.707\,h$$

and

$$h = \frac{t}{0.707} = \frac{5.51 \text{ mm}}{0.707} = 7.8 \text{ mm}$$

Specify an 8.0-mm fillet weld.

REFERENCES

1. *Manual of Steel Construction*, 7th ed., American Institute of Steel Construction, New York, 1970.
2. Johnston, B. G., and F. Lin, *Basic Steel Design*, Prentice-Hall, Inc., Englewood Cliffs, N.J., 1974.
3. *Specifications for Aluminum Structures*, 2nd ed., The Aluminum Association, New York, 1971.

PROBLEMS

16-1   Determine the allowable loads on the joints shown in Figure 16-12. The fasteners are all steel rivets, ASTM A502, Grade 1. The plates are all ASTM A36 steel.

16-2   Determine the allowable loads on the joints shown in Figure 16-13. The fasteners are all steel rivets, ASTM A502, Grade 2. The plates are all ASTM A242 steel.

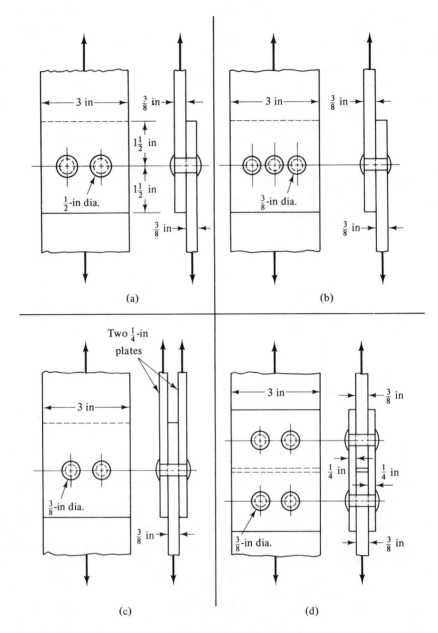

(a)

(b)

Two $\frac{1}{4}$-in plates

(c)

(d)

Figure 16-12

16-3   Determine the allowable loads on the joints shown in Figure 16-12 if the
fasteners are all ASTM A325 steel bolts providing a bearing-type connection
and the plates are all ASTM A441 steel.

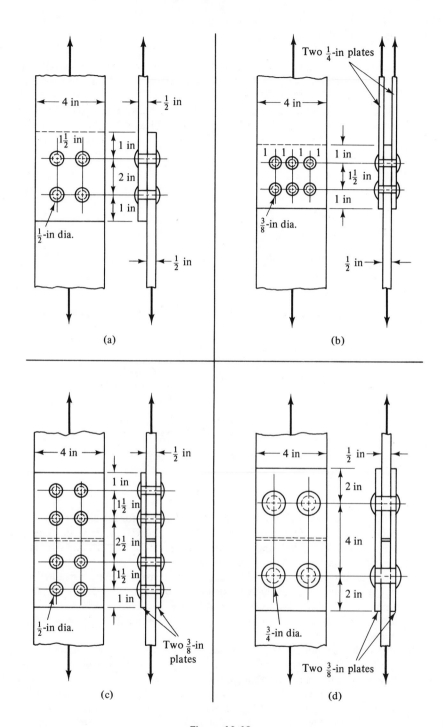

Figure 16-13

16-4   Determine the allowable loads on the joints shown in Figure 16-13. The fasteners are all ASTM A490 steel bolts providing a friction-type joint. The plates are all ASTM A514 steel.

16-5   Determine the required diameter of the bolts used to attach the cantilever beam to the column as shown in Figure 16-14. Use ASTM A325 steel bolts.

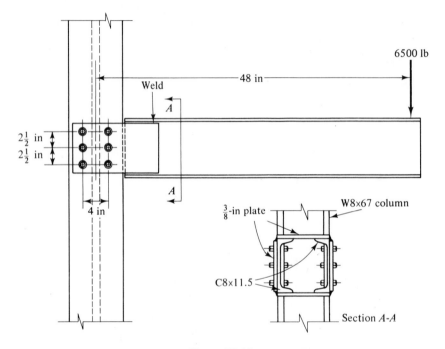

Figure 16-14

16-6   Design the connection of the channel to the column for the hanger shown in Figure 16-15. Both members are A36 steel. Specify the type of fastener (rivet, bolt), the pattern and spacing, the number of fasteners, and the material for the fasteners. Use AISC specifications.

16-7   For the connection shown in Figure 16-12(a), assume that, instead of the two rivets, the two plates were welded across the ends of the 3-in. wide plates using $\frac{5}{16}$-in. welds. The plates are A36 steel and the electric arc welding technique is used with E60 electrodes. Determine the allowable load on the connection.

16-8   Determine the allowable load on the joint shown in Figure 16-13(c) if $\frac{1}{4}$-in. welds using E70 electrodes were placed along both ends of both cover plates. The plates are ASTM A242 steel.

16-9   Design the joint at the top of the straps in Figure 16-1 if the total load in the hopper is 54.4 megagrams (Mg). The beam is a WT12 × 34 made of ASTM A36 steel and has a web thickness of 10.6 mm. The clear vertical height of the web is about 250 mm. Use steel rivets and specify the pattern, number

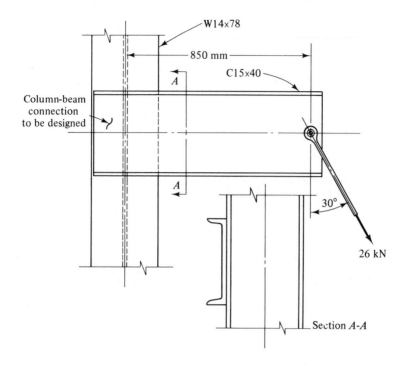

Figure 16-15

of rivets, rivet diameter, rivet material, strap material, and strap dimensions. Specify dimensions, using the preferred metric sizes shown in the Appendix.

16-10   Design the joint at the bottom of the straps in Figure 16-1 if the total lo:d in the hopper is 54.4 Mg. Use steel bolts and a bearing-type connection. Specify the pattern, number of bolts, bolt diameter, bolt material, strap material, and strap dimensions. You may want to coordinate the strap design with the results of Problem 16-9. The design of the tab in Problem 16-11 is also affected by the design of the bolted joint.

16-11   Design the tab to be welded to the hopper for connection to the support straps as shown in Figure 16-1. The hopper load is 54.4 Mg. The material from which the hopper is made is ASTM A36 steel. Specify the width and thickness of the tab and the design of the welded joint. You may want to coordinate the tab design with the bolted connection call:d for in Problem 16-10.

APPENDICES

Pa = N/m²

= 1000 PSI

= 1 mega pascals

N/mm²

TABLE A-1 Properties of Carbon and Alloy Steels*

Material AISI No.	Condition†	Ultimate Strength s_u		Yield Strength s_y		Percent Elongation
		ksi	MPa	ksi	MPa	
1020	Hot-rolled	55	379	30	207	25
1020	Cold-drawn	61	421	51	352	15
1040	Hot-rolled	76	524	42	290	18
1040	Cold-drawn	85	586	71	490	12
1040	WQT 400	113	779	86	593	19
1040	WQT 1300	89	614	62	427	33
1050	Hot-rolled	90	621	49	338	15
1050	Cold-drawn	100	690	84	579	10
1050	OQT 400	143	986	108	745	10
1050	OQT 1300	96	662	61	421	30
1095	Hot-rolled	130	896	66	455	9
1095	OQT 400	188	1300	120	827	10
1095	OQT 1300	90	621	74	510	26
1137	OQT 400	158	1090	138	951	6
1137	OQT 1300	87	600	60	414	28
1141	OQT 400	237	1630	188	1300	7
1141	OQT 1300	94	648	68	469	28
4130	WQT 400	234	1610	197	1360	12
4130	WQT 1300	98	676	89	614	28
4140	OQT 400	290	2000	251	1730	11
4140	OQT 1300	117	807	100	690	23
4150	OQT 400	301	2080	248	1710	10
4150	OQT 1300	128	883	117	807	20
5160	OQT 400	322	2220	260	1790	4
5160	OQT 1300	115	793	106	731	24

*Other properties approximately the same for all carbon and alloy steels:
    Modulus of elasticity in tension = 30 000 000 psi (207 GPa)
    Modulus of elasticity in shear = 11 500 000 psi (80 GPa)
    Density = 0.283 lb/in.³ (7680 kg/m³)
†OQT means oil-quenched and tempered. WQT means water-quenched and tempered.

403

TABLE A-2 Properties of Stainless Steels and Nonferrous Metals

Material and Condition	Ultimate Strength s_u ksi	MPa	Yield Strength s_y ksi	MPa	Percent Elongation	Density lb/in.³	kg/m³	Modulus of Elasticity E psi × 10⁻⁶	GPa
Stainless Steels									
AISI 301 annealed	110	758	40	276	60	0.290	8030	28	193
AISI 301 cold-worked	185	1280	140	965	9	0.290	8030	28	193
AISI 430 annealed	75	517	40	276	25	0.280	7750	29	200
AISI 430 cold-worked	90	621	80	552	15	0.280	7750	29	200
AISI 501 annealed	70	483	30	207	28	0.280	7750	29	200
AISI 501 OQT 1000	175	1210	135	931	15	0.280	7750	29	200
17-4PH cond. H900	195	1340	180	1240	13	0.281	7780	28.5	197
PH 13-8 cond. H1000	215	1480	205	1410	13	0.279	7720	29	200
*Copper and its Alloys**									
CDA 145 Copper, hard	48	331	44	303	8	0.323	8940	16	110
CDA 172 Beryllium Copper, hard	175	1210	240	965	8	0.298	8250	19	131
CDA 220 Bronze, hard	61	421	54	372	5	0.318	8800	17	117
CDA 260 Brass, hard	76	524	63	434	8	0.308	8530	16	110
Magnesium									
ASTM AZ 63A-T6	40	276	19	131	5	0.066	1830	6.5	45
Zinc									
Zn-Cu-Ti alloy, cold-rolled	29	200	—	—	44	0.259	7170	13	90
ASTM AC41A-cast	47†	324	—	—	7	0.240	6650	13	90
Titanium									
6Al-4V, aged at 900°F	170	1170	155	1070	8	0.160	4430	16.5	114

*CDA stands for the Copper Development Association.
†Strength in tension. Compressive ultimate strength is 87 ksi (600 MPa).

Material ASTM No.	Products	Ultimate Strength s_u*		Yield Strength s_y*		Percent Elongation in 2 in.
		ksi	MPa	ksi	MPa	
A36	Shapes and plates	50	345	36	248	23
A242, A440, A441	Shapes and plates	67	462	46	317	24
A514	Plates	115	793	100	690	18
A500	Construction tubing	45	310	33	228	25

*Minimum values; may range higher.

TABLE A-4 Minimum Properties of Aluminum Alloys*

Alloy and Temper	Product†	Ultimate Strength s_u		Yield Strength s_y		Percent Elonga- tion	Shear Strength s_{us}	
		ksi	MPa	ksi	MPa		ksi	MPa
1100-H12	SPTR	14	97	11	76	8	9	62
1100-H112	E	11	76	3	21	—	7	48
2014-T4	S	59	407	35	241	14	35	241
2014-T4	E	50	345	35	241	12	30	207
2014-T4	T	54	372	30	207	12	32	221
2014-T4	R	55	379	32	221	16	33	228
2014-T6	S	66	455	58	400	7	40	276
2014-T5	E	60	414	53	365	7	35	241
2014-T6	TR	65	448	55	379	7	38	262
3003-H12	SPT	17	117	12	83	7	11	76
3003-H18	ST	27	186	24	165	3	15	103
5154-H38	STR	45	310	35	241	4	24	165
5154-H112	PRE	30	207	11	76	12	17	117
6061-O	SPERT	18	124	8	55	30	12	83
6061-T4	SPRT	30	207	16	110	14	20	138
6061-T4	E	26	179	16	110	16	17	117
6061-T6	SPRT	42	290	35	241	8	27	186
6061-T6	E	38	262	35	241	8	24	165
6063-T4	T	22	152	10	69	16	13	90
6063-T4	E	18	124	9	62	14	11	76
6063-T6	T	33	228	28	193	12	21	143
6063-T6	E	30	207	25	172	8	19	131
7075-O	SPRET	32	221	14	97	17	22	152
7075-T6	SRET	76	524	66	455	7	44	304

*Modulus of elasticity E for most aluminum alloys, including 1100, 3003, 6061, and 6063, is 10×10^6 psi (69 GPa). For 2014, $E = 10.6 \times 10^6$ psi (73 GPa). For 5154, $E = 10.2 \times 10^6$ psi (70 GPa). For 7075, $E = 10.4 \times 10^6$ psi (72 GPa). Density of most aluminum alloys is 0.10 lb/in.^{3} (2770 kg/m^3).
†Letters stand for: S-Sheet; P-Plate; T-Drawn Tube; R-Rolled Bar; E-Extruded Shapes.

ultimate compressive strength
ultimate strength
X1000

TABLE A-5 Properties of Cast Iron*

Material Type and Grade	Ultimate Strength						Yield Strength				Modulus of Elasticity E‡		Percent Elonga-tion
	s_u†		s_{uc}‡		s_{us}		s_{yt}†		s_{yc}				
	ksi	MPa	ksi	MPa	ksi	MPa	ksi	MPa	ksi	MPa	psi × 10⁻⁶	GPa	
Gray Iron													
ASTM A48													
Grade 20	20	138	80	552	32	221	—	—	—	—	12.2	84	—
Grade 40	40	276	140	965	57	393	—	—	—	—	19.4	134	—
Grade 60	60	414	170	1170	72	496	—	—	—	—	21.5	148	—
Ductile Iron													
ASTM A536													
60-40-18	60	414	—	—	57	393	40	276	56	386	24	165	18
80-55- 6	80	552	—	—	73	503	55	379	88	607	24	165	6
100-70- 3	100	690	180	1240	—	—	70	483	134	924	24	165	3
120-90- 2	120	827	—	—	—	—	90	621	—	—	23	159	2
ASTM A439													
D2	58	400	180	1240	—	—	30	207	35	241	17	117	8
D2C	58	400	—	—	—	—	28	193	33	228	15	103	20
Malleable Iron													
ASTM A47													
35018	53	365	220	1520	48	331	35	241	—	—	25	172	18
ASTM A220													
400 10	60	414	197	1360	—	—	40	276	63	434	26	179	10
600 04	80	552	240	1650	—	—	60	414	90	621	27	186	4
900 01	105	724	290	2000	—	—	90	621	122	841	28	193	1

*The density of cast iron ranges from 0.25 to 0.27 lb/in.³ (6920 to 7480 kg/m³).
†Minimum values; may range higher.
‡Approximate values; may range higher or lower by about 15%.

TABLE A-6 Mechanical Properties of Wood

| Type and Grade | Bending and Tension | | Horizontal Shear | | Compression | | | | Modulus of Elasticity | |
| | | | | | Perpendicular to Grain | | Parallel to Grain | | | |
	psi	MPa	psi	MPa	psi	MPa	psi	MPa	psi	GPa
Douglas Fir										
Select structural	1900	13	120	0.83	415	2.86	1500	10	1.76×10^6	12.1
Construction	1500	10	120	0.83	390	2.69	1200	8	1.76×10^6	12.1
Standard	1200	8	95	0.66	390	2.69	1000	7	1.76×10^6	12.1
Western Hemlock										
Select structural	1600	11	100	0.69	365	2.52	1200	8	1.54×10^6	10.6
Construction	1500	10	100	0.69	365	2.52	1100	7.6	1.54×10^6	10.6
Standard	1200	8	80	0.55	365	2.52	1000	7	1.54×10^6	10.6
*Southern Pine**										
Dense structural 86	2900	20	150	1.03	455	3.14	2200	15	1.76×10^6	12.1
Dense structural 72	2350	16	135	0.93	455	3.14	1800	12	1.76×10^6	12.1
No. 2 stress rated	1200	8	105	0.72	390	2.69	900	6	1.76×10^6	12.1

*Up to 4 in. thick only.

Type	Tensile Strength		Modulus of Elasticity		Coefficient of Thermal Expansion $\alpha \times 10^5$	
	ksi	MPa	ksi	GPa	°F^{-1}	°C^{-1}
ABS	5.9	41	300	2.1	5.3	9.5
Acrylic sheet	10.5	72	450	3.1	3.9	7.0
TFE fluoro-carbons (glass-reinforced)	3.3	23	208	1.4	6.9	12.4
Nylon type 6/6 (glass-reinforced)	30.0	207	1800	12.4	0.9	1.6
Polypropylene (asbestos-reinforced)	4.8	33	600	4.1	2.1	3.8
Rigid vinyl	7.0	48	430	3.0	3.5	6.3
Melamine (amino) (asbestos-filled)	6.0	41	1950	13.4	1.7	3.0
Molded epoxy (glass-filled)	20	138	3040	21.0	1.1	2.0

*Typical properties only. Variations should be expected depending on fillers used and processing.

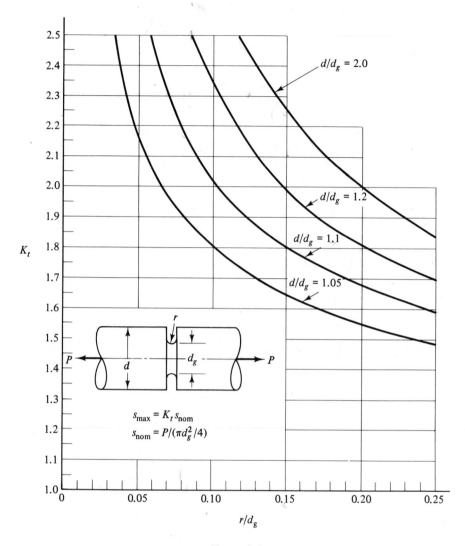

Figure A-1

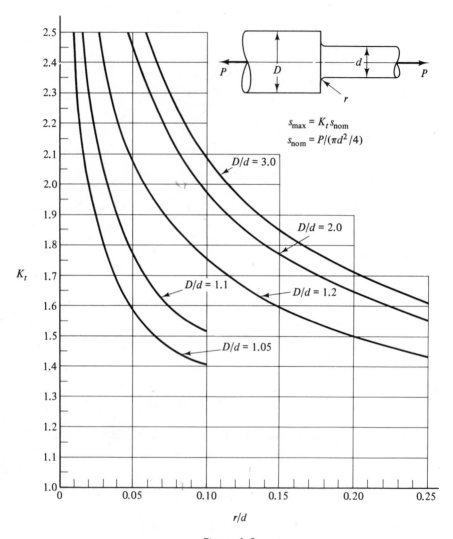

Figure A-2

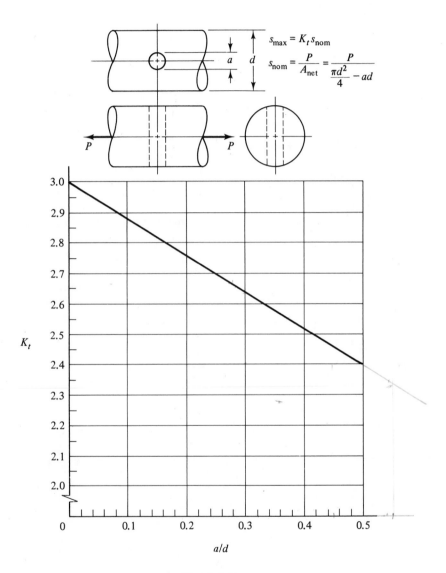

Figure A-3

411

Circle

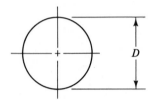

$$A = \pi D^2 /4 \qquad\qquad r = D/4$$

$$I = \pi D^4 /64 \qquad\qquad J = \pi D^4 /32$$

$$Z = \pi D^3/32 \qquad\qquad Z_p = \pi D^3 /16$$

Hollow circle (tube)

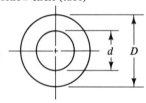

$$A = \pi (D^2 - d^2)/4 \qquad r = \sqrt{D^2 + d^2}/4$$

$$I = \pi (D^4 - d^4)/64 \qquad J' = \pi (D^4 - d^4)/32$$

$$Z = \pi (D^4 - d^4)/32D \quad Z_p = \pi (D^4 - d^4)/16D$$

Square

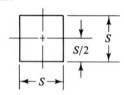

$$A = S^2 \qquad\qquad r = S/\sqrt{12}$$

$$I = S^4 /12$$

$$Z = S^3 /6$$

Rectangle

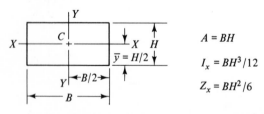

$$A = BH \qquad\qquad r_x = H/\sqrt{12}$$

$$I_x = BH^3 /12 \qquad\qquad r_y = B/\sqrt{12}$$

$$Z_x = BH^2 /6$$

Triangle

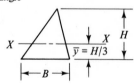

$$A = BH/2 \qquad\qquad r = H/\sqrt{18}$$

$$I = BH^3 /36$$

$$Z = BH^2 /24$$

Semicircle

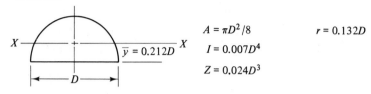

$$A = \pi D^2 /8 \qquad\qquad r = 0.132D$$
$$I = 0.007D^4$$
$$Z = 0.024D^3$$

Regular hexagon

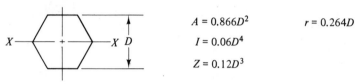

$$A = 0.866D^2 \qquad\qquad r = 0.264D$$
$$I = 0.06D^4$$
$$Z = 0.12D^3$$

*Symbols used are:

A = area
I = moment of inertia
Z = section modulus

r = radius of gyration = $\sqrt{I/A}$
J = polar moment of inertia
Z_p = polar section modulus

TABLE A-9 Preferred Metric Sizes

Nominal Size, mm		Nominal Size, mm		Nominal Size, mm	
First	Second	First	Second	First	Second
1		10		100	
	1.1		11		110
1.2		12		120	
	1.4		14		140
1.6		16		160	
	1.8		18		180
2		20		200	
	2.2		22		220
2.5		25		250	
	2.8		28		280
3		30		300	
	3.5		35		350
4		40		400	
	4.5		45		450
5		50		500	
	5.5		55		550
6		60		600	
	7		70		700
8		80		800	
	9		90		900
				1000	

TABLE A-10 Metric Screw Threads

Thread Designation*	Minimum Minor Dia. D_R, mm	Root Area, mm^2
M1.6 × 0.35	1.063	0.8875
M2 × 0.40	1.394	1.526
M2.5 × 0.45	1.825	2.616
M3 × 0.50	2.256	3.997
M4 × 0.70	2.979	6.970
M5 × 0.80	3.841	11.59
M6 × 1.00	4.563	16.35
M8 × 1.25	6.231	30.49
M10 × 1.50	7.879	48.76
M12 × 1.75	9.543	71.53
M16 × 2.0	13.204	136.9
M20 × 2.5	16.541	214.9
M24 × 3.0	19.855	309.6
M30 × 3.5	25.189	498.3
M36 × 4.0	30.521	731.6

*Designation means, for example:

M36 × 4.0
  └ Pitch in mm
  └ Nominal size (major dia. in mm)
  └ Metric thread

414

TABLE A-11   Unified Standard Screw Threads

Thread Designation*	Minimum Minor Dia. D_R, in.	Root Area, in.^{2}
$\frac{1}{4}$–20	0.185	0.027
$\frac{3}{8}$–16	0.294	0.068
$\frac{1}{2}$–13	0.400	0.126
$\frac{5}{8}$–11	0.507	0.202
$\frac{3}{4}$–10	0.620	0.302
$\frac{7}{8}$– 9	0.731	0.419
1– 8	0.838	0.551
$1\frac{1}{8}$– 7	0.939	0.693
$1\frac{1}{4}$– 7	1.064	0.890
$1\frac{3}{8}$– 6	1.158	1.05
$1\frac{1}{2}$– 6	1.283	1.29
$1\frac{3}{4}$– 5	1.490	1.74
2– $4\frac{1}{2}$	1.711	2.30

*Thread designation gives nominal size (major dia.) followed by number of threads per inch.

W

Wide Flange Shapes

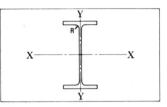

Properties for Designing

Designation and Nominal Size	Weight per Foot	Area	Depth	Flange		Web Thickness	Axis X-X			Axis Y-Y		
				Width	Thickness		I	Z	r	I	Z	r
In.	Lbs.	In.²	In.	In.	In.	In.	In.⁴	In.³	In.	In.⁴	In.³	In.
W24 24 x 12	120	35.4	24.31	12.088	.930	.556	3650	300	10.2	274	45.4	2.78
	110	32.5	24.16	12.042	.855	.510	3330	276	10.1	249	41.4	2.77
	100	29.5	24.00	12.000	.775	.468	3000	250	10.1	223	37.2	2.75
W24 24 x 9	94	27.7	24.29	9.061	.872	.516	2690	221	9.86	108	23.9	1.98
	84	24.7	24.09	9.015	.772	.470	2370	197	9.79	94.5	21.0	1.95
	76	22.4	23.91	8.985	.682	.440	2100	176	9.69	82.6	18.4	1.92
	68	20.0	23.71	8.961	.582	.416	1820	153	9.53	70.0	15.6	1.87
W24 24 x 7	61	18.0	23.72	7.023	.591	.419	1540	130	9.25	34.3	9.76	1.38
	55	16.2	23.55	7.000	.503	.396	1340	114	9.10	28.9	8.25	1.34
W21 21 x 13	142	41.8	21.46	13.132	1.095	.659	3410	317	9.03	414	63.0	3.15
	127	37.4	21.24	13.061	.985	.588	3020	284	8.99	366	56.1	3.13
	112	33.0	21.00	13.000	.865	.527	2620	250	8.92	317	48.8	3.10
W21 21 x 9	96	28.3	21.14	9.038	.935	.575	2100	198	8.61	115	25.5	2.02
	82	24.2	20.86	8.962	.795	.499	1760	169	8.53	95.6	21.3	1.99
W21 21 x 8¼	73	21.5	21.24	8.295	.740	.455	1600	151	8.64	70.6	17.0	1.81
	68	20.0	21.13	8.270	.685	.430	1480	140	8.60	64.7	15.7	1.80
	62	18.3	20.99	8.240	.615	.400	1330	127	8.54	57.5	13.9	1.77
	55	16.2	20.80	8.215	.522	.375	1140	110	8.40	48.3	11.8	1.73
W21 21 x 6½	49	14.4	20.82	6.520	.532	.368	971	93.3	8.21	24.7	7.57	1.31
	44	13.0	20.66	6.500	.451	.348	843	81.6	8.07	20.7	6.38	1.27
W18 18 x 11¾	114	33.5	18.48	11.833	.991	.595	2040	220	7.79	274	46.3	2.86
	105	30.9	18.32	11.792	.911	.554	1850	202	7.75	249	42.3	2.84
	96	28.2	18.16	11.750	.831	.512	1680	185	7.70	225	38.3	2.82
W18 18 x 8¾	85	25.0	18.32	8.838	.911	.526	1440	157	7.57	105	23.8	2.05
	77	22.7	18.16	8.787	.831	.475	1290	142	7.54	94.1	21.4	2.04
	70	20.6	18.00	8.750	.751	.438	1160	129	7.50	84.0	19.2	2.02
	64	18.9	17.87	8.715	.686	.403	1050	118	7.46	75.8	17.4	2.00
W18 18 x 7½	60	17.7	18.25	7.558	.695	.416	986	108	7.47	50.1	13.3	1.68
	55	16.2	18.12	7.532	.630	.390	891	98.4	7.42	45.0	11.9	1.67
	50	14.7	18.00	7.500	.570	.358	802	89.1	7.38	40.2	10.7	1.65
	45	13.2	17.86	7.477	.499	.335	706	79.0	7.30	34.8	9.32	1.62
W18 18 x 6	40	11.8	17.90	6.018	.524	.316	612	68.4	7.21	19.1	6.34	1.27
	35	10.3	17.71	6.000	.429	.298	513	57.9	7.05	15.5	5.16	1.23

W

Wide Flange Shapes

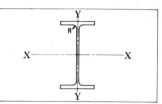

Properties for Designing

Designation and Nominal Size	Weight per Foot	Area	Depth	Flange Width	Flange Thickness	Web Thickness	Axis X-X I	Axis X-X Z	Axis X-X r	Axis Y-Y I	Axis Y-Y Z	Axis Y-Y r
In.	Lbs.	In.²	In.	In.	In.	In.	In.⁴	In.³	In.	In.⁴	In.³	In.
W16 16 x 11½	96	28.2	16.32	11.533	.875	.535	1360	166	6.93	224	38.8	2.82
	88	25.9	16.16	11.502	.795	.504	1220	151	6.87	202	35.1	2.79
W16 16 x 8½	78	23.0	16.32	8.586	.875	.529	1050	128	6.75	92.5	21.6	2.01
	71	20.9	16.16	8.543	.795	.486	941	116	6.71	82.8	19.4	1.99
	64	18.8	16.00	8.500	.715	.443	836	104	6.66	73.3	17.3	1.97
	58	17.1	15.86	8.464	.645	.407	748	94.4	6.62	65.3	15.4	1.96
W16 16 x 7	50	14.7	16.25	7.073	.628	.380	657	80.8	6.68	37.1	10.5	1.59
	45	13.3	16.12	7.039	.563	.346	584	72.5	6.64	32.8	9.32	1.57
	40	11.8	16.00	7.000	.503	.307	517	64.6	6.62	28.8	8.23	1.56
	36	10.6	15.85	6.992	.428	.299	447	56.5	6.50	24.4	6.99	1.52
W16 16 x 5½	31	9.13	15.84	5.525	.442	.275	374	47.2	6.40	12.5	4.51	1.17
	26	7.67	15.65	5.500	.345	.250	300	38.3	6.25	9.59	3.49	1.12
W14 14 x 16	730	215	22.44	17.889	4.910	3.069	14400	1280	8.18	4720	527	4.69
	665	196	21.67	17.646	4.522	2.826	12500	1150	7.99	4170	472	4.62
	605	178	20.94	17.418	4.157	2.598	10900	1040	7.81	3680	423	4.55
	550	162	20.26	17.206	3.818	2.386	9450	933	7.64	3260	378	4.49
	500	147	19.63	17.008	3.501	2.188	8250	840	7.49	2880	339	4.43
	455	134	19.05	16.828	3.213	2.008	7220	758	7.35	2560	304	4.37
	426	125	18.69	16.695	3.033	1.875	6610	707	7.26	2360	283	4.34
	398	117	18.31	16.590	2.843	1.770	6010	657	7.17	2170	262	4.31
	370	109	17.94	16.475	2.658	1.655	5450	608	7.08	1990	241	4.27
	342	101	17.56	16.365	2.468	1.545	4910	559	6.99	1810	221	4.24
	314	92.3	17.19	16.235	2.283	1.415	4400	512	6.90	1630	201	4.20
	287	84.4	16.81	16.130	2.093	1.310	3910	465	6.81	1470	182	4.17
	264	77.6	16.50	16.025	1.938	1.205	3530	427	6.74	1330	166	4.14
	246	72.3	16.25	15.945	1.813	1.125	3230	397	6.68	1230	154	4.12
	237	69.7	16.12	15.910	1.748	1.090	3080	382	6.65	1170	148	4.11
	228	67.1	16.00	15.865	1.688	1.045	2940	368	6.62	1120	142	4.10
	219	64.4	15.87	15.825	1.623	1.005	2800	353	6.59	1070	136	4.08
	211	62.1	15.75	15.800	1.563	.980	2670	339	6.56	1030	130	4.07
	202	59.4	15.63	15.750	1.503	.930	2540	325	6.54	980	124	4.06
	193	56.7	15.50	15.710	1.438	.890	2400	310	6.51	930	118	4.05
	184	54.1	15.38	15.660	1.378	.840	2270	296	6.49	883	113	4.04
	176	51.7	15.25	15.640	1.313	.820	2150	282	6.45	838	107	4.02
	167	49.1	15.12	15.600	1.248	.780	2020	267	6.42	790	101	4.01
	158	46.5	15.00	15.550	1.188	.730	1900	253	6.40	745	95.8	4.00
	150	44.1	14.88	15.515	1.128	.695	1790	240	6.37	703	90.6	3.99
	142	41.8	14.75	15.500	1.063	.680	1670	227	6.32	660	85.2	3.97

W

Wide Flange Shapes

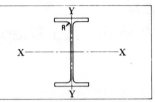

Properties for Designing

Designation and Nominal Size	Weight per Foot	Area	Depth	Flange		Web Thick-ness	Axis X-X			Axis Y-Y		
				Width	Thick-ness		I	Z	r	I	Z	r
In.	Lbs.	In.²	In.	In.	In.	In.	In.⁴	In.³	In.	In.⁴	In.³	In.
W14 14 x 16	320*	94.1	16.81	16.710	2.093	1.890	4140	493	6.63	1640	196	4.17
W14 14 x 14½	136	40.0	14.75	14.740	1.063	.660	1590	216	6.31	568	77.0	3.77
	127	37.3	14.62	14.690	.998	.610	1480	202	6.29	528	71.8	3.76
	119	35.0	14.50	14.650	.938	.570	1370	189	6.26	492	67.1	3.75
	111	32.7	14.37	14.620	.873	.540	1270	176	6.23	455	62.2	3.73
	103	30.3	14.25	14.575	.813	.495	1170	164	6.21	420	57.6	3.72
	95	27.9	14.12	14.545	.748	.465	1060	151	6.17	384	52.8	3.71
	87	25.6	14.00	14.500	.688	.420	967	138	6.15	350	48.2	3.70
W14 14 x 12	84	24.7	14.18	12.023	.778	.451	928	131	6.13	225	37.5	3.02
	78	22.9	14.06	12.000	.718	.428	851	121	6.09	207	34.5	3.00
W14 14 x 10	74	21.8	14.19	10.072	.783	.450	797	112	6.05	133	26.5	2.48
	68	20.0	14.06	10.040	.718	.418	724	103	6.02	121	24.1	2.46
	61	17.9	13.91	10.000	.643	.378	641	92.2	5.98	107	21.5	2.45
W14 14 x 8	53	15.6	13.94	8.062	.658	.370	542	77.8	5.90	57.5	14.3	1.92
	48	14.1	13.81	8.031	.593	.339	485	70.2	5.86	51.3	12.8	1.91
	43	12.6	13.68	8.000	.528	.308	429	62.7	5.82	45.1	11.3	1.89
W14 14 x 6¾	38	11.2	14.12	6.776	.513	.313	386	54.7	5.88	26.6	7.86	1.54
	34	10.0	14.00	6.750	.453	.287	340	48.6	5.83	23.3	6.89	1.52
	30	8.83	13.86	6.733	.383	.270	290	41.9	5.74	19.5	5.80	1.49
W14 14 x 5	26	7.67	13.89	5.025	.418	.255	244	35.1	5.64	8.86	3.53	1.08
	22	6.49	13.72	5.000	.335	.230	198	28.9	5.53	7.00	2.80	1.04
W12 12 x 12	190	55.9	14.38	12.670	1.736	1.060	1890	263	5.82	590	93.1	3.25
	161	47.4	13.88	12.515	1.486	.905	1540	222	5.70	486	77.7	3.20
	133	39.1	13.38	12.365	1.236	.755	1220	183	5.59	390	63.1	3.16
	120	35.3	13.12	12.320	1.106	.710	1070	163	5.51	345	56.0	3.13
	106	31.2	12.88	12.230	.986	.620	931	145	5.46	301	49.2	3.11
	99	29.1	12.75	12.192	.921	.582	859	135	5.43	278	45.7	3.09
	92	27.1	12.62	12.155	.856	.545	789	125	5.40	256	42.2	3.08
	85	25.0	12.50	12.105	.796	.495	723	116	5.38	235	38.9	3.07
	79	23.2	12.38	12.080	.736	.470	663	107	5.34	216	35.8	3.05
	72	21.2	12.25	12.040	.671	.430	597	97.5	5.31	195	32.4	3.04
	65	19.1	12.12	12.000	.606	.390	533	88.0	5.28	175	29.1	3.02
W12 12 x 10	58	17.1	12.19	10.014	.641	.359	476	78.1	5.28	107	21.4	2.51
	53	15.6	12.06	10.000	.576	.345	426	70.7	5.23	96.1	19.2	2.48

*Column core section.

W

Wide Flange Shapes

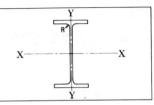

Properties for Designing

Designation and Nominal Size	Weight per Foot	Area	Depth	Flange		Web Thickness	Axis X-X			Axis Y-Y		
				Width	Thickness		I	Z	r	I	Z	r
In.	Lbs.	In.²	In.	In.	In.	In.	In.⁴	In.³	In.	In.⁴	In.³	In.
W12 12 x 8	50	14.7	12.19	8.077	.641	.371	395	64.7	5.18	56.4	14.0	1.96
	45	13.2	12.06	8.042	.576	.336	351	58.2	5.15	50.0	12.4	1.94
	40	11.8	11.94	8.000	.516	.294	310	51.9	5.13	44.1	11.0	1.94
W12 12 x 6½	36	10.6	12.24	6.565	.540	.305	281	46.0	5.15	25.5	7.77	1.55
	31	9.13	12.09	6.525	.465	.265	239	39.5	5.12	21.6	6.61	1.54
	27	7.95	11.96	6.497	.400	.240	204	34.2	5.07	18.3	5.63	1.52
W12 12 x 4	22	6.47	12.31	4.030	.424	.260	156	25.3	4.91	4.64	2.31	.847
	19	5.59	12.16	4.007	.349	.237	130	21.3	4.82	3.76	1.88	.820
	16.5	4.87	12.00	4.000	.269	.230	105	17.6	4.65	2.88	1.44	.770
W12 12 x 4	14	4.12	11.91	3.968	.224	.198	88.0	14.8	4.62	2.34	1.18	.754
W10 10 x 10	112	32.9	11.38	10.415	1.248	.755	719	126	4.67	235	45.2	2.67
	100	29.4	11.12	10.345	1.118	.685	625	112	4.61	207	39.9	2.65
	89	26.2	10.88	10.275	.998	.615	542	99.7	4.55	181	35.2	2.63
	77	22.7	10.62	10.195	.868	.535	457	86.1	4.49	153	30.1	2.60
	72	21.2	10.50	10.170	.808	.510	421	80.1	4.46	142	27.9	2.59
	66	19.4	10.38	10.117	.748	.457	382	73.7	4.44	129	25.5	2.58
	60	17.7	10.25	10.075	.683	.415	344	67.1	4.41	116	23.1	2.57
	54	15.9	10.12	10.028	.618	.368	306	60.4	4.39	104	20.7	2.56
	49	14.4	10.00	10.000	.558	.340	273	54.6	4.35	93.0	18.6	2.54
W10 10 x 8	45	13.2	10.12	8.022	.618	.350	249	49.1	4.33	53.2	13.3	2.00
	39	11.5	9.94	7.990	.528	.318	210	42.2	4.27	44.9	11.2	1.98
	33	9.71	9.75	7.964	.433	.292	171	35.0	4.20	36.5	9.16	1.94
W10 10 x 5¾	29	8.54	10.22	5.799	.500	.289	158	30.8	4.30	16.3	5.61	1.38
	25	7.36	10.08	5.762	.430	.252	133	26.5	4.26	13.7	4.76	1.37
	21	6.20	9.90	5.750	.340	.240	107	21.5	4.15	10.8	3.75	1.32
W10 10 x 4	19	5.61	10.25	4.020	.394	.250	96.3	18.8	4.14	4.28	2.13	.874
	17	4.99	10.12	4.010	.329	.240	81.9	16.2	4.05	3.55	1.77	.844
	15	4.41	10.00	4.000	.269	.230	68.9	13.8	3.95	2.88	1.44	.809
W10 10 x 4	11.5	3.39	9.87	3.950	.204	.180	52.0	10.5	3.92	2.10	1.06	.787

W
Wide Flange Shapes

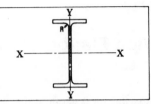

Properties for Designing

Designation and Nominal Size	Weight per Foot	Area	Depth	Flange Width	Flange Thickness	Web Thickness	Axis X-X			Axis Y-Y		
In.	Lbs.	In.²	In.	In.	In.	In.	I In.⁴	Z In.³	r In.	I In.⁴	Z In.³	r In.
W8 8 x 8	67	19.7	9.00	8.287	.933	.575	272	60.4	3.71	88.6	21.4	2.12
	58	17.1	8.75	8.222	.808	.510	227	52.0	3.65	74.9	18.2	2.10
	48	14.1	8.50	8.117	.683	.405	184	43.2	3.61	60.9	15.0	2.08
	40	11.8	8.25	8.077	.558	.365	146	35.5	3.53	49.0	12.1	2.04
	35	10.3	8.12	8.027	.493	.315	126	31.1	3.50	42.5	10.6	2.03
	31	9.12	8.00	8.000	.433	.288	110	27.4	3.47	37.0	9.24	2.01
W8 8 x 6½	28	8.23	8.06	6.540	.463	.285	97.8	24.3	3.45	21.6	6.61	1.62
	24	7.06	7.93	6.500	.398	.245	82.5	20.8	3.42	18.2	5.61	1.61
W8 8 x 5¼	20	5.89	8.14	5.268	.378	.248	69.4	17.0	3.43	9.22	3.50	1.25
	17	5.01	8.00	5.250	.308	.230	56.6	14.1	3.36	7.44	2.83	1.22
W8 8 x 4	15	4.43	8.12	4.015	.314	.245	48.1	11.8	3.29	3.40	1.69	.876
	13	3.83	8.00	4.000	.254	.230	39.6	9.90	3.21	2.72	1.36	.842
W8 8 x 4	10	2.96	7.90	3.940	.204	.170	30.8	7.80	3.23	2.08	1.06	.839
W6 6 x 6	25	7.35	6.37	6.080	.456	.320	53.3	16.7	2.69	17.1	5.62	1.53
	20	5.88	6.20	6.018	.367	.258	41.5	13.4	2.66	13.3	4.43	1.51
	15.5	4.56	6.00	5.995	.269	.235	30.1	10.0	2.57	9.67	3.23	1.46
W6 6 x 4	16	4.72	6.25	4.030	.404	.260	31.7	10.2	2.59	4.42	2.19	.967
	12	3.54	6.00	4.000	.279	.230	21.7	7.25	2.48	2.98	1.49	.918
W6 6 x 4	8.5	2.51	5.83	3.940	.194	.170	14.8	5.08	2.43	1.98	1.01	.889

Source: *Shapes and Plates* (Pittsburgh, Pa. © United States Steel).

TABLE A-13   American Standard Beams

S

American Standard Beams

Properties for Designing

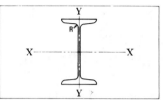

Designation and Nominal Size	Weight per Foot	Area	Depth	Width of Flange	Aver. Flange Thick-ness	Web Thick-ness	Axis X-X			Axis Y-Y		
							I	Z	r	I	Z	r
In.	Lbs.	In.²	In.	In.	In.	In.	In.⁴	In.³	In.	In.⁴	In.³	In.
S18 18 x 6	70	20.6	18.00	6.251	.691	.711	926	103	6.71	24.1	7.72	1.08
	54.7	16.1	18.00	6.001	.691	.461	804	89.4	7.07	20.8	6.94	1.14
S15 15 x 5½	50	14.7	15.00	5.640	.622	.550	486	64.8	5.75	15.7	5.57	1.03
	42.9	12.6	15.00	5.501	.622	.411	447	59.6	5.95	14.4	5.23	1.07
S12 12 x 5¼	50	14.7	12.00	5.477	.659	.687	305	50.8	4.55	15.7	5.74	1.03
	40.8	12.0	12.00	5.252	.659	.462	272	45.4	4.77	13.6	5.16	1.06
S12 12 x 5	35	10.3	12.00	5.078	.544	.428	229	38.2	4.72	9.87	3.89	.980
	31.8	9.35	12.00	5.000	.544	.350	218	36.4	4.83	9.36	3.74	1.00
S10 10 x 4⅝	35	10.3	10.00	4.944	.491	.594	147	29.4	3.78	8.36	3.38	.901
	25.4	7.46	10.00	4.661	.491	.311	124	24.7	4.07	6.79	2.91	.954
S8 8 x 4	23	6.77	8.00	4.171	.425	.441	64.9	16.2	3.10	4.31	2.07	.798
	18.4	5.41	8.00	4.001	.425	.271	57.6	14.4	3.26	3.73	1.86	.831
S7 7 x 3⅝	15.3	4.50	7.00	3.662	.392	.252	36.7	10.5	2.86	2.64	1.44	.766
S6 6 x 3⅜	17.25	5.07	6.00	3.565	.359	.465	26.3	8.77	2.28	2.31	1.30	.675
	12.5	3.67	6.00	3.332	.359	.232	22.1	7.37	2.45	1.82	1.09	.705
S5 5 x 3	10	2.94	5.00	3.004	.326	.214	12.3	4.92	2.05	1.22	.809	.643
S4 4 x 2⅝	7.7	2.26	4.00	2.663	.293	.193	6.08	3.04	1.64	.764	.574	.581
S3 3 x 2⅜	5.7	1.67	3.00	2.330	.260	.170	2.52	1.68	1.23	.455	.390	.522

Source: *Shapes and Plates* (Pittsburgh, Pa. © United States Steel).

TABLE A-14 American Standard Channels

C

American Standard Channels

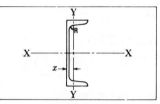

Properties for Designing

Designation and Nominal Size	Weight per Foot	Area	Depth	Width of Flange	Aver. Flange Thickness	Web Thickness	Axis X-X			Axis Y-Y			$\frac{x}{r}$
							I	Z	r	I	Z	r	
In.	Lbs.	In.²	In.	In.	In.	In.	In.⁴	In.³	In.	In.⁴	In.³	In.	In.
C15 15 x 3⅜	50	14.7	15.00	3.716	.650	.716	404	53.8	5.24	11.0	3.78	.867	.799
	40	11.8	15.00	3.520	.650	.520	349	46.5	5.44	9.23	3.36	.886	.778
	33.9	9.96	15.00	3.400	.650	.400	315	42.0	5.62	8.13	3.11	.904	.787
C12 12 x 3	30	8.82	12.00	3.170	.501	.510	162	27.0	4.29	5.14	2.06	.763	.674
	25	7.35	12.00	3.047	.501	.387	144	24.1	4.43	4.47	1.88	.780	.674
	20.7	6.09	12.00	2.942	.501	.282	129	21.5	4.61	3.88	1.73	.799	.698
C10 10 x 2⅝	30	8.82	10.00	3.033	.436	.673	103	20.7	3.42	3.94	1.65	.669	.649
	25	7.35	10.00	2.886	.436	.526	91.2	18.2	3.52	3.36	1.48	.676	.617
	20	5.88	10.00	2.739	.436	.379	78.9	15.8	3.66	2.81	1.32	.691	.606
	15.3	4.49	10.00	2.600	.436	.240	67.4	13.5	3.87	2.28	1.16	.713	.634
C9 9 x 2½	15	4.41	9.00	2.485	.413	.285	51.0	11.3	3.40	1.93	1.01	.661	.586
	13.4	3.94	9.00	2.433	.413	.233	47.9	10.6	3.48	1.76	.962	.668	.601
C8 8 x 2¼	18.75	5.51	8.00	2.527	.390	.487	44.0	11.0	2.82	1.98	1.01	.599	.565
	13.75	4.04	8.00	2.343	.390	.303	36.1	9.03	2.99	1.53	.853	.615	.553
	11.5	3.38	8.00	2.260	.390	.220	32.6	8.14	3.11	1.32	.781	.625	.571
C7 7 x 2¼	12.25	3.60	7.00	2.194	.366	.314	24.2	6.93	2.60	1.17	.702	.571	.525
	9.8	2.87	7.00	2.090	.366	.210	21.3	6.08	2.72	.968	.625	.581	.541
C6 6 x 2	13	3.83	6.00	2.157	.343	.437	17.4	5.80	2.13	1.05	.642	.525	.514
	10.5	3.09	6.00	2.034	.343	.314	15.2	5.06	2.22	.865	.564	.529	.500
	8.2	2.40	6.00	1.920	.343	.200	13.1	4.38	2.34	.692	.492	.537	.512
C5 5 x 1¾	9	2.64	5.00	1.885	.320	.325	8.90	3.56	1.83	.632	.449	.489	.478
	6.7	1.97	5.00	1.750	.320	.190	7.49	3.00	1.95	.478	.378	.493	.484
C4 4 x 1⅝	7.25	2.13	4.00	1.721	.296	.321	4.59	2.29	1.47	.432	.343	.450	.459
	5.4	1.59	4.00	1.584	.296	.184	3.85	1.93	1.56	.319	.283	.449	.458
C3 3 x 1½	5	1.47	3.00	1.498	.273	.258	1.85	1.24	1.12	.247	.233	.410	.438
	4.1	1.21	3.00	1.410	.273	.170	1.66	1.10	1.17	.197	.202	.404	.437

Source: *Shapes and Plates* (Pittsburgh, Pa. © United States Steel).

TABLE A-15 Angles—Equal Leg

L

Angles
Equal Leg

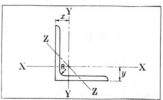

Properties for Designing

Designation and Nominal Size	Thickness	Weight per Foot	Area	Axis X-X and Axis Y-Y					Fillet Radius R
				I	Z	r	x or y	r_{min}	
In.	In.	Lbs.	In.^{2}	In.^{4}	In.^{3}	In.	In.	In.	In.
L8x8	1⅛	56.9	16.7	98.0	17.5	2.42	2.41	1.56	
	1	51.0	15.0	89.0	15.8	2.44	2.37	1.56	
	⅞	45.0	13.2	79.6	14.0	2.45	2.32	1.57	⅝
	¾	38.9	11.4	69.7	12.2	2.47	2.28	1.58	
	⅝	32.7	9.61	59.4	10.3	2.49	2.23	1.58	
	½	26.4	7.75	48.6	8.36	2.50	2.19	1.59	
L6x6	1	37.4	11.0	35.5	8.57	1.80	1.86	1.17	
	⅞	33.1	9.73	31.9	7.63	1.81	1.82	1.17	
	¾	28.7	8.44	28.2	6.66	1.83	1.78	1.17	
	⅝	24.2	7.11	24.2	5.66	1.84	1.73	1.18	½
	½	19.6	5.75	19.9	4.61	1.86	1.68	1.18	
	⅜	14.9	4.36	15.4	3.53	1.88	1.64	1.19	
L5x5	⅞	27.2	7.98	17.8	5.17	1.49	1.57	.973	
	¾	23.6	6.94	15.7	4.53	1.51	1.52	.975	
	½	16.2	4.75	11.3	3.16	1.54	1.43	.983	½
	⅜	12.3	3.61	8.74	2.42	1.56	1.39	.990	
	⁵⁄₁₆	10.3	3.03	7.42	2.04	1.57	1.37	.994	
L4x4	¾	18.5	5.44	7.67	2.81	1.19	1.27	.778	
	⅝	15.7	4.61	6.66	2.40	1.20	1.23	.779	
	½	12.8	3.75	5.56	1.97	1.22	1.18	.782	
	⅜	9.8	2.86	4.36	1.52	1.23	1.14	.788	⅜
	⁵⁄₁₆	8.2	2.40	3.71	1.29	1.24	1.12	.791	
	¼	6.6	1.94	3.04	1.05	1.25	1.09	.795	
L3½x3½	⅜	8.5	2.48	2.87	1.15	1.07	1.01	.687	
	⁵⁄₁₆	7.2	2.09	2.45	.976	1.08	.990	.690	⅜
	¼	5.8	1.69	2.01	.794	1.09	.968	.694	
L3x3	½	9.4	2.75	2.22	1.07	.898	.932	.584	
	⅜	7.2	2.11	1.76	.833	.913	.888	.587	
	⁵⁄₁₆	6.1	1.78	1.51	.707	.922	.865	.589	⁵⁄₁₆
	¼	4.9	1.44	1.24	.577	.930	.842	.592	
	³⁄₁₆	3.71	1.09	.962	.441	.939	.820	.596	

TABLE A-15 (*Cont.*)

L

Bar Size Angles Equal Legs

Dimensions and Properties for Designing

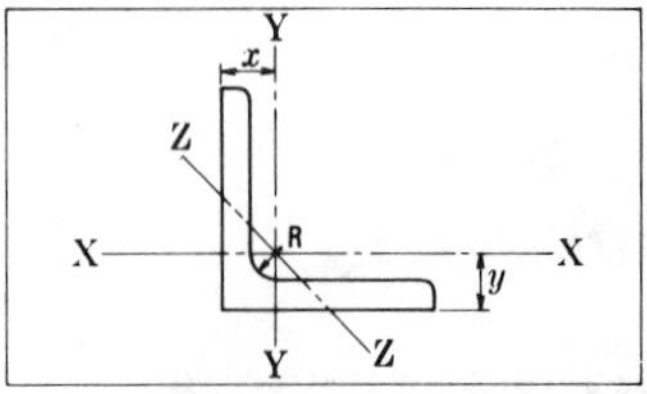

Designation and Nominal Size	Thickness	Weight per Foot	Area	Axis X-X and Axis Y-Y				Axis Z-Z	Fillet Radius R
				I	Z	r	x or y	$r_{min.}$	
In.	In.	Lbs.	In.²	In.⁴	In.³	In.	In.	In.	In.
L2½x2½	½	7.7	2.25	1.23	.724	.739	.806	.487	5/16*
	⅜	5.9	1.73	.984	.566	.753	.762	.487	
	5/16	5.0	1.46	.849	.482	.761	.740	.489	
	¼	4.1	1.19	.703	.394	.769	.717	.491	
	3/16	3.07	.902	.547	.303	.778	.694	.495	
	⅛	2.07	.609	.378	.207	.788	.671	.499	3/16
L2x2	⅜	4.7	1.36	.479	.351	.594	.636	.389	¼*
	5/16	3.92	1.15	.416	.300	.601	.614	.390	
	¼	3.19	.938	.348	.247	.609	.592	.391	
	.205	2.65	.778	.294	.207	.615	.575	.393	
	3/16	2.44	.715	.272	.190	.617	.569	.394	
	⅛	1.65	.484	.190	.131	.626	.546	.398	
L1¾x1¾	5/16	3.39	1.00	.271	.226	.521	.551	.341	¼*
	¼	2.77	.813	.227	.186	.529	.529	.341	
	3/16	2.12	.621	.179	.144	.537	.506	.343	
	⅛	1.44	.422	.126	.099	.546	.484	.347	
L1½x1½	5/16	2.86	.840	.164	.162	.442	.488	.292	3/16*
	¼	2.34	.688	.139	.134	.449	.466	.292	
	3/16	1.80	.527	.110	.104	.457	.444	.293	
	.165	1.59	.468	.099	.093	.460	.436	.294	
	5/32	1.52	.444	.094	.088	.461	.433	.295	
	⅛	1.23	.359	.078	.072	.465	.421	.296	
L1¼x1¼	¼	1.92	.563	.077	.091	.369	.403	.243	3/16*
	3/16	1.48	.434	.061	.071	.377	.381	.244	
	⅛	1.01	.297	.044	.049	.385	.359	.246	
L1⅛x1⅛	⅛	.90	.266	.032	.040	.345	.327	.221	3/32
L1x1	¼	1.49	.438	.037	.056	.290	.339	.196	⅛*
	3/16	1.16	.340	.030	.044	.297	.318	.195	⅛*
	⅛	.80	.234	.022	.031	.304	.296	.196	⅛*
L⅞x⅞	⅛	.70	.203	.014	.023	.264	.264	.171	⅛*
L¾x¾	⅛	.59	.172	.009	.017	.224	.233	.146	⅛*
L⅝x⅝	⅛	.48	.141	.005	.011	.185	.201	.122	5/64
L½x½	⅛	.38	.109	.002	.007	.145	.170	.098	1/32

*Angles are produced with various size fillet radii depending on where they are rolled. The maximum radii produced are those shown.

Source: *Shapes and Plates* (Pittsburgh, Pa. © United States Steel).

TABLE A-16 Angles—Unequal Leg

L

Angles Unequal Leg

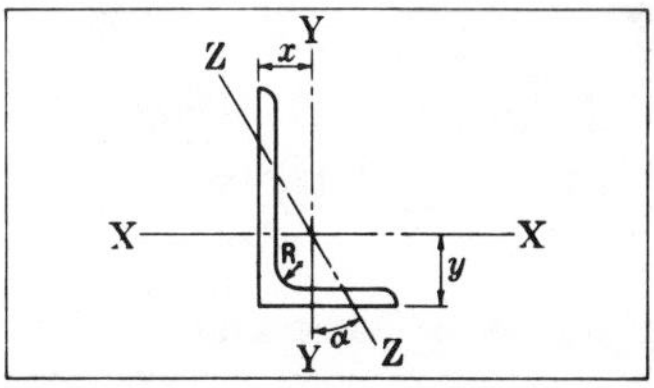

Properties for Designing

Designation and Nominal Size	Thickness	Weight per Foot	Area	Axis X-X I	Axis X-X Z	Axis X-X r	Axis X-X y	Axis Y-Y I	Axis Y-Y Z	Axis Y-Y r	Axis Y-Y x	Axis Z-Z r_{min}	Axis Z-Z Tan α	Fillet Radius R
In.	In.	Lbs.	In.^{2}	In.^{4}	In.^{3}	In.	In.	In.^{4}	In.^{3}	In.	In.	In.		In.
L8x6	1	44.2	13.0	80.8	15.1	2.49	2.65	38.8	8.92	1.73	1.65	1.28	.543	½
	¾	33.8	9.94	63.4	11.7	2.53	2.56	30.7	6.92	1.76	1.56	1.29	.551	
	½	23.0	6.75	44.3	8.02	2.56	2.47	21.7	4.79	1.79	1.47	1.30	.558	
L8x4	1	37.4	11.0	69.6	14.1	2.52	3.05	11.6	3.94	1.03	1.05	.846	.247	½
	¾	28.7	8.44	54.9	10.9	2.55	2.95	9.36	3.07	1.05	.953	.852	.258	
	½	19.6	5.75	38.5	7.49	2.59	2.86	6.74	2.15	1.08	.859	.865	.267	
L7x4	¾	26.2	7.69	37.8	8.42	2.22	2.51	9.05	3.03	1.09	1.01	.860	.324	½
	½	17.9	5.25	26.7	5.81	2.25	2.42	6.53	2.12	1.11	.917	.872	.335	
	⅜	13.6	3.98	20.6	4.44	2.27	2.37	5.10	1.63	1.13	.870	.880	.340	
L6x4	¾	23.6	6.94	24.5	6.25	1.88	2.08	8.68	2.97	1.12	1.08	.860	.428	½
	½	16.2	4.75	17.4	4.33	1.91	1.99	6.27	2.08	1.15	.987	.870	.440	
	⅜	12.3	3.61	13.5	3.32	1.93	1.94	4.90	1.60	1.17	.941	.877	.446	
L6x3½	⅜	11.7	3.42	12.9	3.24	1.94	2.04	3.34	1.23	.980	.787	.767	.350	½
	5/16	9.8	2.87	10.9	2.73	1.95	2.01	2.85	1.04	.996	.763	.772	.352	
L5x3½	¾	19.8	5.81	13.9	4.28	1.55	1.75	5.55	2.22	.977	.996	.748	.464	7/16
	½	13.6	4.00	9.99	2.99	1.58	1.66	4.05	1.56	1.01	.906	.755	.479	
	⅜	10.4	3.05	7.78	2.29	1.60	1.61	3.18	1.21	1.02	.861	.762	.486	
	5/16	8.7	2.56	6.60	1.94	1.61	1.59	2.72	1.02	1.03	.838	.766	.489	
L5x3	⅝	15.7	4.61	11.4	3.55	1.57	1.80	3.06	1.39	.815	.796	.644	.349	⅜
	½	12.8	3.75	9.45	2.91	1.59	1.75	2.58	1.15	.829	.750	.648	.357	
	⅜	9.8	2.86	7.37	2.24	1.61	1.70	2.04	.888	.845	.704	.654	.364	
	5/16	8.2	2.40	6.26	1.89	1.61	1.68	1.75	.753	.853	.681	.658	.368	
	¼	6.6	1.94	5.11	1.53	1.62	1.66	1.44	.614	.861	.657	.663	.371	
L4x3½	½	11.9	3.50	5.32	1.94	1.23	1.25	3.79	1.52	1.04	1.00	.722	.750	⅜
	⅜	9.1	2.67	4.18	1.49	1.25	1.21	2.95	1.17	1.06	.955	.727	.755	
	5/16	7.7	2.25	3.56	1.26	1.26	1.18	2.55	.994	1.07	.932	.730	.757	
	¼	6.2	1.81	2.91	1.03	1.27	1.16	2.09	.808	1.07	.909	.734	.759	
L4x3	½	11.1	3.25	5.05	1.89	1.25	1.33	2.42	1.12	.864	.827	.639	.543	⅜
	⅜	8.5	2.48	3.96	1.46	1.26	1.28	1.92	.866	.879	.782	.644	.551	
	5/16	7.2	2.09	3.38	1.23	1.27	1.26	1.65	.734	.887	.759	.647	.554	
	¼	5.8	1.69	2.77	1.00	1.28	1.24	1.36	.599	.896	.736	.651	.558	
L3½x3	⅜	7.9	2.30	2.72	1.13	1.09	1.08	1.85	.851	.897	.830	.625	.721	⅜
	5/16	6.6	1.93	2.33	.954	1.10	1.06	1.58	.722	.905	.808	.627	.724	
	¼	5.4	1.56	1.91	.776	1.11	1.04	1.30	.589	.914	.785	.631	.727	
L3½x2½	⅜	7.2	2.11	2.56	1.09	1.10	1.16	1.09	.592	.719	.660	.537	.496	5/16
	5/16	6.1	1.78	2.19	.927	1.11	1.14	.939	.504	.727	.637	.540	.501	
	¼	4.9	1.44	1.80	.755	1.12	1.11	.777	.412	.735	.614	.544	.506	
L3x2½	⅜	6.6	1.92	1.66	.810	.928	.956	1.04	.581	.736	.706	.522	.676	5/16
	5/16	5.6	1.62	1.42	.688	.937	.933	.898	.494	.744	.683	.525	.680	
	¼	4.5	1.31	1.17	.561	.945	.911	.743	.404	.753	.661	.528	.684	
L3x2	⅜	5.9	1.73	1.53	.781	.940	1.04	.543	.371	.559	.539	.430	.428	5/16
	5/16	5.0	1.46	1.32	.664	.948	1.02	.470	.317	.567	.516	.432	.435	
	¼	4.1	1.19	1.09	.542	.957	.993	.392	.260	.574	.493	.435	.440	

L

Bar Size Angles Unequal Legs

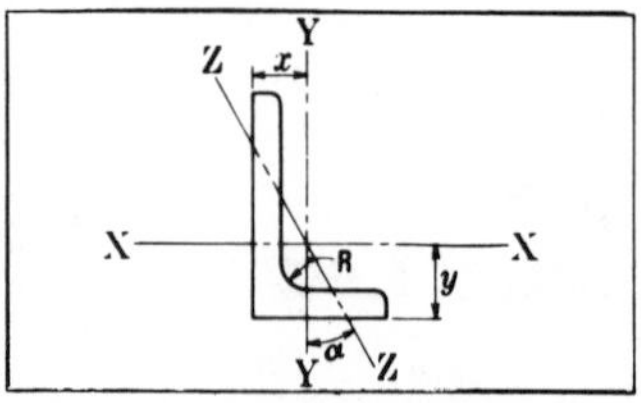

Properties for Designing

Designation and Nominal Size	Thickness	Weight per Foot	Area	Axis X-X				Axis Y-Y				Axis Z-Z		Fillet Radius R
				I	Z	r	y	I	Z	r	x	$r_{min.}$	Tan α	
In.	In.	Lbs.	In.^{2}	In.^{4}	In.^{3}	In.	In.	In.^{4}	In.^{3}	In.	In.	In.		In.
L2½x2	⅜	5.3	1.55	.912	.547	.768	.831	.514	.363	.577	.581	.420	.614	
	5/16	4.5	1.31	.788	.466	.776	.809	.446	.310	.584	.559	.422	.620	3/16*
	¼	3.62	1.06	.654	.381	.784	.787	.372	.254	.592	.537	.424	.626	
	3/16	2.75	.809	.509	.293	.793	.764	.291	.196	.600	.514	.427	.631	
L2½x1½	5/16	3.92	1.15	.711	.444	.785	.898	.191	.174	.408	.398	.322	.349	
	¼	3.19	.938	.591	.364	.794	.875	.161	.143	.415	.375	.324	.357	3/16*
	3/16	2.44	.715	.461	.279	.803	.852	.127	.111	.422	.352	.327	.364	
L2¼x1½	3/16	2.28	.668	.344	.229	.718	.745	.124	.110	.431	.370	.326	.440	⅛
L2x1½	¼	2.77	.813	.316	.236	.623	.663	.151	.139	.432	.413	.320	.543	
	3/16	2.12	.621	.248	.182	.632	.641	.120	.108	.440	.391	.322	.551	3/16*
	⅛	1.44	.422	.173	.125	.641	.618	.085	.075	.448	.368	.326	.558	
L2x1¼	5/16	3.12	.918	.353	.278	.620	.731	.104	.117	.337	.356	.268	.367	
	¼	2.55	.750	.296	.229	.628	.703	.089	.097	.344	.333	.269	.378	¼*
	3/16	1.96	.574	.232	.177	.636	.686	.071	.075	.351	.311	.271	.387	
	⅛	1.33	.391	.163	.122	.645	.663	.050	.052	.359	.237	.274	.396	
L2x1	3/16	1.81	.527	.214	.170	.638	.738	.037	.048	.263	.238	.213	.258	¼
L1¾x1¼	¼	2.34	.688	.202	.176	.543	.602	.085	.095	.352	.352	.267	.486	
	3/16	1.80	.527	.160	.137	.551	.580	.068	.074	.359	.330	.269	.496	3/16*
	⅛	1.23	.359	.113	.094	.560	.557	.049	.051	.368	.307	.272	.506	
L1½x1¼	¼	2.13	.625	.130	.130	.456	.500	.081	.093	.361	.375	.260	.667	
	3/16	1.64	.480	.104	.104	.464	.478	.065	.073	.368	.353	.261	.676	⅛
L1½x¾	⅛	.91	.266	.061	.064	.480	.548	.011	.018	.199	.173	.160	.261	3/32
L1⅜x⅞	3/16	1.32	.387	.071	.081	.429	.490	.022	.035	.240	.240	.188	.387	
	⅛	.91	.266	.051	.056	.438	.467	.016	.025	.247	.217	.190	.401	⅛*
L1x¾	⅛	.70	.203	.020	.030	.312	.332	.009	.017	.216	.207	.160	.543	1/16
L1x⅝	⅛	.64	.188	.019	.029	.314	.354	.006	.012	.172	.167	.134	.378	⅛*

*Angles are produced with various size fillet radii depending on where they are rolled. The maximum radii produced are those shown.

Source: *Shapes and Plates* (Pittsburgh, Pa. © United States Steel).

TS
F36 Construction Pipe

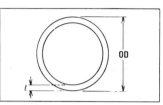

Properties for Designing

Nominal Size ((OD))	Wall Thickness t	Weight per Foot	Area	I	Z	r
In.	In.	Lbs.	In.²			
TS½ (.840)	.109	.85	.250	.017	.041	.261
	.147	1.09	.320	.020	.048	.250
TS¾ (1.050)	.113	1.13	.333	.037	.071	.334
	.154	1.47	.433	.045	.085	.321
TS1 (1.315)	.133	1.68	.494	.087	.133	.421
	.179	2.17	.639	.106	.161	.407
TS1¼ (1.660)	.140	2.27	.669	.195	.235	.540
	.191	3.00	.881	.242	.291	.524
TS1½ (1.900)	.145	2.72	.799	.310	.326	.623
	.200	3.63	1.07	.391	.412	.605
TS2 (2.375)	.154	3.65	1.07	.666	.561	.787
	.218	5.02	1.48	.868	.731	.766
TS2½ (2.875)	.203	5.79	1.70	1.53	1.06	.947
	.276	7.66	2.25	1.92	1.34	.924
TS3 (3.500)	.156	5.57	1.64	2.30	1.31	1.18
	.188	6.65	1.96	2.69	1.54	1.17
	.216	7.58	2.23	3.02	1.72	1.16
	.300	10.25	3.02	3.89	2.23	1.14
TS3½ (4.000)	.188	7.65	2.25	4.10	2.05	1.35
	.226	9.11	2.68	4.79	2.39	1.34
	.318	12.50	3.68	6.28	3.14	1.31
TS4 (4.500)	.156	7.24	2.13	5.03	2.23	1.54
	.188	8.66	2.55	5.93	2.64	1.53
	.219	10.01	2.95	6.77	3.01	1.52
	.237	10.79	3.17	7.23	3.21	1.51
	.337	14.98	4.41	9.61	4.27	1.48

F36 pipe is for general structural use and not intended for applications involving internal pressure.

Source: *Shapes and Plates* (Pittsburgh, Pa. © United States Steel).

TABLE A-18 Aluminum Association Standard Channels—Dimensions, Areas, Weights and Section Properties

Aluminum Association Standard Channels—
Dimensions, Areas, Weights and Section Properties

| Size | | Area① | Weight② | Flange Thick-ness | Web Thick-ness | Fillet Radius | Section Properties⑧ | | | | | | |
| Depth A in. | Width B in. | in.² | lb/ft | t₁ in. | t in. | R in. | Axis X-X | | | Axis Y-Y | | | |
							I in.⁴	Z in.³	r in.	I in.⁴	Z in.³	r in.	x in.
2.00	1.00	0.491	0.577	0.13	0.13	0.10	0.288	0.288	0.766	0.045	0.064	0.303	0.298
2.00	1.25	0.911	1.071	0.26	0.17	0.15	0.546	0.546	0.774	0.139	0.178	0.391	0.471
3.00	1.50	0.965	1.135	0.20	0.13	0.25	1.41	0.94	1.21	0.22	0.22	0.47	0.49
3.00	1.75	1.358	1.597	0.26	0.17	0.25	1.97	1.31	1.20	0.42	0.37	0.55	0.62
4.00	2.00	1.478	1.738	0.23	0.15	0.25	3.91	1.95	1.63	0.60	0.45	0.64	0.65
4.00	2.25	1.982	2.331	0.29	0.19	0.25	5.21	2.60	1.62	1.02	0.69	0.72	0.78
5.00	2.25	1.881	2.212	0.26	0.15	0.30	7.88	3.15	2.05	0.98	0.64	0.72	0.73
5.00	2.75	2.627	3.089	0.32	0.19	0.30	11.14	4.45	2.06	2.05	1.14	0.88	0.95
6.00	2.50	2.410	2.834	0.29	0.17	0.30	14.35	4.78	2.44	1.53	0.90	0.80	0.79
6.00	3.25	3.427	4.030	0.35	0.21	0.30	21.04	7.01	2.48	3.76	1.76	1.05	1.12
7.00	2.75	2.725	3.205	0.29	0.17	0.30	22.09	6.31	2.85	2.10	1.10	0.88	0.84
7.00	3.5	4.009	4.715	0.38	0.21	0.30	33.79	9.65	2.90	5.13	2.23	1.13	1.20
8.00	3.00	3.526	4.147	0.35	0.19	0.30	37.40	9.35	3.26	3.25	1.57	0.96	0.93
8.00	3.75	4.923	5.789	0.41	0.25	0.35	52.69	13.17	3.27	7.13	2.82	1.20	1.22
9.00	3.25	4.237	4.983	0.35	0.23	0.35	54.41	12.09	3.58	4.40	1.89	1.02	0.93
9.00	4.00	5.927	6.970	0.44	0.29	0.35	78.31	17.40	3.63	9.61	3.49	1.27	1.25
10.00	3.50	5.218	6.136	0.41	0.25	0.35	83.22	16.64	3.99	6.33	2.56	1.10	1.02
10.00	4.25	7.109	8.360	0.50	0.31	0.40	116.15	23.23	4.04	13.02	4.47	1.35	1.34
12.00	4.00	7.036	8.274	0.47	0.29	0.40	159.76	26.63	4.77	11.03	3.86	1.25	1.14
12.00	5.00	10.053	11.822	0.62	0.35	0.45	239.69	39.95	4.88	25.74	7.60	1.60	1.61

TABLE A-18 *(Cont.)*

Aluminum Association Standard I-Beams — Dimensions, Areas, Weights and Section Properties

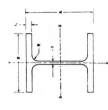

Size							Section Properties③					
Depth A in.	Width B in.	Area① in.²	Weight② lb/ft	Flange Thickness t₁ in.	Web Thickness t in.	Fillet Radius R in.	Axis X-X			Axis Y-Y		
							I in.⁴	Z in.³	r in.	I in.⁴	Z in.³	r in.
3.00	2.50	1.392	1.637	0.20	0.13	0.25	2.24	1.49	1.27	0.52	0.42	0.61
3.00	2.50	1.726	2.030	0.26	0.15	0.25	2.71	1.81	1.25	0.68	0.54	0.63
4.00	3.00	1.965	2.311	0.23	0.15	0.25	5.62	2.81	1.69	1.04	0.69	0.73
4.00	3.00	2.375	2.793	0.29	0.17	0.25	6.71	3.36	1.68	1.31	0.87	0.74
5.00	3.50	3.146	3.700	0.32	0.19	0.30	13.94	5.58	2.11	2.29	1.31	0.85
6.00	4.00	3.427	4.030	0.29	0.19	0.30	21.99	7.33	2.53	3.10	1.55	0.95
6.00	4.00	3.990	4.692	0.35	0.21	0.30	25.50	8.50	2.53	3.74	1.87	0.97
7.00	4.50	4.932	5.800	0.38	0.23	0.30	42.89	12.25	2.95	5.78	2.57	1.08
8.00	5.00	5.255	6.181	0.35	0.23	0.30	59.69	14.92	3.37	7.30	2.92	1.18
8.00	5.00	5.972	7.023	0.41	0.25	0.30	67.78	16.94	3.37	8.55	3.42	1.20
9.00	5.50	7.110	8.361	0.44	0.27	0.30	102.02	22.67	3.79	12.22	4.44	1.31
10.00	6.00	7.352	8.646	0.41	0.25	0.40	132.09	26.42	4.24	14.78	4.93	1.42
10.00	6.00	8.747	10.286	0.50	0.29	0.40	155.79	31.16	4.22	18.03	6.01	1.44
12.00	7.00	9.925	11.672	0.47	0.29	0.40	255.57	42.60	5.07	26.90	7.69	1.65
12.00	7.00	12.153	14.292	0.62	0.31	0.40	317.33	52.89	5.11	35.48	10.14	1.71

① Areas listed are based on nominal dimensions.
② Weights per foot are based on nominal dimensions and a density of 0.098 pound per cubic inch which is the density of alloy 6061.
③ I=moment of inertia; Z=section modulus; r=radius of gyration.

Source: *Aluminum Standards and Data*, 5th ed. (New York, N.Y.: The Aluminum Association, © 1976) p. 180.

Type of Beam	Stresses	
	General Formula for Stress at any Point	Stresses at Critical Points
Case 1. — Supported at Both Ends, Uniform Load TOTAL LOAD W	$s = -\dfrac{W}{2Zl}\, x\,(l-x)$	Stress at center, $-\dfrac{Wl}{8Z}$ If cross-section is constant, this is the maximum stress.
Case 2. — Supported at Both Ends, Load at Center	Between each support and load, $s = -\dfrac{Wx}{2Z}$	Stress at center, $-\dfrac{Wl}{4Z}$ If cross-section is constant, this is the maximum stress.
Case 3. — Supported at Both Ends, Load at any Point $a+b=l$	For segment of length a, $s = -\dfrac{Wbx}{Zl}$ For segment of length b, $s = -\dfrac{Wav}{Zl}$	Stress at load, $-\dfrac{Wab}{Zl}$ If cross-section is constant, this is the maximum stress.
Case 4. — Supported at Both Ends, Two Symmetrical Loads	Between each support and adjacent load, $s = -\dfrac{Wx}{Z}$ Between loads, $s = -\dfrac{Wa}{Z}$	Stress at each load, and at all points between, $-\dfrac{Wa}{Z}$
Case 5. — Both Ends Overhanging Supports Symmetrically, Uniform Load TOTAL LOAD W $L = l + 2c$	Between each support and adjacent end, $s = \dfrac{W}{2ZL}\,(c-u)^2$ Between supports, $s = \dfrac{W}{2ZL}[c^2 - x\,(l-x)]$	Stress at each support, $\dfrac{Wc^2}{2ZL}$ Stress at center, $\dfrac{W}{2ZL}(c^2 - \tfrac{1}{4}l^2)$ If cross-section is constant, the greater of these is the maximum stress. If l is greater than $2c$, the stress is zero at points $\sqrt{\tfrac{1}{4}l^2 - c^2}$ on both sides of the center. If cross-section is constant and if $l = 2.828\,c$, the stresses at supports and center are equal and opposite, and are $\pm \dfrac{WL}{46.62\,Z}$

430

Deflections (*See footnote*)	
General Formula for Deflection at any Point	Deflections at Critical Points
$y = \dfrac{-Wx\,(l-x)}{24\,EIl}\,[l^2 + x\,(l-x)]$	Maximum deflection, at center, $$\frac{-5}{384}\ \frac{Wl^3}{EI}$$
Between each support and load, $$y = \frac{-Wx}{48\,EI}\,(3\,l^2 - 4\,x^2)$$	Maximum deflection, at load, $\dfrac{-Wl^3}{48\,EI}$
For segment of length a, $$y = \frac{-Wbx}{6\,EIl}\,(l^2 - x^2 - b^2)$$ For segment of length b, $$y = \frac{-Wav}{6\,EIl}\,(l^2 - v^2 - a^2)$$	Deflection at load, $\dfrac{-Wa^2b^2}{3\,EIl}$ Let a be the length of the shorter segment and b of the longer one. The maximum deflection is in the longer segment, at $$v = b\sqrt{\frac{1}{3}+\frac{2\,a}{3\,b}} = v_1,\text{ and is }\frac{-Wav_1^3}{3\,EIl}$$
Between each support and adjacent load, $$y = \frac{-Wx}{6\,EI}\,[3\,a\,(l-a) - x^2]$$ Between loads, $$y = \frac{-Wa}{6\,EI}\,[3\,v\,(l-v) - a^2]$$	Maximum deflection at center, $$\frac{-Wa}{24\,EI}\,(3\,l^2 - 4\,a^2)$$ Deflection at loads $\dfrac{-Wa^2}{6\,EI}\,(3\,l - 4\,a)$
Between each support and adjacent end, $$y = \frac{-Wu}{24\,EIL}\,[6\,c^2\,(l+u) - u^2\,(4\,c - u) - l^2]$$ Between supports, $$y = \frac{-Wx\,(l-x)}{24\,EIL}\,[x\,(l-x) + l^2 - 6\,c^2]$$	Deflection at ends, $$\frac{-Wc}{24\,EIL}[3\,c^2\,(c+2\,l) - l^2]$$ Deflection at center, $$\frac{-Wl^2}{384\,EIL}\,(5\,l^2 - 24c^2)$$ If l is between $2\,c$ and $2.449\,c$, there are maximum upward deflections at points $\sqrt{3\,(\tfrac{1}{4}\,l^2 - c^2)}$ on both sides of the center, which are, $+\dfrac{W}{96\,EIL}\,(6\,c^2 - l^2)^2$

The deflections apply only to cases where the cross-section of the beam is constant for its entire length

Type of Beam	Stresses	
	General Formula for Stress at any Point	Stresses at Critical Points
Case 6. — Both Ends Overhanging Supports Unsymmetrically, Uniform Load TOTAL LOAD W $\frac{W}{2l}(l-d+c)$ $\frac{W}{2l}(l+d-c)$	For overhanging end of length c, $s = \dfrac{W}{2ZL}(c-u)^2$ Between supports, $s = \dfrac{W}{2ZL}\left\{ c^2\left(\dfrac{l-x}{l}\right)\right.$ $\left. + d^2\dfrac{x}{l} - x(l-x)\right\}$ For overhanging end of length d, $s = \dfrac{W}{2ZL}(d-w)^2$	Stress at support next end of length c, $\dfrac{Wc^2}{2ZL}$ Critical stress between supports is at $x = \dfrac{l^2+c^2-d^2}{2l} = x_1$ and is $\dfrac{W}{2ZL}(c^2-x_1^2)$ Stress at support next end of length d, $\dfrac{Wd^2}{2ZL}$ If cross-section is constant, the greatest of these three is the maximum stress. If $x_1 > c$, the stress is zero at points $\sqrt{x_1^2-c^2}$ on both sides of $x = x_1$.
Case 7. — Both Ends Overhanging Supports, Load at any Point Between $\dfrac{Wb}{l}$ $(a+b=l)$ $\dfrac{Wa}{l}$	Between supports: For segment of length a, $s = -\dfrac{Wbx}{Zl}$ For segment of length b, $s = -\dfrac{Wav}{Zl}$ Beyond supports $s = 0$.	Stress at load, $-\dfrac{Wab}{Zl}$ If cross-section is constant, this is the maximum stress.
Case 8. — Both Ends Overhanging Supports, Single Overhanging Load $\dfrac{W(a+l)}{l}$ $-\dfrac{Wa}{l}$	Between load and adjacent support, $s = \dfrac{W}{Z}(c-u)$ Between supports, $s = \dfrac{Wc}{Zl}(l-x)$ Between unloaded end and adjacent support, $s = 0$.	Stress at support adjacent to load, $\dfrac{Wc}{Z}$ If cross-section is constant, this is the maximum stress. Stress is zero at other support.
Case 9. — Both Ends Overhanging Supports, Symmetrical Overhanging Loads 	Between each load and adjacent support, $s = \dfrac{W}{Z}(c-u)$ Between supports, $s = \dfrac{Wc}{Z}$	Stress at supports and at all points between, $\dfrac{Wc}{Z}$ If cross-section is constant, this is the maximum stress.

Deflections *(See footnote at beginning of Table)*	
General Formula for Deflections at any Point	Deflections at Critical Points
For overhanging end of length c, $$y = \frac{-Wu}{24\,EIL}\,[2\,l\,(d^2 + 2\,c^2)$$ $$+\,6\,c^2u - u^2\,(4\,c - u) - l^2]$$ Between supports, $$y = \frac{-Wx\,(l-x)}{24\,EIL}\left\{ x\,(l-x) + l^2 - 2\,(d^2 + c^2) \right.$$ $$\left. -\,\frac{2}{l}\,[d^2x + c^2\,(l-x)] \right\}$$ For overhanging end of length d, $$y = \frac{-Ww}{24\,EIL}\,[2\,l\,(c^2 + 2\,d^2)$$ $$+\,6\,d^2w - w^2\,(4\,d - w) - l^2]$$	Deflection at end c, $$\frac{-Wc}{24\,EIL}\,[2\,l\,(d^2 + 2\,c^2) + 3\,c^3 - l^3]$$ Deflection at end d, $$\frac{-Wd}{24\,EIL}\,[2\,l\,(c^2 + 2\,d^2) + 3\,d^3 - l^3]$$ This case is so complicated that convenient general expressions for the critical deflections between supports cannot be obtained.
Between supports, same as Case 3. For overhanging end of length c, $$y = +\,\frac{Wabu}{6\,EIl}\,(l+b)$$ For overhanging end of length d, $$y = +\,\frac{Wabw}{6\,EIl}\,(l+a)$$	Between supports, same as Case 3. Deflection at end c, $+\,\dfrac{Wabc}{6\,EIl}\,(l+b)$ Deflection at end d, $+\,\dfrac{Wabd}{6\,EIl}\,(l+a)$
Between load and adjacent support, $$y = \frac{-Wu}{6\,EI}\,(3\,cu - u^2 + 2\,cl)$$ Between supports, $$y = +\,\frac{Wcx}{6\,EIl}\,(l-x)(2\,l-x)$$ Between unloaded end and adjacent support, $y = \dfrac{-Wclw}{6\,EI}$	Deflection at load, $\dfrac{-Wc^2}{3\,EI}\,(c+l)$ Maximum upward deflection is at $x = 0.42265\,l$, and is $+\,\dfrac{Wcl^2}{15.55\,EI}$ Deflection at unloaded end, $\dfrac{-Wcld}{6\,EI}$
Between each load and adjacent support, $y = \dfrac{-Wu}{6\,EI}\,[3\,c\,(l+u) - u^2]$ Between supports, $y = +\,\dfrac{Wcx}{2\,EI}\,(l-x)$	Deflections at loads, $\dfrac{-Wc^2}{6\,EI}\,(2\,c+3\,l)$ Deflection at center, $+\,\dfrac{Wcl^2}{8\,EI}$

The above expressions involve the usual approximations of the theory of flexure, and hold only for small deflections. Exact expressions for deflections of any magnitude are as follows:

Between supports the curve is a circle of radius $r = \dfrac{EI}{Wc}$; $y = \sqrt{r^2 - \tfrac{1}{4}\,l^2} - \sqrt{r^2 - (\tfrac{1}{2}\,l - x)^2}$

Deflection at center, $\sqrt{r^2 - \tfrac{1}{4}\,l^2} - r$

Type of Beam	Stresses	
	General Formula for Stress at any Point	Stresses at Critical Points
Case 10. — Fixed at One End, Uniform Load	$s = \dfrac{W}{2\,Zl}(l-x)^2$	Stress at support, $$\dfrac{Wl}{2\,Z}$$ If cross-section is constant, this is the maximum stress.
Case 11. — Fixed at One End, Load at Other	$s = \dfrac{W}{Z}(l-x)$	Stress at support, $\dfrac{Wl}{Z}$ If cross-section is constant, this is the maximum stress.
Case 12. — Fixed at One End, Intermediate Load	Between support and load, $$s = \dfrac{W}{Z}(l-x)$$ Beyond load, $s = $ o.	Stress at support, $\dfrac{Wl}{Z}$ If cross-section is constant, this is the maximum stress.
Case 13. — Fixed at Both Ends, Uniform Load	$s = \dfrac{Wl}{2\,Z}\left\{\dfrac{1}{6}-\dfrac{x}{l}+\left(\dfrac{x}{l}\right)^2\right\}$	Maximum stress, at ends, $\dfrac{Wl}{12\,Z}$ Stress is zero at $x = $ o.7887 l and at $x = $ o.2113 l Greatest negative stress, at center, $- \dfrac{Wl}{24\,Z}$
Case 14. — Fixed at Both Ends, Load at Center	Between each end and load, $$s = \dfrac{W}{2\,Z}(\tfrac{1}{4}\,l - x)$$	Stress at ends $\dfrac{Wl}{8\,Z}$; at load $-\dfrac{Wl}{8\,Z}$ These are the maximum stresses and are equal and opposite. Stress is zero at $x = \tfrac{1}{4}\,l$
Case 15. — Fixed at Both Ends, Load at any Point	For segment of length a, $$s = \dfrac{Wb^2}{Zl^3}[al - x\,(l + 2\,a)]$$ For segment of length b, $$s = \dfrac{Wa^2}{Zl^3}[bl - v\,(l + 2\,b)]$$	Stress at end next segment of length a, $\dfrac{Wab^2}{Zl^2}$ Stress at end next segment of length b, $\dfrac{Wa^2b}{Zl^2}$ Maximum stress is at end next shorter segment. Stress is zero for $x = \dfrac{al}{l+2\,a}$ and $v = \dfrac{bl}{l+2\,b}$ Greatest negative stress, at load, $-\dfrac{2\,Wa^2b^2}{Zl^3}$

Deflections *(See footnote at beginning of Table)*	
General Formula for Deflection at any Point	Deflections at Critical Points
$y = \dfrac{-Wx^2}{24\,EIl}\,[2\,l^2 + (2\,l - x)^2]$	Maximum deflection, at end, $\dfrac{-Wl^3}{8\,EI}$
$y = \dfrac{-Wx^2}{6\,EI}\,(3\,l - x)$	Maximum deflection, at end, $\dfrac{-Wl^3}{3\,EI}$
Between support and load, $y = \dfrac{-Wx^2}{6\,EI}\,(3\,l - x)$ Beyond load, $y = \dfrac{-Wl^2}{6\,EI}\,(3\,v - l)$	Deflection at load, $\dfrac{-Wl^3}{3\,EI}$ Maximum deflection, at end, $\dfrac{-Wl^2}{6\,EI}\,(2\,l + 3\,b)$
$y = \dfrac{-Wx^2}{24\,EIl}\,(l - x)^2$	Maximum deflection, at center, $\dfrac{-Wl^3}{384\,EI}$
$y = \dfrac{-Wx^2}{48\,EI}\,(3\,l - 4\,x)$	Maximum deflection, at load, $\dfrac{-Wl^3}{192\,EI}$
For segment of length a, $y = \dfrac{-Wx^2b^2}{6\,EIl^3}\,[2\,a\,(l - x) + l\,(a - x)]$ For segment of length b, $y = \dfrac{-Wv^2a^2}{6\,EIl^3}\,[2\,b\,(l - v) + l\,(b - v)]$	Deflection at load, $\dfrac{-Wa^3b^3}{3\,EIl^3}$ Let b be the length of the longer segment and a of the shorter one. The maximum deflection is in the longer segment, at $v = \dfrac{2\,bl}{l + 2\,b}$, and is $\dfrac{-2\,Wa^3b^3}{3\,EI\,(l + 2\,b)^2}$

Source: Erik Oberg, Franklin D. Jones, and Holbrook L. Horton, *Machinery Handbook*, 20th ed. (New York, N.Y.: Industrial Press, Inc., © 1975). Reprinted with permission.

TABLE A-20   Properties of American National Standard Schedule 40 Welded and Seamless Wrought Steel Pipe

Nominal	Diameter, in. Actual Inside	Actual Outside	Wall Thickness, in.	Cross-Sectional Area of Metal, in.^{2}	I Moment of Inertia, in.^{4}	Radius of Gyration, in.	Section Modulus Z, in.^{3}	Polar Section Modulus Z_p, in.^{3}
$\frac{1}{8}$	0.269	0.405	0.068	0.072	0.00106	0.122	0.00525	0.01050
$\frac{1}{4}$	0.364	0.540	0.088	0.125	0.00331	0.163	0.01227	0.02454
$\frac{3}{8}$	0.493	0.675	0.091	0.167	0.00729	0.209	0.02160	0.04320
$\frac{1}{2}$	0.622	0.840	0.109	0.250	0.01709	0.261	0.04070	0.08140
$\frac{3}{4}$	0.824	1.050	0.113	0.333	0.03704	0.334	0.07055	0.1411
1	1.049	1.315	0.133	0.494	0.08734	0.421	0.1328	0.2656
$1\frac{1}{4}$	1.380	1.660	0.140	0.669	0.1947	0.539	0.2346	0.4692
$1\frac{1}{2}$	1.610	1.900	0.145	0.799	0.3099	0.623	0.3262	0.6524
2	2.067	2.375	0.154	1.075	0.6658	0.787	0.5607	1.121
$2\frac{1}{2}$	2.469	2.875	0.203	1.704	1.530	0.947	1.064	2.128
3	3.068	3.500	0.216	2.228	3.017	1.163	1.724	3.448
$3\frac{1}{2}$	3.548	4.000	0.226	2.680	4.788	1.337	2.394	4.788
4	4.026	4.500	0.237	3.174	7.233	1.510	3.215	6.430
5	5.047	5.563	0.258	4.300	15.16	1.878	5.451	10.90
6	6.065	6.625	0.280	5.581	28.14	2.245	8.496	16.99
8	7.981	8.625	0.322	8.399	72.49	2.938	16.81	33.62
10	10.020	10.750	0.365	11.91	160.7	3.674	29.91	59.82
12	11.938	12.750	0.406	15.74	300.2	4.364	47.09	94.18
16	15.000	16.000	0.500	24.35	732.0	5.484	91.50	183.0
18	16.876	18.000	0.562	30.79	1172.	6.168	130.2	260.4

TABLE A-21 Conversion Factors

Quantity	English Unit	SI Unit	English to SI	SI to English
			Multiply Given Value by Factor to Convert from:	
Area	in.^{2}	mm^2	645.16	1.550×10^{-3}
	ft^2	m^2	0.0929	10.76
Loading	lb/in.^{2}	kPa	6.895	0.1450
	lb/ft^2	kPa	0.0479	20.89
Bending moment	lb·in.	N·m	0.1130	8.851
	lb·ft	N·m	1.356	0.7376
Density	lb$_m$/in.^{3}	kg/m^3	2.768×10^4	3.613×10^{-5}
	lb$_m$/ft^3	kg/m^3	16.02	0.0624
Force	lb	N	4.448	0.2248
	kip	kN	4.448	0.2248
Length	in.	mm	25.4	0.03937
	ft	m	0.3048	3.281
Mass	lb$_m$	kg	0.454	2.205
Torque	lb·in.	N·m	0.1130	8.851
Power	hp	kW	0.7457	1.341
Stress or pressure	psi	kPa	6.895	0.1450
	ksi	MPa	6.895	0.1450
	psi	GPa	6.895×10^{-6}	1.450×10^5
	ksi	GPa	6.895×10^{-3}	145.0
Section modulus	in.^{3}	mm^3	1.639×10^4	6.102×10^{-5}
Moment of inertia	in.^{4}	mm^4	4.162×10^5	2.403×10^{-6}

ANSWERS
TO SELECTED
PROBLEMS

Chapter 1

1-7 7.84 kN, front
11.76 kN, rear

1-9 54.4 mm

1-13 1762 lb, front
2644 lb, rear

1-15 55.1 lb
25.7 lb/in.
2.14 in.

1-17 8300 kPa

1-19 97 MPa to 524 MPa

1-21 9100 mm²

1-23 Area: 324 in.²
209×10^3 mm²
Volume: 3890 in.³
2.25 ft³
63.7×10^6 mm³
63.7×10^{-3} m³

Chapter 3

3-1 40.7 MPa

3-3 5380 psi

3-5 990 psi

3-7 32.9 mm minimum

3-9 1390 psi

3-11 54.7 MPa

3-13 15.5 kN

3-15 Required $s_u = 326$ MPa

3-17 Required $s_u = 560$ MPa

3-19 18 900 lb

3-21 11 790 psi
$N = 6.8$

3-23 63.6 MPa

Chapter 4

4-1 0.051 in.

4-3 2357 lb
3655 psi

4-5 (a) 10.6 mm; mass = 0.430 kg
(b) 10.6 mm; mass = 0.430 kg
(c) 18.4 mm; mass = 0.465 kg

4-7 (a) 0.857 mm
(b) 0.488 mm

4-9 0.00403 in. tension
0.00045 in. compression

4-11 10.8 mm

4-13 1.10 mm

4-15 1.41 in.

4-17 0.50 in.

4-19 349 MPa

4-21 0.59 in.

4-23 222 000 lb

4-25 Steel: 150 MPa
Aluminum: 50 MPa

Chapter 5

5-1 72.9 MPa

5-3 0.438 in.

5-5 Required s_u = 2149 MPa

5-7 19.4 kN

5-9 Required s_u = 716 MPa

5-11 251 kN

5-13 47 200 lb

5-15 24.7 MPa

5-17 19.2 mm

5-19 1.79 in.

5-21 38 540 psi

5-23 183 MPa

Chapter 6

6-1 178 MPa

6-3 4036 psi

6-5 83.8 MPa

6-7 15 800 psi

6-9 2207 psi

6-11 116.7 N·m

6-13 12.74 in.
Ratio = 1.96

6-15 130 MPa
0.135 rad

6-17 0.0378 rad

6-19 1.04 in.

6-21 6.40 mm
96.8 MPa; N = 1.24 (low)

6-23 164 MPa
N = 3.26

6-25 43.9 mm
82.2 MPa

6-27 137 MPa

6-29 At A: 112 MPa
At B: 31.5 MPa
At C: 115 MPa
At D: 115 MPa
At E: 41.3 MPa
At F: 122 MPa

6-31 Shaft 1: 10.5 mm
Shaft 2: 16.3 mm
Shaft 3: 27.9 mm

Chapter 7

Note: The following answers refer to Figures 7-35 to 7-48. For reactions, R_1 is the left; R_2 is the right. V and M refer to the maximum absolute values for shear and bending moment, respectively.

7-35(a) $R_1 = R_2 = 325$ lb
$V = 325$ lb
$M = 4550$ lb·in.
 (c) $R_1 = 11.43$ K; $R_2 = 4.57$ K
$V = 11.43$ K
$M = 45.7$ K·ft

7-36(a) $R_1 = 575$ N; $R_2 = 325$ N
$V = 575$ N
$M = 195$ N·m
 (c) $R_1 = R_2 = 50$ kN
$V = 50$ kN
$M = 85$ kN·m

7-37(a) $R_1 = 1557$ lb
$R_2 = 1743$ lb
$V = 1557$ lb
$M = 6228$ lb·in.
 (c) $R_1 = 7.5$ K
$R_2 = 37.5$ K
$V = 20$ K
$M = 60$ K·ft

7-38(a) $R_1 = R_2 = 250$ N
$V = 850$ N
$M = 362.5$ N·m
 (c) $R_1 = 37.4$ kN (down)
$R_2 = 38.3$ kN (up)
$V = 24.9$ kN
$M = 50$ kN·m

7-39(a) $R = 120$ lb
$V = 120$ lb
$M = 960$ lb·ft
 (c) $R = 24$ K
$V = 24$ K
$M = 168$ K·ft

7-40(a) $R = 1800$ N
$V = 1800$ N
$M = 1020$ N·m
 (c) $R = 120$ kN
$V = 120$ kN
$M = 240$ kN·m

7-41(a) $R_1 = R_2 = 180$ lb
$V = 180$ lb
$M = 810$ lb·in.
 (c) $R_1 = R_2 = 12$ K
$V = 12$ K
$M = 84$ K·ft

7-42(a) $R_1 = 99.2$ N;
$R_2 = 65.8$ N
$V = 99.2$ N
$M = 9.9$ N·m
 (c) $R_1 = R_2 = 40$ kN
$V = 40$ kN
$M = 140$ kN·m

7-43(a) $R_1 = R_2 = 440$ lb
$V = 240$ lb
$M = 360$ lb·in.
 (c) $R_1 = 1456$ N;
$R_2 = 644$ N
$V = 956$ N
$M = 125$ N·m
 (e) $R_1 = R_2 = 54$ kN
$V = 30$ kN
$M = 24$ kN·m

7-44(a) $R = 360$ lb
$V = 360$ lb
$M = 1620$ lb·in.
 (c) $R = 600$ N
$V = 600$ N
$M = 200$ N·m

7-45(a) $R_1 = R_2 = 330$ lb
$V = 330$ lb
$M = 4200$ lb·in.
 (c) $R_1 = 36.6$ K; $R_2 = 30.4$ K
$V = 36.6$ K
$M = 183.2$ K·ft

7-46(a) $R_1 = R_2 = 450$ N
$V = 450$ N
$M = 172.5$ N·m
 (c) $R_1 = R_2 = 175$ kN
$V = 175$ kN
$M = 600$ kN·m

7-47(a) $R_1 = 636$ lb; $R_2 = 1344$ lb
$V = 804$ lb
$M = 2528$ lb·in.
 (c) $R_1 = 4950$ N;
$R_2 = 3100$ N
$V = 2950$ N
$M = 3350$ N·m

7-48(a) $R = 236$ lb
$V = 236$ lb
$M = 1504$ lb·in.

(c) $R = 1130$ N
     $V = 1130$ N
     $M = 709$ N·m
(e) $R = 230$ kN
     $V = 230$ kN
     $M = 430$ kN·m

Chapter 8

Note: The following answers refer to Figures 8-14 to 8-20. The first number is the distance to the centroid from the bottom of the section. The second number is the moment of inertia with respect to the centroidal axis.

8-14(a) 0.900 in.; 0.366 in.^{4}
   (c) 4.00 in.; 166 in.^{4}

8-15(a) 152.5 mm; 4.64×10^7 mm^4
   (c) 25.33 mm; 3.38×10^5 mm^4

8-16(a) 12.5 mm; 6.167×10^4 mm^4
   (c) 20 mm; 5.36×10^4 mm^4

8-17(a) 7.50 in.; 1729 in.^{4}
   (c) 7.40 in.; 423.5 in.^{4}

8-18(a) 3.50 in.; 88.94 in.^{4}
   (c) 3.00 in. from center of either pipe; 22.91 in.^{4}

8-19(a) 4.25 in.; 151.4 in.^{4}
   (c) 2.25 in.; 107.2 in.^{4}

8-20(a) 30 mm; 2.64×10^5 mm^4
   (c) 12.39 mm; 16 956 mm^4

Chapter 9

9-1 94.4 MPa

9-3 (a) 20 620 psi
         (b) 41 240 psi

9-5 26.9 mm

9-7 Maximum stress = 92.0 MPa
         $s_{sd} = 103.5$ MPa — (OK)

9-9 Required $Z = 0.294$ in.^{3}
         $1\frac{1}{2}$-in. Schedule 40 pipe

9-11 Stress = 504 psi — (OK)

9-13 16 744 psi

9-15 6882 psi tensile
          12 970 psi compressive

9-17 39.7 MPa — (OK)

9-19 47.4 MPa (6880 psi)

9-21 (a) $D = 112.8$ mm
              $A = 9998$ mm^2
          (b) $S = 94.6$ mm
              $A = S^2 = 8945$ mm^2
          (c) $b = 37.5$ mm
              $h = 150$ mm
              $A = 5625$ mm^2
          (d) S7×15.3
              $A = 2903$ mm^2
              (lightest)

9-23 Required $Z = 1.35$ in.^{3}
          3-in. Schedule 40 pipe

9-25 24 kN·m
          W8×13

9-27 5.36 ft

9-29 398 MPa

9-31 $e = 0.345$ in.

Chapter 10

10-1 12.8 MPa

10-3 131 psi

10-5 28.1 MPa

10-7 1.69 MPa

10-9 944 lb

10-11 1125 lb

10-13 4210 psi

10-15 1976 psi

10-17 1168 psi

10-19 $s_s = 3.40$ MPa
           $s_d = 37.3$ MPa

10-21 1.37 MPa

10-23 106.2 lb

10-25 2.62 in./row

10-27 2.76 in./row (70 mm/row)

Chapter 11

11-1 10 510 psi compressive

11-3 At N: 6428 psi tension
At M: 5008 psi compressive

11-5 At N: 9426 psi tension
At M: 10 846 psi compressive

11-7 92.4 MPa compressive at load

11-9 433 N

11-11 727 MPa

11-13 23 985 psi tension
18 375 psi compressive

11-15 Load = 7934 N
Mass = 809 kg

11-17 51.6 MPa

11-19 s_s = 9903 psi
Required s_y = 118 800 psi

11-21 79.5 MPa

11-23 $L5 \times 3 \times \frac{5}{16}$

Chapter 12

12-1 -2.01 mm

12-3 -0.503 mm

12-5 -5.40 mm

12-7 At loads: -0.242 in.
At center: -0.371 in.

12-9 -0.424 in.

12-11 -0.230 in.

12-13 $+0.079$-in. deflection at $x = 50.7$ in.

12-15 At 500-N load: -16.18 mm
At 400-N load: -21.63 mm

12-17 -0.0078 in.

12-19 -0.869 mm

12-21 -3.997 mm

12-23 64.8 mm

12-25 Required $I = 1.66 \times 10^8$ mm^4
W18 × 35 beam

12-27 0.020 in.

12-29 $y = -0.0078$ in.
at $x = 8.56$ in.

12-31 $y = -3.79$ mm

12-33 $D = 69.2$ mm

12-35 17 × 5.800 aluminum beam
$y = -5.37$ mm

12-37 $D = 109$ mm

Chapter 13

13-1(a) $R_A = R_E = 371$ lb
$R_C = 858$ lb
$V = 429$ lb
$M = 1113$ lb·ft

(c) $R_A = R_E = 1873$ lb
$R_C = 5854$ lb
$V = 2927$ lb
$M = 5016$ lb·ft

13-2(a) $R_A = 22.5$ kN
$R_B = 13.5$ kN
$V = 22.5$ kN
$M = 8.1$ kN·m

(c) $R_A = 127.4$ kN
$R_C = 76.6$ kN
$V = 127.4$ kN
$M = 103.4$ kN·m

13-3(a) Same as 13-1(a)
(c) Same as 13-1(c)

13-4(a) $R_A = 8.90$ kN
$R_C = 34.1$ kN
$R_D = 6.00$ kN
$V = 18$ kN
$M = 8.9$ kN·m

(c) $R_A = 15.8$ kN
$R_C = 68.9$ kN
$R_E = 7.3$ kN
$V = 62.9$ kN
$M = 31.7$ kN·m

13-5(a) $R_A = R_D = 21$ kN
$R_B = R_C = 44$ kN
$V = 29$ kN
$M = 22.05$ kN·m

(c) $R_A = 52.56$ kN
$R_C = 156.44$ kN
$R_D = 35.05$ kN
$R_F = 25.95$ kN
$V = 107.44$ kN
$M = 154.4$ kN·m

Chapter 14

14-1 24.2 kN

14-3 8.35 kN

14-5 25.3 kN

14-7 110 kN

14-9 10 columns

14-11 499 kN

14-13 65 300 lb

14-15 36.8 mm

14-17 $P_{cr} = 2.60 \times 10^5$ N
Design factor = 8.37

14-19 15.7 kN

14-21 $P_{cr} = 17\ 860$ lb
$N = 3.57 -$ (OK)

14-23 $P_{cr} = 3130$ lb
$N = 2.09$, marginal

Chapter 15

15-1 115 MPa

15-3 2.94 mm

15-5 2134 psi

15-7 1.32 mm

15-9 N = 3.78

15-11 Hoop: 212 MPa
Radial: −70 MPa
(compressive)

15-13
Radius	Stress
110 mm	166 MPa
120 mm	149 MPa
130 mm	136 MPa
140 mm	125 MPa
150 mm	116 MPa

Chapter 16

16-1(a) $F_s = 5890$ lb (allowable)
$F_b = 18\ 225$ lb
$F_t = 14\ 175$ lb

(c) $F_s = 6627$ lb (allowable)
$F_b = 13\ 669$ lb
$F_t = 16\ 200$ lb

16-2(a) $F_s = 15\ 708$ lb (allowable)
$F_b = 62\ 100$ lb
$F_t = 37\ 950$ lb

(c) $F_s = 31\ 416$ lb (allowable)
$F_b = 62\ 100$ lb
$F_t = 37\ 950$ lb

16-3(a) $F_s = 8639$ lb (allowable)
$F_b = 23\ 288$ lb
$F_t = 18\ 113$ lb

(c) $F_s = 9719$ lb (allowable)
$F_b = 17\ 466$ lb
$F_t = 20\ 700$ lb

16-4(a) $F_s = 15\ 708$ lb (allowable)
$F_t = 82\ 500$ lb

(c) $F_s = 31\ 416$ lb (allowable)
$F_t = 82\ 500$ lb

16-5 Required $D = 1.34$ in.;
use $1\text{-}\frac{3}{8}$ in.

16-7 23 860 lb on welds

INDEX